Volterra and Functional Differential Equations

LECTURE NOTES
IN PURE AND APPLIED MATHEMATICS

1. *N. Jacobson*, Exceptional Lie Algebras
2. *L.-Å. Lindahl and F. Poulsen*, Thin Sets in Harmonic Analysis
3. *I. Satake*, Classification Theory of Semi-Simple Algebraic Groups
4. *F. Hirzebruch, W. D. Newmann, and S. S. Koh*, Differentiable Manifolds and Quadratic Forms (out of print)
5. *I. Chavel*, Riemannian Symmetric Spaces of Rank One (out of print)
6. *R. B. Burckel*, Characterization of C(X) Among Its Subalgebras
7. *B. R. McDonald, A. R. Magid, and K. C. Smith*, Ring Theory: Proceedings of the Oklahoma Conference
8. *Y.-T. Siu*, Techniques of Extension of Analytic Objects
9. *S. R. Caradus, W. E. Pfaffenberger, and B. Yood*, Calkin Algebras and Algebras of Operators on Banach Spaces
10. *E. O. Roxin, P.-T. Liu, and R. L. Sternberg*, Differential Games and Control Theory
11. *M. Orzech and C. Small*, The Brauer Group of Commutative Rings
12. *S. Thomeier*, Topology and Its Applications
13. *J. M. López and K. A. Ross*, Sidon Sets
14. *W. W. Comfort and S. Negrepontis*, Continuous Pseudometrics
15. *K. McKennon and J. M. Robertson*, Locally Convex Spaces
16. *M. Carmeli and S. Malin*, Representations of the Rotation and Lorentz Groups: An Introduction
17. *G. B. Seligman*, Rational Methods in Lie Algebras
18. *D. G. de Figueiredo*, Functional Analysis: Proceedings of the Brazilian Mathematical Society Symposium
19. *L. Cesari, R. Kannan, and J. D. Schuur*, Nonlinear Functional Analysis and Differential Equations: Proceedings of the Michigan State University Conference
20. *J. J. Schäffer*, Geometry of Spheres in Normed Spaces
21. *K. Yano and M. Kon*, Anti-Invariant Submanifolds
22. *W. V. Vasconcelos*, The Rings of Dimension Two
23. *R. E. Chandler*, Hausdorff Compactifications
24. *S. P. Franklin and B. V. S. Thomas*, Topology: Proceedings of the Memphis State University Conference
25. *S. K. Jain*, Ring Theory: Proceedings of the Ohio University Conference
26. *B. R. McDonald and R. A. Morris*, Ring Theory II: Proceedings of the Second Oklahoma Conference
27. *R. B. Mura and A. Rhemtulla*, Orderable Groups
28. *J. R. Graef*, Stability of Dynamical Systems: Theory and Applications
29. *H.-C. Wang*, Homogeneous Banach Algebras
30. *E. O. Roxin, P.-T. Liu, and R. L. Sternberg*, Differential Games and Control Theory II
31. *R. D. Porter*, Introduction to Fibre Bundles
32. *M. Altman*, Contractors and Contractor Directions Theory and Applications
33. *J. S. Golan*, Decomposition and Dimension in Module Categories
34. *G. Fairweather*, Finite Element Galerkin Methods for Differential Equations
35. *J. D. Sally*, Numbers of Generators of Ideals in Local Rings
36. *S. S. Miller*, Complex Analysis: Proceedings of the S.U.N.Y. Brockport Conference
37. *R. Gordon*, Representation Theory of Algebras: Proceedings of the Philadelphia Conference
38. *M. Goto and F. D. Grosshans*, Semisimple Lie Algebras
39. *A. I. Arruda, N. C. A. da Costa, and R. Chuaqui*, Mathematical Logic: Proceedings of the First Brazilian Conference

40. *F. Van Oystaeyen*, Ring Theory: Proceedings of the 1977 Antwerp Conference
41. *F. Van Oystaeyen and A. Verschoren*, Reflectors and Localization: Application to Sheaf Theory
42. *M. Satyanarayana*, Positively Ordered Semigroups
43. *D. L. Russell*, Mathematics of Finite-Dimensional Control Systems
44. *P.-T. Liu and E. Roxin*, Differential Games and Control Theory III: Proceedings of the Third Kingston Conference, Part A
45. *A. Geramita and J. Seberry*, Orthogonal Designs: Quadratic Forms and Hadamard Matrices
46. *J. Cigler, V. Losert, and P. Michor*, Banach Modules and Functors on Categories of Banach Spaces
47. *P.-T. Liu and J. G. Sutinen*, Control Theory in Mathematical Economics: Proceedings of the Third Kingston Conference, Part B
48. *C. Byrnes*, Partial Differential Equations and Geometry
49. *G. Klambauer*, Problems and Propositions in Analysis
50. *J. Knopfmacher*, Analytic Arithmetic of Algebraic Function Fields
51. *F. Van Oystaeyen*, Ring Theory: Proceedings of the 1978 Antwerp Conference
52. *B. Kedem*, Binary Time Series
53. *J. Barros-Neto and R. A. Artino*, Hypoelliptic Boundary-Value Problems
54. *R. L. Sternberg, A. J. Kalinowski, and J. S. Papadakis*, Nonlinear Partial Differential Equations in Engineering and Applied Science
55. *B. R. McDonald*, Ring Theory and Algebra III: Proceedings of The Third Oklahoma Conference
56. *J. S. Golan*, Structure Sheaves over a Noncommutative Ring
57. *T. V. Narayana, J. G. Williams, and R. M. Mathsen*, Combinatorics, Representation Theory and Statistical Methods in Groups: YOUNG DAY Proceedings
58. *T. A. Burton*, Modeling and Differential Equations in Biology
59. *K. H. Kim and F. W. Roush*, Introduction to Mathematical Consensus Theory
60. *J. Banas and K. Goebel*, Measures of Noncompactness in Banach Spaces
61. *O. A. Nielsen*, Direct Integral Theory
62. *J. E. Smith, G. O. Kenny, and R. N. Ball*, Ordered Groups: Proceedings of the Boise State Conference
63. *J. Cronin*, Mathematics of Cell Electrophysiology
64. *J. W. Brewer*, Power Series Over Commutative Rings
65. *P. K. Kamthan and M. Gupta*, Sequence Spaces and Series
66. *T. G. McLaughlin*, Regressive Sets and the Theory of Isols
67. *T. L. Herdman, S. M. Rankin, III, and H. W. Stech*, Integral and Functional Differential Equations
68. *R. Draper*, Commutative Algebra: Analytic Methods
69. *W. G. McKay and J. Patera*, Tables of Dimensions, Indices, and Branching Rules for Representations of Simple Lie Algebras
70. *R. L. Devaney and Z. H. Nitecki*, Classical Mechanics and Dynamical Systems
71. *J. Van Geel*, Places and Valuations in Noncommutative Ring Theory
72. *C. Faith*, Injective Modules and Injective Quotient Rings
73. *A. Fiacco*, Mathematical Programming with Data Perturbations I
74. *P. Schultz, C. Praeger, and R. Sullivan*, Algebraic Structures and Applications: Proceedings of the First Western Australian Conference on Algebra
75. *L. Bican, T. Kepka, and P. Němec*, Rings, Modules, and Preradicals
76. *D. C. Kay and M. Breen*, Convexity and Related Combinatorial Geometry: Proceedings of the Second University of Oklahoma Conference

77. *P. Fletcher and W. F. Lindgren*, Quasi-Uniform Spaces
78. *C.-C. Yang*, Factorization Theory of Meromorphic Functions
79. *O. Taussky*, Ternary Quadratic Forms and Norms
80. *S. P. Singh and J. H. Burry*, Nonlinear Analysis and Applications
81. *K. B. Hannsgen, T. L. Herdman, H. W. Stech and R. L. Wheeler*, Volterra and Functional Differential Equations

Other Volumes in Preparation

Volterra and Functional Differential Equations

edited by

KENNETH B. HANNSGEN
TERRY L. HERDMAN
HARLAN W. STECH
ROBERT L. WHEELER

Department of Mathematics
Virginia Polytechnic Institute
and State University
Blacksburg, Virginia

MARCEL DEKKER, INC. New York and Basel

Library of Congress Cataloging in Publication Data
Main entry under title:

Volterra and functional differential equations.

(Lecture notes in pure and applied mathematics ; v. 81)
Proceedings of the Conference on Volterra and Functional Differential Equations, held June 10-13, 1981, at Virginia Polytechnic Institute and State University in Blacksburg, Virginia.
1. Volterra equations--Congresses. 2. Functional differential equations--Congresses. I. Hannsgen, Kenneth B., [date]. II. Conference on Volterra and Functional Differential Equations (1981 : Virginia Polytechnic Institute and State University)
QA431.V63 1982 515.4'5 82-13998
ISBN 0-8247-1721-X

MARCEL DEKKER, INC.
270 Madison Avenue, New York, New York 10016

Current printing (last digit):
10 9 8 7 6 5 4 3 2 1

PRINTED IN THE UNITED STATES OF AMERICA

PREFACE

The Conference on Volterra and Functional Differential Equations was held June 10 - 13, 1981 at Virginia Polytechnic Institute and State University in Blacksburg, Virginia. Over sixty - five participants, including mathematicians from thirty - three academic institutions and research laboratories, attended lectures and informal sessions in McBryde Hall on the Virginia Tech campus. Kenneth B. Hannsgen, Terry L. Herdman and Robert L. Wheeler constituted the Organizing Committee; John A. Burns and Harlan W. Stech served as advisors.

This volume represents the contents of the twenty - four invited lectures presented at the conference. Papers based on the eleven one - hour lectures appear in Part One. Part Two contains the thirteen half - hour lectures.

The conference was made possible by the National Science Foundation through grant No. MCS - 8023198.

In addition, the editors gratefully acknowledge the assistance and support of many individuals and organizations. The Virginia Tech Department of Mathematics and its Head, C. Wayne Patty, generously supplied facilities and personnel both for the administration of the conference and the preparation of this proceedings; the Donaldson Brown Center for Continuing Education handled the housing arrangements; Susan Anderson, Robert Powers and Leslie Ratliff contributed in the day - to - day running of the conference; and a considerable amount of the necessary paperwork was managed with the invaluable assistance of Sharon Irvin and Debbie Wardinski. Finally, with special thanks, we acknowledge Cynthia Duncan for her skillful preparation of the final manuscript for this volume.

Kenneth B. Hannsgen
Terry L. Herdman
Harlan W. Stech
Robert L. Wheeler

CONTENTS

CONTRIBUTORS

H. THOMAS BANKS, Division of Applied Mathematics, Brown University, Providence, Rhode Island.

DENNIS W. BREWER, Department of Mathematics, University of Arkansas, Fayetteville, Arkansas.

STAVROS N. BUSENBERG, Department of Mathematics, Harvey Mudd College, Claremont, California.

RALPH W. CARR, Department of Mathematics, St. Cloud State University, St. Cloud, Minnesota.

GOONG CHEN, Department of Mathematics, Pennsylvania State University, University Park, Pennsylvania.

CONSTANTIN CORDUNEANU, Department of Mathematics, University of Texas at Arlington, Arlington, Texas.

JAMES M. GREENBERG, Division of Mathematical and Computer Sciences, National Science Foundation, Washington, D.C.

RONALD GRIMMER, Department of Mathematics, Southern Illinois University, Carbondale, Illinois.

JACK K. HALE, Division of Applied Mathematics, Brown University, Providence, Rhode Island.

LING HSIAO, Division of Applied Mathematics, Brown University, Providence, Rhode Island.

ETTORE F. INFANTE*, Division of Applied Mathematics, Brown University, Providence, Rhode Island.

G. SAMUEL JORDAN, Department of Mathematics, University of Tennessee, Knoxville, Tennessee.

THOMAS R. KIFFE, Department of Mathematics, Texas A & M University, College Station, Texas.

JACOB J. LEVIN, Department of Mathematics, University of Wisconsin, Madison, Wisconsin.

STIG-OLOF LONDEN, Institute of Mathematics, Helsinki University of Technology, Otaniemi, Finland.

RICHARD C. MacCAMY, Department of Mathematics, Carnegie - Mellon University, Pittsburgh, Pennsylvania.

LUIS MAGALHÃES, Division of Applied Mathematics, Brown University, Providence, Rhode Island.

REZA MALEK-MADANI[†], Department of Mathematics, University of Wisconsin, Madison, Wisconsin.

PAUL MASSATT, Department of Mathematics, University of Oklahoma, Norman, Oklahoma.

ANTHONY N. MICHEL, Department of Electrical Engineering, Iowa State University, Ames, Iowa.

RICHARD K. MILLER, Department of Mathematics, Iowa State University, Ames, Iowa.

JOHN A. NOHEL, Mathematics Research Center, University of Wisconsin - Madison, Madison, Wisconsin.

SAMUEL M. RANKIN, III, Department of Mathematics, West Virginia University, Morgantown, West Virginia.

MICHAEL RENARDY, Mathematics Research Center, University of Wisconsin - Madison, Madison, Wisconsin.

DAVID L. RUSSELL, Department of Mathematics, University of Wisconsin, Madison, Wisconsin.

ROBERT J. SACKER, Department of Mathematics, University of Southern California, Los Angeles, California.

GEORGE R. SELL, School of Mathematics, University of Minnesota, Minneapolis, Minnesota.

*Current affiliation: Division of Mathematical and Computer Sciences, National Science Foundation, Washington, D.C.

[†]Current affiliation: Department of Mathematics, Virginia Polytechnic Institute and State University, Blacksburg, Virginia.

SHERWIN J. SKAR, Department of Mathematics, Oklahoma State University, Stillwater, Oklahoma.

OLOF J. STAFFANS, Institute of Mathematics, Helsinki University of Technology, Otaniemi, Finland.

CURTIS C. TRAVIS, Oak Ridge National Laboratory, Health and Safety Research Division, Oak Ridge, Tennessee.

LUTHER W. WHITE, Department of Mathematics, University of Oklahoma, Norman, Oklahoma.

CONFERENCE PARTICIPANTS

Joseph Ball
Virginia Tech

H. Thomas Banks
Brown University

John B. Bennett
Arkansas State University

Steve Bins
Clemson University

Steven Black
Clemson University

Dennis Brewer
University of Arkansas

John A. Burns
Virginia Tech

Theodore A. Burton
Southern Illinois University

Stavros Busenberg
Harvey Mudd College

Dean A. Carlson
University of Delaware

Ralph W. Carr
St. Cloud State University

Goong Chen
Pennsylvania State University

Eugene M. Cliff
Virginia Tech

Kenneth L. Cooke
Pomona College

Constantin Corduneanu
University of Texas at Arlington

Martin Day
Virginia Tech

Rodney Driver
University of Rhode Island

Robert Fennell
Clemson University

William Grasman
Virginia Tech

David Green, Jr.
General Motors Institute

James M. Greenberg
National Science Foundation

George Hagedorn
Virginia Tech

Jack K. Hale
Brown University

Kenneth B. Hannsgen
Virginia Tech

Terry L. Herdman
Virginia Tech

Herbert Hethcote
University of Iowa

Ettore F. Infante
Brown University

Harry Johnson
Virginia Tech

G. Samuel Jordan
University of Tennessee

Thomas Kiffe
Texas A & M University

Werner Kohler
Virginia Tech

Jacob J. Levin
University of Wisconsin

James Lightbourne, III
West Virginia University

Chih-Bing Ling
Virginia Tech

Stig-Olof Londen
Helsinki University of Technology

Michael Lyons
North Carolina State University

Richard C. MacCamy
Carnegie - Mellon University

Joe Mahaffy
North Carolina State

Reza Malek-Madani
University of Wisconsin

Robert Martin
North Carolina State University

Paul Massatt
University of Oklahoma

Jim Mosely
West Virginia University

Beny Neta
Texas Tech University

John A. Nohel
University of Wisconsin

Mary E. Parrott
University of South Florida

Raymond Plaut
Virginia Tech

Carl Prather
Virginia Tech

T. Gilbert Proctor
Clemson University

Samuel M. Rankin, III
West Virginia University

Russell M. Reid
University of Missouri

Michael Renardy
University of Wisconsin

Charles Rennolet
Rose - Hulman Institute of Technology

Dave Reynolds
Carnegie - Mellon University

David L. Russell
University of Wisconsin

Stephen Saperstone
George Mason University

George R. Sell
University of Minnesota

J. Kenneth Shaw
Virginia Tech

Henry C. Simpson
University of Tennessee

Sherwin J. Skar
Oklahoma State University

Olof J. Staffans
Helsinki University of Technology

Harlan W. Stech
Virginia Tech

Kenneth E. Swick
Queens College

Curtis C. Travis
Oak Ridge National Laboratory

Robert L. Wheeler
Virginia Tech

Luther White
University of Oklahoma

Robert White
Bell Laboratories

Donald F. Young
Agnes Scott College

PART ONE

A SURVEY OF SOME PROBLEMS AND RECENT RESULTS FOR PARAMETER ESTIMATION AND OPTIMAL CONTROL IN DELAY AND DISTRIBUTED PARAMETER SYSTEMS

H. T. Banks

Lefschetz Center for Dynamical Systems
Division of Applied Mathematics
Brown University
Providence, Rhode Island

1. INTRODUCTION

In this lecture we shall first present a brief account of several areas of applications which have motivated our recent efforts, both theoretical and numerical, on approximation methods for estimation and control of infinite dimensional systems. We then shall sketch the general theoretical ideas we have employed to establish convergence results for related iterative schemes. Finally we return to two of the applications and illustrate the use of these ideas by explaining in more detail our investigations for these problems.
As we shall make clear, our efforts on many of the problems mentioned below involve joint endeavors with colleagues and students. In addition to a well-deserved thank you to Richard Ambrasino, James Crowley, Patti Daniel, Mary Garrett, Karl Kunisch, and Gary Rosen, we would also like to publicly acknowledge E. Armstrong (NASA Langley Research Center), R. Ewing and G. Moeckel (Mobil Research and Development Corp.), P. Kareiva (Brown University), J. P. Kernevez (Université de Technologie de Compiègne), W. T. Kyner (University of New Mexico), and G. A. Rosenberg (V. A. Medical Center, U. N. M. School of Medicine) for numerous stimulating discussions and suggestions which have substantially affected the investigations of our group at Brown University.

Our discussions here focus on a general class of systems including nonlinear delay systems

$$\dot{x}(t) = f(\alpha,t,x(t),x_t,x(t-\tau_1),\ldots,x(t-\tau_\nu)) + g(t)$$

$$x(\theta) = \phi(\theta), \quad -\tau_\nu \le \theta \le 0 \tag{1}$$

$$q = (\alpha,\tau_1,\ldots,\tau_\nu),$$

nonlinear distributed parameter systems

$$\frac{\partial^2 u}{\partial t^2} = q_1 \frac{\partial^2 u}{\partial x^2} + q_2 \frac{\partial u}{\partial t} + q_3 u + f(q_6,t,x,u)$$

$$u(0,x) = q_4\phi(x), \quad \frac{\partial u}{\partial t}(0,x) = q_5\psi(x) \tag{2}$$

$$u(t,0) = g_1(t), \quad u(t,1) = g_2(t)$$

of hyperbolic type, and parabolic systems of the form

$$\frac{\partial u}{\partial t} = \frac{q_1}{k(x)} \frac{\partial}{\partial x}\left(p(x)\frac{\partial u}{\partial x}\right) + q_2 u + f(q_4,t,x,u)$$

$$u(0,x) = q_3\phi(x) \tag{3}$$

$$\begin{bmatrix} \alpha_{11} & \alpha_{12} & \alpha_{13} & \alpha_{14} \\ \alpha_{21} & \alpha_{22} & \alpha_{23} & \alpha_{24} \end{bmatrix} \left[u(t,0),\ \frac{\partial u}{\partial x}(t,0),\ u(t,1),\ \frac{\partial u}{\partial x}(t,1)\right]^T = \begin{bmatrix} g_1(t) \\ g_2(t) \end{bmatrix}.$$

A typical estimation problem consists of the inverse problem of finding the vector parameter q, given observations $\{\xi_j\}$ of the state (or components of the state) corresponding to known inputs g or g_i. A typical control problem (for fixed parameter values q) might consist of minimizing a given payoff or cost functional subject to (1), (2), or (3), over some admissible class of control functions g or g_i.

2. MOTIVATING EXAMPLES

2.1. The LN_2 Wind Tunnel

The liquid nitrogen wind tunnel (National Transonic Facility) currently being constructed at NASA Langley Research Center is a cryogenic wind tunnel for which the cost of liquid nitrogen alone is estimated at $\$6.5 \times 10^6$ per year of operation. The tunnel represents the latest advances in technology in that essentially independent control of Mach number and Reynolds number (~ temperature) is an anticipated feature. Schematically, the tunnel can be represented as in Figure 1.

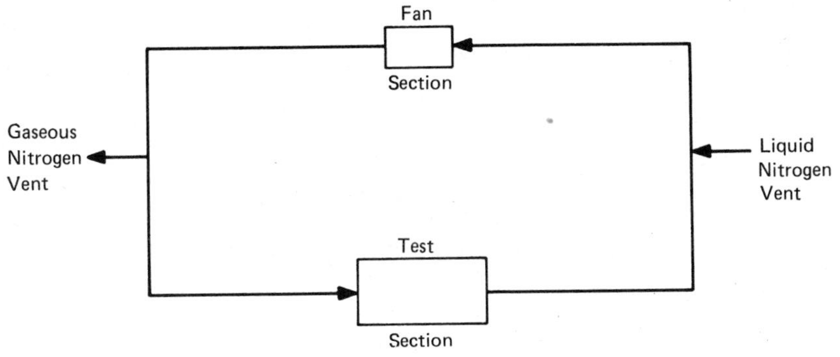

Figure 1

The basic physical model relating states such as Reynolds number, pressure and Mach number to controls such as LN_2 input, GN_2 bleed, and fan operation involves a formidable set of partial differential equations (Navier - Stokes) to describe fluid flow in the tunnel and test chamber. This model has, not surprisingly, proved to be very unwieldly computationally and probably cannot be used directly in design of sophisticated control laws. (Both open loop and feedback controllers are needed for efficient operation of the tunnel - - and this is clearly a desirable goal. Given the current estimates of costs of operation, the funds from only a 1 or 2% savings in operation costs would support a nontrivial amount of related research by scientists and engineers!)

In view of the schematic in Figure 1, it is not surprising that engineers (e.g., see [18]) have proposed lumped parameter models (the variables representing values of states and controllers at various discrete locations in the tunnel) with transport delays to represent flow times in sections of the tunnel. A specific example is the model (see [1]) for the Mach number (in the test chamber) loop which to first order is controlled by the fan guide vane angle setting (in the fan section) - - i.e., $M(t) \sim GVA(t-\tau)$ where τ represents a transport time from the fan section to the test section.

In addition to the design of both open loop and closed loop controllers, parameter estimation techniques will be useful once data from the completed tunnel is available (current investigations involve use of data from a 1/3 meter scale model of the tunnel).

2.2. Enzyme Tubular Reactors

Column reactors in which enzyme mediated chemical reactions take place to produce a desirable product (or products) from a given substrate (or substrates) are of some importance because of the numerous potential applications in commercial production (e.g., purification of fruit juices, proteolytic treatment of beer, synthesis of antibiotics and steroids). Research on the operation of such reactors has been carried out in the laboratory of D. Thomas at Université de Technologie de Compiègne for several years. Mathematical models for these processes (which involve reaction, diffusion, and convective transport) range from simple plug-flow (PF) models to full-fledged diffusion-convection-reaction (DCR) models [16], [19]. In an attempt to formulate models with the desirable accuracy exhibited by the DCR models (which are computationally expensive and unwieldy to use, especially on small computers) but which approach the simplicity of the PF models (which in these applications prove too inaccurate in their representation of qualitative phenomena to be of practical use), J. P. Kernevez and his colleagues have proposed lumped parameter models with delays. In these models there are several delays representing convective transport and a number of diffusive transport mechanisms. One version of such models, which are nonlinear due to certain reaction velocity terms, is discussed in some detail by P. Daniel in [12] where additional references may also be found. To investigate the accuracy and potential usefulness of these models, efficient methods for parameter estimation (unknown parameters include several delays as well as kinetic constants) and control techniques for nonlinear delay systems are essential.

2.3. Gas and Oil Exploration and Recovery

a) Reservoir Engineering Problems

The importance of inverse or parameter estimation problems in the gas and oil industry is rather well documented. One class of problems [8], [15], [26] involves use of the flow equations in a porous medium (a reservoir or oil/gas field) to determine the field porosity ϕ (the ratio of pore volume to total volume) and field permeability function k. A greatly simplified model would be based on an equation (derived from conservation of mass and Darcy's law - - see [11], [20]) for the pressure $p = p(t,x,y)$ in a vertically homogeneous field of depth h, say

$$\phi ch \frac{\partial p}{\partial t} = \frac{\partial}{\partial x} \left(\frac{hk}{\mu} \frac{\partial p}{\partial x} \right) + f$$

where μ = fluid viscosity, c = fluid compressibility, and f is a general sink/source term. The field usually contains a number of wells (for production or observation or both) and a typical problem is to estimate ϕ and k or, alternatively, the total pore volume $\Phi \equiv \iint \phi h$, from observations of p at the well heads.

More realistic models involve several miscible fluids and a coupled set of partial differential equations [14]; the fundamental inverse problem is similar, but, of course, much more complicated.

b) Seismic Exploration

A second class of inverse problems concerns determination of the elastic properties of an inhomogeneous medium via surface observations after perturbing "shocks" have produced waves in the medium. Models usually involve the equations of elasticity [2], [17]; for example, in the 1-dimensional problem one might consider

$$\rho(z) \frac{\partial^2 u}{\partial t^2} = \frac{\partial}{\partial z} \left(E(z) \frac{\partial u}{\partial z} \right)$$

where ρ is the mass density and $E = \lambda + 2\mu$ for compressional or P-waves, $E = \mu$ for shear or S-waves with λ, μ the Lamé parameters. The boundary conditions at $z = 0$ (here z is the vertical distance from the surface) include excitation or perturbation of the medium (often this source input itself is a quantity to be "identified".) From observations at the surface z = 0 (these observations usually involve the unknown source input and a velocity term $\frac{\partial u}{\partial t}$ for particle displacement), one wishes to determine the unknown functions ρ and E and, in addition, the source term if it is unknown.

2.4. Large Space Structures

Another class of control and identification problems for which the models are based on the equations for elastic structures are those dealing with large space antennas. One such antenna that is currently being developed by NASA is the Maypole Hoop/Column antenna which is depicted in Figure 2.

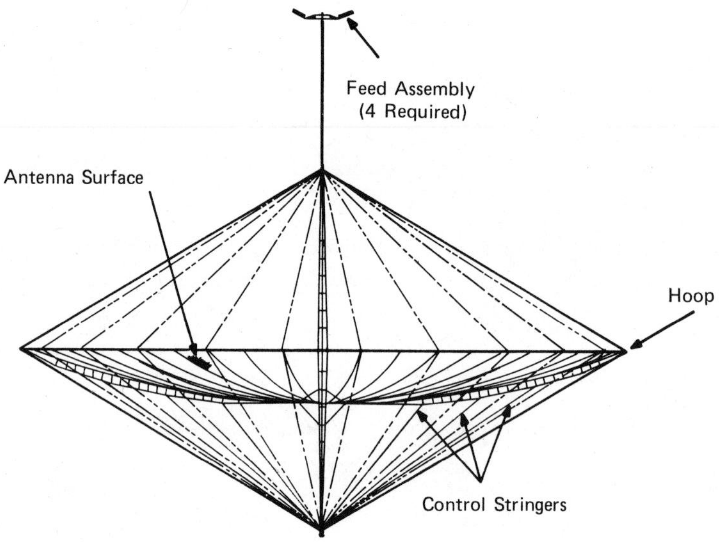

Figure 2

This antenna, which when fully deployed somewhat resembles an inverted umbrella (100 m. in diameter), consists of a membranous surface of gold-plated molybdenum reflective mesh, a collapsible hoop or ring on which the surface is stretched, and a telescoping column to which the antenna surface is anchored and on which feed assemblies are mounted. The antenna in collapsed configuration (similar to the popular "travel umbrellas" that collapse to fit into a briefcase or small suitcase) is to be transported into space in the space shuttle; it is then deployed for use as a communication antenna. The antenna surface itself is flexible and its shape (and hence focusing properties) can be changed via control stringers attached to 48 equally spaced radial teflon coated graphite "cords" (4 control stringers per radial cord or "gore" edge). In addition to the dynamic identification and control problems associated with initial deployment of the Maypole Hoop/Column, it is anticipated that after long periods of operation, the reflector surface will (due to changes in elastic properties and forces) require adjustment. Thus a static problem of interest consists of the following: Determine from observations (through sensing devices placed on the gore edges on the surface) the present configuration of the antenna surface and then effect the desired configuration or "displacement" through adjustment of the control stringers.

A typical problem then might involve a partial differential equation for the displacement of a circular membrane or thin plate, say $\mathcal{P}(D,q)u = f$, where D represents spatial differential operators, q represents elastic parameters to be estimated, and f entails applied forces. A simple example might be

$$\frac{\partial}{\partial r}\left(r\mathcal{E}\,\frac{\partial u}{\partial r}\right) + \frac{\partial}{\partial \theta}\left(\frac{\mathcal{E}}{r}\,\frac{\partial u}{\partial \theta}\right) = f$$

where $\mathcal{E} = \mathcal{E}(r,\theta)$.

2.5. Dispersion Models in Ecology

An important problem to population ecologists [21], [22] concerns the movement of insects (or, more generally, herbivores) through vegetation patches. Outbreaks or cyclical population explosions of some insects are observed and it is believed that the nature of the transport mechanisms for the insects affect the occurrence (or lack thereof) of outbreaks and their magnitude and periodicity. Typical equations to describe movement of the insects involve both diffusive and advective (convective) terms, e.g., for 1-dimensional models

$$\frac{\partial N}{\partial t} + \frac{\partial}{\partial x}(VN) = \frac{\partial}{\partial x}\left(D\,\frac{\partial N}{\partial x}\right) + f(N)$$

in addition to the usual sink/source terms f. Depending on the species involved, it is generally expected that D and/or V can depend on N, the population level, and/or x, the spatial variable. The diffusion coefficient D and the convective velocity V may alternatively, or, in addition, depend on temperature or time (e.g., as in seasonal migration of pests.)

There are numerous estimation and control problems of importance in the context of ecological investigations. Typically, one wishes to determine the coefficient functions D and V from observations of N; once this is done, one might wish to estimate the optimal vegetation density in a patch in order to hold population levels in the patch to a minimum, or at least below some given level.

2.6. Transport Models in Physiology

In physiology a great deal of research is devoted to questions concerning transport mechanisms such as simple (passive) diffusion, bulk flow or convective transport, facilitated diffusion, and active transport. An example is the effort [9], [10], [23], [24] devoted in recent years to the controversy involving bulk flow vs. molecular diffusion of brain interstitial fluid in gray and white matter. The mathematical models again are based on the convection-diffusion equation

$$\frac{\partial u}{\partial t} + V \frac{\partial u}{\partial x} = D \frac{\partial^2 u}{\partial x^2}$$

for the concentration u of a labeled substance such as sucrose in brain tissue. From experimental data (for u at various times and locations in the tissue samples) one seeks to estimate values for V and D in gray and white matter and contrast the transport properties of each type of tissue. We shall, in a subsequent section of this presentation, discuss in some detail an application to these transport problems of some of the methods that are the focus of attention in this lecture.

3. THEORETICAL FOUNDATIONS

We turn now to a discussion of the theoretical techniques that one can employ to establish convergence results for certain approximation schemes for nonlinear systems such as (1), (2) or (3). For the sake of brevity, we shall restrict our considerations to parameter estimation problems. A discussion of the use of the ideas presented here in control problems can be found in [12] in the case of nonlinear delay systems while the case of distributed parameter control problems is considered briefly in [6].

For the purposes of illustration, we shall use a least squares formulation (for a discussion of maximum likelihood estimator ideas, see [3] and the references therein) of the parameter estimation problem. In particular, one seeks to minimize

$$J(q) = \frac{1}{2}\sum_{i=1}^{m} |y(t_i;q) - \xi_i|^2$$

over a given set Q of admissible parameters. Here ξ_i is an observation for the output $t \to y(t;q)$ at t_i with $y(t) = \mathcal{C}x(t)$ in the case of (1), $y(t) = \mathrm{col}(\mathcal{C}u(t,x_1),\ldots,\mathcal{C}u(t,x_p), \mathcal{D}u_t(t,x_1),\ldots,\mathcal{D}u_t(t,x_p))$ in the case of (2), and $y(t) = \mathrm{col}(\mathcal{C}u(t,x_1),\ldots,\mathcal{C}u(t,x_p))$ in the case of (3), where $\mathcal{C}$ and $\mathcal{D}$ are matrix operators of appropriate dimension and rank.

Our approach entails rewriting (1), (2), or (3) as an abstract equation

$$\begin{aligned} \dot{z}(t) &= \mathcal{A}(q)z(t) + G(t) \\ z(0) &= z_0 \end{aligned} \tag{4}$$

in an appropriately chosen Hilbert space Z. The operator $\mathcal{A}$ may be linear or nonlinear and depends on the unknown parameters q. We reformulate the estimation problems as ones of minimizing

$$J(q) = \frac{1}{2}\sum_{i=1}^{m} |\Gamma(z(t_i;q)) - \xi_i|^2$$

where $y(t;q) = \Gamma(z(t;q))$ is an appropriately defined output.

We take a classical Ritz-Galerkin type approach to reducing these infinite-dimensional state space problems to a sequence of approximating finite dimensional state space problems that are readily solved numerically. For a given sequence Z^N of "subspaces" of Z with "projections" $P^N: Z \to Z^N$, we minimize

$$J^N(q) = \frac{1}{2}\sum_{i=1}^{m} |\Gamma(z^N(t_i;q)) - \xi_i|^2$$

over Q where z^N is the solution of an approximating system

$$\begin{aligned} \dot{z}^N(t) &= A^N(q)z^N(t) + P^N G(t) \\ z^N(0) &= P^N z_0. \end{aligned} \tag{5}$$

In the methods discussed here, we always take $A^N(q) = P^N A(q) P^N$ and obtain a state convergence $z^N(t;q) \to z(t;q)$. The ultimate goal, of course, is to insure convergence of some sequence $\{\bar{q}^N\}$ of solutions of approximating estimation problems involving (5) to a solution $\bar{q}$ of the problems involving (4). This objective can be attained in the cases of the "modal" and "spline" schemes we have developed and tested numerically in [4], [5], [6], [7], and [12].

To date we have employed two different theories to establish state and parameter convergence. For distributed parameter systems, both modal [6] and spline [7] schemes have been investigated using an abstract semigroup formulation and Trotter - Kato type theorems. Briefly, one establishes that the linear operators A and A^N (we suppress the q dependence here) satisfy a uniform (in q and N) dissipativeness condition and generate C_0-semigroups $T(t)$ and $T^N(t)$ respectively. Then treating the nonlinearities $G(\sigma) = F(q,\sigma,z(\sigma))$ (F is defined in an appropriate manner using f from (2) or (3)) as perturbations, one considers in place of (4) and (5) the implicit equations

$$z(t) = T(t)z_0 + \int_0^t T(t-\sigma)F(q,\sigma,z(\sigma))d\sigma \tag{6}$$

and

$$z^N(t) = T^N(t)P^N z_0 + \int_0^t T^N(t-\sigma)P^N F(q,\sigma,z^N(\sigma))d\sigma. \tag{7}$$

The basic tool then is a Trotter - Kato type result which, under the conditions:

$|T^N(t)| \leq Me^{\omega t}$ for some M and ω independent of N; (8i)

there exists a set $\mathcal{D} \subset \mathrm{Dom}(A)$, $\mathcal{D}$ dense in Z, such that $(\lambda_0 - A)\mathcal{D}$ is dense in Z for some $\lambda_0 > 0$; (8ii)

$|A^N z - Az| \to 0$ for $z \in \mathcal{D}$; (8iii)

guarantees the convergence

$T^N(t)z \to T(t)z$ for $z \in Z$, uniformly in t. (8iv)

The convergence in (8iv), along with (6), (7) and $P^N \to I$ strongly, can be used to argue state convergence $z^N(t) \to z(t)$. This in turn can be used to establish a desired parameter convergence. (For the rather technical details - - which are nontrivial when the full dependence of the operators, projections, semigroups and, in some cases, the subspaces, on the unknown parameters q is taken into account - - one should consult [6] and [7].)

A somewhat different approach to spline methods for delay systems (1) has been taken in [4], [5], [12] where the nonlinearity f is treated directly as part of a nonlinear operator $A = A(t)$ (which is now possibly time dependent). In this case one uses the implicit equations

$$z(t) = z_0 + \int_0^t \{A(\sigma)z(\sigma) + G(\sigma)\}d\sigma$$

$$z^N(t) = P^N z_0 + \int_0^t \{A^N(\sigma)z^N(\sigma) + P^N G(\sigma)\}d\sigma$$

in place of (4) and (5). Under reasonable conditions on f, one can establish dissipative type inequalities

$$\langle\langle A^N(\sigma)z - A^N(\sigma)w, z - w\rangle\rangle \leq \omega(\sigma)\langle\langle z - w, z - w\rangle\rangle,$$

where $\langle\langle , \rangle\rangle$ is a specially defined inner product on Z. With some elementary analysis and use of a Gronwall inequality, one then obtains estimates for

$|z^N(t) - z(t)|$ in terms of integrals of $\{A^N(\sigma) - A(\sigma)\}z(\sigma)$. Desired convergence results then follow from convergence properties of A^N. Again the technical details become quite involved when one treats general nonlinear delay systems with multiple unknown delays. These can be found in [5], [12].

We remark that one need not have Z^N a subspace of Z in carrying out the above theories. Indeed in both cases (delay systems with unknown delays, distributed parameter systems with unknown coefficients) outlined above, one finds that the appropriate P^N, Z^N, and Z all depend on the unknown parameters q (through the domain of the function space in the case of unknown delays and through the inner product for Z and Z^N in the case of some distributed parameter examples as well as in the unknown delays problems) which of course vary as one iterates through the sequence of approximating problems (i.e., on N.) This feature results in interesting difficulties from both a conceptual and computational viewpoint.

4. AN APPLICATION TO TRANSPORT IN BRAIN TISSUE

We return to the example 2.6 of Section 2 involving the transport of labeled sucrose in gray and white matter. A complete description of the experimental procedures and the questions being investigated can be found in [24]. Briefly, cats are anesthetized and experiments of either 1, 2, or 4 hours duration are carried out. Labeled sucrose is perfused into the lateral ventricle. At the end of the perfusion period, the animals are sacrificed and their brains are removed and frozen. Well-stained areas of gray and white matter perpendicular to the ventricular surface (along the x-axis in our notation below) are sectioned and analyzed. This yields data corresponding to a fixed time t_i for a maximum of 4 spatial locations x_j, $j=1,\ldots,4$, in gray matter and 8 spatial locations x_j in white matter. From this data $\{\hat{u}(t_i,x_j)\}$ for the concentration u, one wishes to compare transport in gray matter with that in white matter. The primary questions pertain to transport via molecular diffusion alone vs. transport via diffusion and bulk (convective) flow. In particular, the mathematical problems reduce to those of estimating D, V, and C_0 in

$$\frac{\partial u}{\partial t} + V\frac{\partial u}{\partial x} = D\frac{\partial^2 u}{\partial x^2}$$

$$u(t,0) = C_0.$$

In the early experimental work, data for only one t_i (1, 2, or 4 hours) and for anywhere from 4 to 8 spatial locations x_j were available. A substantial concern is whether one can develop accurate methods for estimation of the three parameters in question from such limited data.

We have successfully applied the modal methods of Example 4.4 of [6] to investigate these problems. We first summarize briefly the pertinent ideas behind the methods. For the purpose of illustration, consider the example

$$u_t = q_1 u_{xx} + q_2 u_x$$

$$u(t,0) = u(t,1) = 0$$

$$u(0,x) = \phi(x),$$

which can be reformulated in the form (4) in $Z = L_2(0,1)$ by choosing $\mathcal{A}(q) = q_1 D^2 + q_2 D$ (here D is the differential operator in $L_2(0,1)$) with $\mathrm{Dom}(\mathcal{A}(q)) = H^2 \cap H_0^1$. It can be argued that $\mathcal{A}(q)$ is uniformly maximal dissipative (although it is not self-adjoint). For the development of modal approximation schemes, general spectral results given in [13] can be employed. It can be seen that $\mathcal{A}(q)$ is a relatively bounded perturbation of a discrete spectral operator and is itself a discrete spectral operator with a resolution of identity and associated eigenmanifolds and projections. The eigenvalues are found to be $\lambda_j(q) = -j^2\pi^2 q_1 - q_2^2/2q_1$ with associated eigenfunctions $\psi_j(q) = \exp\{-q_2 x/2q_1\}\sin j\pi x$. The natural modes or eigenfunctions form a complete (but not orthogonal) set in $Z = L_2(0,1)$. However, a choice of $\tilde{Z}^N = \mathrm{span}\{\psi_1(q),\ldots,\psi_N(q)\}$, while desirable theoretically, is not useful in parameter estimation algorithms since the basis elements are then dependent upon the unknown parameters (and thus change with each new estimate of the q's). One can use instead the near-modal functions $\Phi_j(x) = \sqrt{2}\sin j\pi x$ and take $Z^N = \mathrm{span}\{\Phi_1,\ldots,\Phi_N\}$ with, of course, $\mathcal{A}^N = P^N \mathcal{A} P^N$ where P^N is the canonical projection of Z onto Z^N.

Convergence can be argued using the Trotter - Kato formulation of (8i) - (8iv) above. The stability condition (8i) follows immediately from the uniform dissipativeness. Choosing $\mathcal{D} = \bigcup_{N=1}^{\infty} \tilde{Z}^N(\bar{q})$, where $\bar{q}$ is a limit of the sequence of estimates $\bar{q}^N$, the spectral results yield (8ii) trivially while one must work somewhat more to establish (8iii).

With regard to implementation, the scheme offers some nice computational features since the matrix realizations of the operators $A^N(q)$ are given by

$$[A^N(q)]_{ij} = \begin{cases} -q_1 i^2 \pi^2 & i = j \\ 0 & i \neq j, \quad i + j \quad \text{even} \\ 2jq_2 [\dfrac{2i}{i^2 - j^2}] & i \neq j, \quad i + j \quad \text{odd.} \end{cases}$$

Turning to our investigation of these methods for possible use in the brain transport questions, we first tested the methods with an example for which the solution was "known." (J. Crowley and M. Garrett carried out the computations for this problem. J. Saltzman supplied a "known" solution technique involving an infinite series which was used to generate "data" corresponding to fixed parameter values in the equation. This technique is totally unrelated to the methods we were testing.) The example used was

$$u_t - q_2 u_x = q_1 u_{xx}$$

$$u(t,0) = q_3, \quad u(t,1) = 0$$

$$u(0,x) = \phi(x)$$

where $\phi(x) = a_0 x^2 + a_1 x + a_2$ is a quadratic satisfying $\phi(0) = 1$, $\phi(1) = 0$, and max $\phi = \phi(1/4)$. "Data" were generated corresponding to true values $q_1^* = .3$, $q_2^* = 1.75$, and $q_3^* = 1.0$. A number of numerical trials with the above described "modal" scheme were conducted in which the inverse problem for varying amounts of "data" was "solved". We summarize briefly some of our findings. In the examples presented here, the notation I = k, J = p in example indicates that the data set for this test consisted of values $u(t_i, x_j)$, i=1,...,k, j=1,...,p (i.e., k × p "observations" were employed in the inverse problem.)

EXAMPLE 1. It was assumed that q_3 was known and an attempt to fit the data was made by searching for q_1 and q_2. Initial guesses (for each value of N tried) were $q_1^{N,0} = 1.0$, $q_2^{N,0} = 0.0$. The "converged" values (corresponding to N = 8 or 16 in this and all examples presented here) were:

$$I = 1,\ J = 3: \quad \bar{q}_1 = 1.82 \quad \bar{q}_2 = .83$$

$$I = 2,\ J = 3: \quad \bar{q}_1 = .2979 \quad \bar{q}_2 = 1.7557.$$

EXAMPLE 2. This example was exactly the same as Example 1 except initial guesses $q_1^{N,0} = .75$, $q_2^{N,0} = 1.0$ (somewhat closer to the "true" values than those used in Example 1) were used. For I = 1, J = 3, the results were $\bar{q}_1 = .3009$, $\bar{q}_2 = 1.7529$, quite acceptable in this case.

EXAMPLE 3. We investigated the effect of using increasingly more spatial points in our data grid (i.e., I = 1 with J = 4,5,6). For initial guesses $q_1^{N,0} = .8$, $q_2^{N,0} = .9$, increasingly better estimates were obtained as the number of spatial grid points increased. We obtained:

$$I = 1,\ J = 3: \quad \bar{q}_1 = .6115 \quad \bar{q}_2 = 1.4903$$

$$I = 1,\ J = 4: \quad \bar{q}_1 = .3018 \quad \bar{q}_2 = 1.7468$$

$$I = 1,\ J = 5: \quad \bar{q}_1 = .2978 \quad \bar{q}_2 = 1.7492$$

$$I = 1,\ J = 6: \quad \bar{q}_1 = .2984 \quad \bar{q}_2 = 1.7493.$$

It is clear that 4 points in the spatial grid yields an adequate amount of data for this example.

EXAMPLE 4. In this case we wished to estimate q_1, q_2 and the boundary concentration q_3. Initial guesses were $q_1^{N,0} = .8$, $q_2^{N,0} = .9$, $q_3^{N,0} = .5$. The converged values were:

$$I = 1,\ J = 6: \quad \bar{q}_1 = .2990 \quad \bar{q}_2 = 1.6983 \quad \bar{q}_3 = 1.0356$$

$$I = 2,\ J = 6: \quad \bar{q}_1 = .2997 \quad \bar{q}_2 = 1.7469 \quad \bar{q}_3 = 1.0012.$$

EXAMPLE 5. As a final test, we modified the initial function ϕ used in the above examples to $\phi(0) = 1$, $\phi(x) = 0$ for $x \neq 0$. This represents the type of problem (one with a discontinuity in the boundary-initial data) that one encounters when using the actual data collected in the experiments with cats described above. Again the results obtained were encouraging. With $I = 2$, $J = 6$ and initial guesses $q_1^{N,0} = .8$, $q_2^{N,0} = .9$, converged values of $\bar{q}_1 = .3019$, $\bar{q}_2 = 1.7635$ were found with a residual sum of squares of 4.3×10^{-3}.

In summary, the numerical tests reveal that it is probably unreasonable to expect to solve with the "modal" methods the inverse problems for $I = 1$, $J = 3$ in most cases. However, problems with $I = 2$, $J = 3$ correspond to a reasonable number of spatial observations for the method in some cases. (By changing labels during the perfusion period, Rosenberg, Kyner, and colleagues are now collecting data with two time grid points.) For data from white matter (where $J = 6$ is feasible), the methods should prove useful in estimating D, V, and C_0 in the transport models.

We have, in fact, used the methods with actual data sets ($I = 1$, $J = 8$) for white matter supplied by Kyner and his associates. The "modal" methods appear to consistently perform in an acceptable manner. Typical values obtained in solving the inverse problems are $D = 2.7 \times 10^{-6}$ cm^2/sec., $V = -5.99$ μm/min., $C_0 = 128.5$, values that are consistent with expectations based on values obtained by Kyner and associates using other techniques.

We anticipate that extensions of these methods (or perhaps the spline methods developed in [7]) will prove useful in future investigations of the channeling structure in white matter (in these problems the velocity coefficient V will be spatially dependent as will also, in some cases, the coefficient of diffusion D).

5. ESTIMATION OF FUNCTION SPACE PARAMETERS

The theory developed in [6] and [7] deals with estimation of parameters in Euclidean space sets. The framework is, however, general enough to allow one to treat problems in which unknown function space parameters must be estimated. In this section of our presentation we shall give a brief sketch of how one further develops such a theory. At the same time we shall illustrate some of the ideas fundamental to spline methods as opposed to the "modal" methods discussed earlier.

In order to demonstrate the ideas we shall consider an equation of the porous media type (see Section 2.3.a); that is, we consider

$$q_1(x)\,\frac{\partial u}{\partial t} = \frac{\partial}{\partial x}\,(q_2(x)\,\frac{\partial u}{\partial x}\,) + f \tag{9}$$

with homogeneous boundary conditions $u(t,0) = u(t,1) = 0$. In relating this to the porous media application (then u = pressure), one might consider large fields for which the boundary terms are either constant or slowly varying in time. In either case, such nonhomogeneous boundary problems can be transformed to problems with homogeneous boundary conditions in a quite standard manner. With certain smoothness assumptions on q_1, q_2, the operator in equation (9) can be viewed as a standard Sturm - Liouville operator (i.e., identify $q_1 \sim k$, $q_2 \sim p$ in the usual notation for the coefficient functions - - see (3) above and p. 40-42 of [6]). For our discussions here we shall assume that $q = (q_1,q_2)$ is to be chosen from a parameter space $Q \subset L_2(0,1) \times L_2(0,1)$ satisfying $Q \subset \{(q_1,q_2) \in H^2 \times H^3 | q_2 > 0,\ 0 < m \le q_1 \le M\}$. (The smoothness hypothesized will guarantee certain smoothness properties for the eigenfunctions to be discussed momentarily.)

We rewrite (9) as an equation

$$\dot{z}(t) = A(q)z(t) + F(q,f) \tag{10}$$

in the state space $Z = X(q) = L_2(0,1)$ where we take as inner product $\langle\phi,\psi\rangle_q = \int_0^1 \phi\psi q_1$. (Here the spaces *do* depend on the unknown parameters, a complicating possibility we mentioned earlier.) The operators in (10) are given by $F = \frac{1}{q_1} f$ and $A(q)\psi = \frac{1}{q_1} D(q_2 D\psi)$, where $\mathrm{Dom}(A(q)) = H^2 \cap H_0^1$.

Simple integration by parts yields $\langle A(q)z,z\rangle_q \le 0$ so that $A(q)$ is uniformly dissipative in $X(q)$. In fact $A(q)$ is maximal dissipative and generates a C_0-semigroup, and we are thus in a position to consider (6), (7) and the Trotter - Kato approach to approximation schemes.

To describe the spline methods we need to recall the definition of some standard cubic spline basis elements. For any positive integer N, we let $t_j^N = j/N$, $j = -3,\ldots,N+3$, and let $\tilde{B}_j^N$, $j = -1,\ldots,N+1$, be the cubic spline that vanishes outside (t_{j-2}^N, t_{j+2}^N), has value 4 and slope 0 at t_j^N, value 1 and slope 3N at t_{j-1}^N, and value 1 and slope -3N at t_{j+1}^N. (See [25], p. 73 -- and note that our elements here differ from those of Schultz only by a multiplicative factor of 24.)

For our modified basis elements B_j^N we take the restriction to [0,1] of the following:

$$B_0^N = \tilde{B}_0^N - 4\tilde{B}_{-1}^N$$

$$B_1^N = \tilde{B}_0^N - 4\tilde{B}_1^N$$

$$B_j^N = \tilde{B}_j^N, \quad j = 2,\ldots,N-2$$

$$B_{N-1}^N = \tilde{B}_N^N - 4\tilde{B}_{N-1}^N$$

$$B_N^N = \tilde{B}_N^N - 4\tilde{B}_{N+1}^N.$$

We note that these elements are in Dom($A(q)$).

We define our approximation subspaces $X^N(q) \subset X(q)$ by $X^N(q) = \text{span}\{B_0^N, \ldots, B_N^N\}$ and let $P^N(q)$ be the canonical projection of $X(q)$ onto $X^N(q)$, i.e., $P^N(q)\psi = \sum_{j=0}^{N} \langle\psi, B_j^N\rangle_q B_j^N$. Finally, as usual, we take $A^N = A^N(q) = P^N(q)A(q)P^N(q)$.

Under an assumption that Q is compact in $H^0 \times H^0$, one can argue in this case that solutions to the estimation problems for (5) (or (7)) do exist. We in fact assume that Q is compact in the $C \times H^1$ topology so that we henceforth assume without loss of generality (possibly by taking a subsequence) that we have a sequence $\{q^N\}$ of solutions to the estimation problems satisfying $q^N \to \bar{q}$ in $C \times H^1$ with $\bar{q} \in Q$.

We briefly indicate the steps to verify (8i) - (8iii) to ensure convergence of the semigroups generated by $A^N(q^N)$ to the semigroup generated by $A(\bar{q})$. (As we have noted before, this is the fundamental convergence result needed for both state and parameter convergence.) The stability requirement (8i) follows from

$$\langle A^N(q^N)z, z\rangle_{q^N} = \langle A(q^N)P^N(q^N)z, P^N(q^N)z\rangle_{q^N} \le 0,$$

the inequality being a result of the uniform dissipativeness of $A(q^N)$.

The operator $A(\bar{q})$ has, by the usual spectral results, a complete orthonormal set of eigenfunctions $\{\Psi_j(\bar{q})\}$. In (8ii), we take

$$\mathcal{D} = \bigcup_{N=1}^{\infty} \text{span}\{\Psi_1(\bar{q}),\ldots,\Psi_N(\bar{q})\}.$$

It is then easily seen that the conditions of (8ii) follow from the completeness of the Ψ_j and the relationship $(\lambda_0 - \mathcal{A}(\bar{q}))\Psi_j(\bar{q}) = (\lambda_0 - \lambda_j)\Psi_j(\bar{q})$.

We finally consider (8iii) and note that the spaces $X(q)$, $q \in Q$, are all equivalent (recall $0 < m \le q_1 \le M$), a fact which plays a fundamental role in the basic theory developed in [6]. Indeed we may, in considering any convergence results, equivalently use the L_2 topology. Thus, to establish (8iii), it suffices to argue

$$A^N(q^N)\Psi_j \to \mathcal{A}(\bar{q})\Psi_j \tag{11}$$

in L_2. From the smoothness assumptions on Q (and hence $\bar{q}$), it is easily seen that $\Psi_j \in H^4$. Since $\Psi_j \in \text{Dom}(\mathcal{A}(\bar{q}))$, we also have $\Psi_j \in H_0^1$. Hence, it suffices to fix $\psi = \Psi_j(\bar{q})$ in $H^4 \cap H_0^1$ and argue that (11) holds whenever $q^N \to \bar{q}$ in $C \times H^1$.

Estimates similar to those we need can be found in Theorem 6.13, p. 82 of [25]. However we cannot use those estimates directly since our projections $P^N(q^N)$ (onto $X^N(q^N)$) are *not* the same as the standard projections (of Theorem 6.13) of L_2 onto $S(N) = \text{span}\{\tilde{B}_{-1}^N, \tilde{B}_0^N, \ldots, \tilde{B}_{N+1}^N\}$. But, using fundamental ideas similar to those found in [25] (e.g., the Schmidt inequality and estimates for the appropriate interpolating splines), one can establish:

For $\psi \in H^4 \cap H_0^1$,

$$|\psi - P^N\psi| \le \frac{K_1}{N^4} |D^4\psi|$$

$$|D(\psi - P^N\psi)| \le \frac{K_2}{N^3} |D^4\psi|$$

$$|D^2(\psi - P^N\psi)| \le \frac{K_3}{N^2} |D^4\psi|$$

where the norms are the usual L_2 norm, $P^N = P^N(q^N)$ as defined earlier, and the constants K_1, K_2, K_3 are independent of N and ψ.

We thus have $D^2\psi^N \to D^2\psi$ in L_2, $D\psi^N \to D\psi$ in C, where $\psi^N = P^N\psi$. Furthermore, we know that $q_1^N \to \bar{q}_1$ in C, $q_2^N \to \bar{q}_2$ in H^1. Since $P^N \to I$ it thus follows from elementary arguments that

$$A^N(q^N)\psi = P^N\{(1/q_1^N)Dq_2^N D\psi^N + (q_2^N/q_1^N)D^2\psi^N\}$$

converges in L_2 to

$$A(\bar{q})\psi = (1/\bar{q}_1)D\bar{q}_2 D\psi + (\bar{q}_2/\bar{q}_1)D^2\psi.$$

The assumptions on Q made above are not unreasonable for many applications. We have successfully used these spline methods in computational packages for function space parameter estimation in models for insect dispersion (see Section 2.5). In those applications, the smoothness and compactness assumptions listed above are satisfied when one formulates the problems and parameterizes Q in a way that is natural and consistent with the experimental and theoretical efforts of population ecologists.

ACKNOWLEDGEMENTS

Parts of the research discussed here were carried out while the author was a visitor at the Institute for Computer Applications in Science and Engineering, NASA Langley Research Center, Hampton, VA, which is operated under NASA contracts No. NAS1-15810 and No. NAS1-16394. Additional research and the manuscript were completed while the author was a visitor in the Mathematics Department, University of Utah.

Work reported herein was supported in part by the Air Force Office of Scientific Research under contract AFOSR 76-3092D, in part by the National Science Foundation under grant NSF-MCS 7905774-02, and in part by the U.S. Army Research Office under contract ARO-DAAG29-79-C-0161.

REFERENCES

1. E. A. Armstrong and J. S. Tripp, An application of multivariable design techniques to the control of the National Transonic Facility, NASA Tech. Paper 1887, NASA-LRC, August, 1981.

2. A. Bamberger, G. Chavent and P. Lailly, About the stability of the inverse problem in 1-D wave equations -- application to the interpretation of seismic profiles, *Appl. Math. Optim.* 5 (1979), 1-47.

3. H. T. Banks, Parameter identification techniques for physiological control systems, in *Mathematical Aspects of Physiology* (F. Hoppensteadt, ed.) Vol. 19, Lect. in Applied Math, Amer. Math. Soc., 1981, pp. 361-383.

4. H. T. Banks, Identification of nonlinear delay systems using spline methods, in *Proc. Intl. Conf. on Nonlinear Phenomena in Math. Sciences* (V. Lakshmikantham, ed.), Academic Press, 1981.

5. H. T. Banks and P. L. Daniel, Estimation of delays and other parameters in nonlinear functional differential equations, to appear.

6. H. T. Banks and K. Kunisch, An approximation theory for nonlinear partial differential equations with applications to identification and control, LCDS Tech. Rep. 81-7, Brown University, April, 1981.

7. H. T. Banks, J. M. Crowley, and K. Kunisch, Cubic spline approximation techniques for parameter estimation in distributed systems, to appear.

8. G. Chavent, M. Dupuy and P. Lemonnier, History matching by use of optimal control theory, *Soc. Petroleum Eng. J.* (1975), 76-86.

9. H. F. Cserr, D. N. Cooper and T. H. Milhorat, Flow of cerebral interstitial fluid as indicated by the removal of extracellular markers from rat caudate nucleus, *Exp. Eye Res. Suppl.* 25 (1977), 461-473.

10. H. F. Cserr, D. N. Cooper, P. K. Suri, and C. S. Patlak, Efflux of radiolabeled polyethylene glycols and albumin from rat brain, *Am. J. Physiol.* 240 (*Renal Fluid Electrolyte Physiol.* 9), 1981, F319-F328.

11. L. P. Dake, *Fundamentals of Reservoir Engineering*, Elsevier Scientific Publ. Co., New York, 1978.

12. P. L. Daniel, Spline-based approximation methods for identification and control of nonlinear functional differential equations, Ph.D. Thesis, Brown University, 1981.

13. N. Dunford and J. T. Schwartz, *Linear Operators: Part III, Spectral Operators*, Wiley-Interscience, New York, 1971.

14. R. E. Ewing and M. F. Wheeler, Galerkin methods for miscible displacement problems in porous media, *SIAM J. Num. Anal.* 17 (1980), 351-365.

15. G. R. Gavalas and J. H. Seinfeld, Reservoirs with spatially varying properties: estimation of volume from late transient pressure data, *Soc. Petroleum Eng. J.* (1973), 335-342.

16. G. Gellf and J. Henry, Experimental and theoretical study of diffusion, convection and reaction phenomena for immobilized enzyme systems, in *Analysis and Control of Immobilized Enzyme Systems*, (D. Thomas and J. P. Kernevez, eds.) North-Holland/American Elsevier, New York, 1976, pp. 253-274.

17. F. Grant and G. West, *Interpretation Theory in Applied Geophysics*, McGraw-Hill, 1965.

18. G. Gumas, The dynamic modeling of a slotted test section, NASA CR-159069, 1979, Penn State University.

19. J. Henry, Contrôle d'un réacteur enzymatique à l'aide de modèles à paramètres distribués: quelques problèmes de contrôlabilité de systemes paraboliques, Thèses d'État, Université Paris VI, 1978.

20. M. K. Hubbert, Darcy's law and the field equation of the flow of underground fluids, *Petroleum Transactions, AIME*, 207 (1956), 222-239.

21. Peter Kareiva, The application of diffusion to herbivore models, Ecological Monograph, to appear.

22. A. Okubo, *Diffusion and Ecological Problems: Mathematical Models*, Biomathematics, Vol. 10, Springer, New York, 1980.

23. G. A. Rosenberg and W. T. Kyner, Gray and white matter brain-blood transfer constants by steady-state tissue clearance in cat, *Brain Res.* 193 (1980), 59-66.

24. G. A. Rosenberg, W. T. Kyner, and E. Estrada, Bulk flow of brain interstitial fluid under normal and hyperosmolar conditions, *Am. J. Physiol.* 238 (*Renal Fluid Electrolyte Physiol.* 7), 1980, F42-F49.

25. M. H. Schultz, *Spline Analysis*, Prentice-Hall, Englewood Cliffs, 1973.

26. P. C. Shah, G. R. Gavalas and J. H. Seinfeld, Error analysis in history matching: the optimum level of parametrization, *Soc. Petroleum Eng. J.*, (1978), 219-228.

A MODEL RIEMANN PROBLEM FOR VOLTERRA EQUATIONS

J. M. Greenberg

Division of Mathematical and Computer Sciences
National Science Foundation
Washington, D.C.
and
Department of Mathematics
SUNY - Buffalo
Buffalo, New York

Ling Hsiao

Division of Applied Mathematics
Brown University
Providence, Rhode Island
and
Department of Mathematics
Academia - Sinica of China
Taipei, Taiwan

R. C. MacCamy

Department of Mathematics
Carnegie - Mellon University
Pittsburgh, Pennsylvania

1. INTRODUCTION

We seek functions $u(x,t) = \sigma(x,t)$, x and t in $\mathbb{R}$, which satisfy the conditions:

$$\frac{d}{dt}\int_{x_1}^{x_2} u(x,t)dx = -(\sigma(x_2,t) - \sigma(x_1,t)), \quad t > 0 \tag{B}$$

$$\sigma(x,t) = a(0)f(u(x,t)) + \int_0^\infty \dot{a}(\tau)f(u(x,t-\tau))d\tau \tag{C}$$

$$u(x,t) = \phi(x,-t) \quad \text{for} \quad t < 0, \quad u(x,0^+) = u_0(x). \tag{I}$$

The above problem is to serve as a model for a class of hereditary evolution equations in mechanics as described below. In this analogy (B) represents a *balance law*, (C) a *constitutive assumption* and (I) the *initial history*. When $a(t) \equiv a(0)$ we say that there is *no memory*.

There are two situations in continuum mechanics which are being modeled. These are:

$$\begin{aligned} &\frac{d}{dt}\int_{x_1}^{x_2} u_t(x,t)dx = -(\sigma(x_2,t) - \sigma(x_1,t)) \\ &\sigma(x,t) = -a(0)f(u_x(x,t)) - \int_0^\infty \dot{a}(\tau)f(u_x(x,t-\tau))d\tau \end{aligned} \tag{V}$$

$$\begin{aligned} &\frac{d}{dt}\int_{x_1}^{x_2} u(x,t)dx = -(\sigma(x_2,t) - \sigma(x_1,t)) \\ &\sigma(x,t) = -\int_0^\infty a(\tau)f'(u_x(x,t-\tau))d\tau, \quad f'(\zeta) < 0. \end{aligned} \tag{H}$$

As described in [7], (V) and (H) represent models for viscoelasticity and heat flow, respectively, in materials with memory. The balance laws represent balance of momentum and heat, respectively. For (H) condition (I) still holds while for (V) one needs to add the condition $u_t(x,0^+) = u_1(x)$.

We observe that if we differentiate (H) with respect to t we obtain (V). There is, however, an important technical difference, as described in [7]. For (V) the function a satisfies

$$(-1)^k a^{(k)}(t) \geq 0, \tag{1.1}$$

with $a(t) \to a_\infty > 0$ as $t \to \infty$. For (H) condition (1.1) is still reasonable but one must have $a(t) \to 0$ as $t \to \infty$.

Interest in the models (V) and (H) derives from the fact that in the no memory case, $a(t) \equiv a(0)$, there can be global smooth solutions only for very special (and physically unrealistic) initial data. The general situation is that second derivatives of solutions become infinite in finite time ([1], [5] and [8]). In contrast, (V) and (H) will have global smooth solutions, under hypothesis (1.1), provided the initial data are small [7]. It is conjectured, but not quite proved, that if the data become too large, solutions of (V) and (H) will also break down. (See [10] for a related result.)

For the no memory case of (V) and (H) there has been developed a theory of generalized solutions of shock type, see [2] and [6]. A central feature of this theory is the treatment of the *Riemann problem* which consists of initial data which are piecewise constant with a single jump.

Many of the ideas of the no memory version of (V) and (H) are modeled by the corresponding no memory version of the problem studied in this paper; see [6]. Our object here is to continue the analogy to the equation with memory.

We list the hypotheses under which we operate. The function f is to satisfy

$$f'(\zeta) > 0, \quad f''(\zeta) > 0, \quad f(0) = 0. \tag{F}$$

We suppose the initial data fall into one of two categories:

$$\phi(x,0^+) = u_0(x) \tag{C}$$

$$\phi(x,\tau) \equiv 0, \tag{J}$$

which we term the *continuous* and *jump* situations.

The conditions on a are taken from [7] and are:

$$a \in C^{(2)}(0,\infty), \quad (-1)^j a^{(j)}(t) \geq 0, \quad j = 0,1,2, \tag{A_1}$$

$$a(t) = a_\infty + A(t), \quad \text{where } A,\dot{A} \in L_1(0,\infty)$$
$$\text{and } a_\infty > 0 \quad \text{or} \quad a_\infty = 0. \tag{A_2}$$

We denote the cases $a_\infty > 0$ and $a_\infty = 0$ by (E) and (H) to conform with the analogy to elasticity and heat flow in the introduction.

In order that (C) make sense under condition (A_1), we require that the initial history ϕ satisfy

$$\phi(x,\cdot) \in L_\infty(0,\infty) \quad \text{for each } x. \tag{ϕ}$$

We also need a technical hypothesis on a. Let $L_{\dot{a}}$ denote the linear Volterra operator

$$L_{\dot{a}}[\zeta](t) = \int_0^t \dot{a}(t-\tau)\zeta(\tau)d\tau. \tag{1.2}$$

For convenience we assume $a(0) = 1$. Then the operator $I + L_{\dot{a}}$ has an inverse of the form

$$(I+L_{\dot{a}})^{-1} = I + L_k \tag{1.3}$$

where k is the resolvent of $-\dot{a}$. We require that

$$k(t) \geq 0, \quad \dot{k}(t) \leq 0. \tag{A_3}$$

It is not difficult to verify that conditions (A_1) imply the first of conditions (A_3) automatically. The second of conditions (A_3) is not automatic. It can be verified that part two of (A_3) will hold if a satisfies the condition

$$\ddot{a}(t) - \dot{a}(0)\dot{a}(t) \geq 0. \tag{A_3'}$$

Under condition (A_2) a more complete analysis of k is possible by means of Laplace transforms as described in [7]. One result from [7] is this. Assuming that not only are A and $\dot{A}$ in L_1 but that a certain number of their moments are finite, one can show that k has the form

$$\begin{gathered} k(t) = k_\infty + K, \quad K,\dot{K} \in L_1(0,\infty) \\ k_\infty > 0 \text{ if } a_\infty = 0 \text{ and } k_\infty = 0 \text{ if } a_\infty > 0. \end{gathered} \tag{1.4}$$

2. SMOOTH SOLUTIONS

We consider first solutions in which ϕ_x and u_0 are continuous and u is of class $C^{(1)}$ on $t > 0$. In this case (B) implies the *local balance law*

$$u_t + \sigma_x = 0 \quad \text{on} \quad t > 0. \tag{LB}$$

If we put

$$\Phi(x,t) = \int_0^\infty \dot{a}(t+\tau)f(\phi(x,\tau))d\tau, \tag{2.1}$$

then our problem is equivalent to

$$u_t + (I+L_a\cdot)[f(u)_x] = -\Phi_x, \quad u(x,0^+) = u_0(x). \tag{P}$$

The corresponding no memory problem is

$$u_t + f(u)_x = 0, \quad u(x,0^+) = u_0(x). \tag{P_0}$$

We recall the well known results for (P_0) (see [6]).

THEOREM 1. (i) (P_0) has a unique solution locally in t.

(ii) If $u_0'(x) \geq 0$, then (P_0) has a unique solution globally in t.

(iii) If $u_0'(x_0) < 0$ for some x_0, then there exists a time $T < \infty$, and an x_1 such that the local solution exists for $t < T$ but $u_t(x,t) \to \infty$ as $(x,t) \to (x_1,T)$.

We indicate now a result for the memory case which illustrates our remarks in the introduction.

THEOREM 2. (i) (P) has a unique solution locally in t.

(ii) If $u_0'(x) \geq 0$, then (P) - ($\mathcal{J}$) has a unique solution globally in t.

(iii) Suppose f satisfies the condition

$$\lim_{u\to 0} \frac{f''(u)}{f'(u)} = \gamma, \quad 0 < \gamma < \infty. \tag{F_1}$$

Then for (P) - (J) we have:

(a) If $\|u_0^{(j)}\|_{L_\infty(-\infty,\infty)}$, $j = 0,1$ are sufficiently small, then (P) - (J) has a unique solution globally in t.

(b) If $\|u_0\|_{L_\infty(-\infty,\infty)}$ is sufficiently small, then there is an N such that if $u_0'(x_0) < -N$ for some x_0, there is a T and an x_1 such that the local solution of (P) - (J) exists for $t < T$ but $u_t(x,t) \to \infty$ as $(x,t) \to (x_1,T)$.

A complete proof of this result will be given elsewhere but we can indicate the essential idea. We use a device from [7]. We apply (1.3) to (P). With an integration by parts the result is

$$u_t + f(u)_x + k(0)u = \Psi - L_{\dot{k}}[u], \quad u(x,0^+) = u_0(x) \tag{$\tilde{P}$}$$

$$\Psi(x,t) = k(t)u_0(x) - (I+L_k)[\Phi_x(x,\cdot)](t). \tag{2.2}$$

Problem (P_0) has the property that the characteristic curves are given by $\frac{dx}{dt} = f'(u)$ and along these u is constant. When $u_0' \geq 0$ everywhere, the characteristics are straight lines with slopes that increase as x increases and one gets a global solution. On the other hand, if u_0' is ever negative, characteristics will cross and this corresponds to breakdown. The situation for $(\tilde{P})$ is modified. The characteristics are still given by $\frac{dx}{dt} = f'(u)$, but the presence of the term $k(0)u$ on the left side of $(\tilde{P})$ tends to produce exponential decay of u. The right side is controllable with small data and hence one can prevent the crossing of characteristics. This yields (ii) of Theorem 2. For large negative u_0' one can again force the characteristics to cross and this yields (iii).

3. WEAK SOLUTIONS

Theorem 2 shows that we cannot expect to have global smooth solutions of (P) for all data and thus we are led to the idea of weak solutions. For these we allow u and σ to be piecewise continuous and require (B) rather than (LB). In this connection it is important to observe that in physical situations it is the global balance laws which have meaning.

Let us describe precisely the functions we allow. u and σ are to have the property that in any compact x-t set they are continuously differentiable except along a finite number of differentiable curves $x = s(t)$ across which

u, σ and their x and t derivatives can jump. For any function $\chi(x,t)$ which jumps across $x = s(t)$ we write $[\chi](t) = \chi(s(t)^+,t) - \chi(s(t)^-,t)$.* There are three types of discontinuities with which we deal.

(1) *Acceleration Waves.* Here we have u and σ continuous across $x = s(t)$ but $\nabla u, \nabla\sigma$ discontinuous. We borrow the name from (V) where u represents velocity.

(2) *Shocks.* Here u and σ jump across $x = s(t)$. It follows easily from (B) that the *Rankine-Hugoniot* condition

$$\dot{s}(t)[u](t) = [\sigma](t) \equiv [f(u)](t) \tag{3.1}$$

must hold. It is well known in hyperbolic equation theory that meaningful shocks must satisfy an additional condition - the analog of the entropy condition in gas dynamics. The same is true here. We say a shock is *admissible* if

$$f'(u(s(t)^-,t)) > \dot{s}(t) > f'(u(s(t)^+,t)) \qquad (3.2)^+$$

(3) *Contact Discontinuities.* Here we have a jump in u across $x = s(t)$ with σ remaining continuous. It can be verified that this can occur only on vertical lines $s(t) \equiv$ constant.

We study the problem (B), (C), (I) under the following hypotheses:

$$\left\{\begin{array}{l}\phi(x,\tau) \text{ piecewise continuously differentiable with} \\ \text{discontinuities only along vertical lines and } u_0(x) \\ \text{piecewise continuous.}\end{array}\right. \tag{3.3}$$

We say u is a *weak solution* if it has only acceleration waves, shocks and contact discontinuities, satisfies (B) and (C) for $t > 0$ and satisfies (I) at points of continuity of u_0. We say it is *admissible* if all shocks are admissible.

$^*\chi(s^+(t),t) \equiv \lim_{\substack{x\to s(t)\\ x>s(t)}} \chi(x,t)$ and $\chi(s^-(t),t) \equiv \lim_{\substack{x\to s(t)\\ x<s(t)}} \chi(x,t)$.

$^+$This condition for admissibility of a shock is only correct for f's which are convex, as we are assuming. If f were not convex, a more complicated condition would be required.

PROPOSITION. Let u be a shock solution. Then:

(i) u satisfies (LB) at points of continuity of u_t and u_x.

(ii) For case (*J*), u has no contact discontinuities.

(iii) For case (*C*), u has contact discontinuities only at the discontinuities of ϕ.

Proof. (i) is immediate. To verify (ii) and (iii) we need only observe that (C) implies that on a line x = constant

$$[\sigma] = (I+L_{a}\cdot)[f(u)] + \int_0^\infty \dot{a}(t+\tau)[f(\phi(\cdot,\tau))]d\tau. \tag{3.4}$$

Hence in case (*J*) $[\sigma] \equiv 0$ implies $[f(u)] \equiv 0$, that is, $[u] \equiv 0$. In case (*C*), however, we can make $[\sigma] \equiv 0$ along lines where $[f(\phi)] \neq 0$ without having $[u] \equiv 0$.

REMARK. We observe that the conditions for shocks are exactly the same as in the no memory case; see [6]. On the other hand, there are no contact discontinuities in the no memory case.

The role of admissibility of shocks is the same here as in the no memory case; namely, it yields a stability and uniqueness result.

THEOREM 3. Let u^i be admissible shock solutions for initial data (u_0^i, ϕ^i), i = 1,2. Suppose that $|f(\phi^1(x,t)) - f(\phi^2(x,t))| \le \omega(x)$ for $t > 0$, $\overline{\omega} = ||\omega||_{L_1(-\infty,\infty)} < \infty$. Put $\zeta(t) = ||u^1(\cdot,t) - u^2(\cdot,t)||_{L_1(-\infty,\infty)}$. Then

$$\zeta(t) + \int_0^t k(t-\tau)\zeta(\tau)d\tau \le \zeta(0)\{1 + ||K||_{L_1(0,\infty)} + k_\infty t\}$$

where A and K are as in (A_2) and (1.4).

COROLLARY. There exists at most one admissible shock solution.

This theorem is an extension of the one for no memory in [6]. In that case $A \equiv 0$ and $k \equiv 0$ and the conclusion is that $\zeta(t) \le \zeta(0)$. This was the conclusion in [6]. The proof is a variation of that in [6] and will be given elsewhere.

There are two known existence results which bear on our work. In [3] Greenberg considered our problem under conditions which in our language correspond to the case (*J*), (*E*). He showed that there exists a generalized solution in a different sense than ours. His solution is much weaker than ours and consequently does not meet the hypotheses of our uniqueness theorem. In [1] Dafermos and Hsiao discuss generalized solutions of a much more general class of hereditary conservation laws. Their results would specialize to our case (*J*), (*E*) and would again give a weaker solution than ours so that our uniqueness theorem would not apply.

We see that the situation for our problem, at least for special cases, is the same as for nonlinear hyperbolic systems. One has uniqueness in a small class and existence in a large class. In the next section we discuss a special class of problems where one has both existence and uniqueness.

4. THE RIEMANN PROBLEM

Here we consider the special problem in which the initial data are piecewise constant with a single jump. That is, we assume

$$\begin{aligned} \phi(x,t) &= \phi^{\ell} \quad \text{for} \quad x < 0, \quad \phi^{r} \quad \text{for} \quad x > 0 \\ u_0(x) &= u^{\ell} \quad \text{for} \quad x < 0, \quad u^{r} \quad \text{for} \quad x > 0. \end{aligned} \tag{4.1}$$

Let us recall the solution of this problem in the no memory case, $u_t + f(u)_x = 0$. For this we need the idea of a *centered simple wave*. This is a solution of the form $u = U(x/t)$ where $f'(U(\zeta)) = \xi$. Then the unique admissible shock solution has the following form:

CASE 1. $u^{\ell} < u^{r}$

$$u(x,t) \equiv u^{\ell} \quad \text{for} \quad x < f'(u^{\ell})t$$

$$u(x,t) = U(x/t) \quad \text{for} \quad f'(u^{\ell})t < x < f'(u^{r})t$$

$$u(x,t) \equiv u^{r} \quad \text{for} \quad x > f'(u^{r})t.$$

CASE 2. $u^{\ell} > u^{r}$

$$u(x,t) \equiv u^{\ell} \quad \text{for} \quad x < \frac{f(u^{r}) - f(u^{\ell})}{u^{r} - u^{\ell}}\, t$$

$$u(x,t) \equiv u^r \quad \text{for} \quad x > \frac{f(u^r) - f(u^\ell)}{u^r - u^\ell}\, t.$$

Note that Case 1 consists of acceleration waves while Case 2 contains a shock.

Let us turn to the memory situation. There is one easy case.

THEOREM 4. For the case (*C*) - (*H*) the solution is $u \equiv \phi^\ell$ for $x < 0$, $u \equiv \phi^r$ for $x > 0$.

Proof. Suppose u is so chosen. Then by (3.4) we have

$$[\sigma](t) = \{\int_0^t \dot{a}(t-\tau)d\tau + \int_0^\infty \dot{a}(t+\tau)d\tau\}(f(\phi^r) - f(\phi^\ell))$$

$$= (-a(t) + a(t) - a_\infty)(f(\phi^r) - f(\phi^\ell)).$$

In case (*H*), a_∞ is zero so this quantity is zero. Thus the proposed formula gives a solution which is unique by Theorem 3.

The analysis of other cases is not complete. We can, however, give two examples which we feel illustrate the essential ideas.

EXAMPLE I. $a(t) = e^{-\alpha t}$, (*J*).

The exponential kernel is the prototype for the heat flow problem. What makes this case easy to handle and instructive is the very simple form which the transformed problem $(\tilde{P})$ assumes. It is easy to verify that in this case the resolvent $k(t) \equiv \alpha$. Hence with no initial history $(\tilde{P})$ becomes

$$u_t + f(u)_x + \alpha u = \alpha u_0, \quad u(x,0) = u_0(x)$$
$$u_0(x) = u^\ell \quad \text{for} \quad x < 0, \quad u_0(x) = u^r \quad \text{for} \quad x > 0. \tag{4.2}$$

The essential thing to note here is the presence of the term αu_0. This is the product of the memory effect and, since u_0 is discontinuous, we can expect it to have an influence on the nature of the solution.

We proceed to analyze this problem. For simplicity we restrict ourselves to the situation where u_0^ℓ and u_0^r are both positive.

We make two preliminary observations. First, for $x < 0$ a solution of (4.2) is $u(x,t) = u^{\ell}$. If the right side remained equal to αu^{ℓ} as we cross $x = 0$, this would continue to be the solution for $x < f'(u^{\ell})t$. But the right side changes as we cross $x = 0$ so something must happen at $x = 0$ -- a memory effect. Second, a solution for $x > f'(u^r)t$ is $u(x,t) \equiv u^r$. Our proposed solution then will have $u(x,t) \equiv u^{\ell}$ in $x < 0$ and $u(x,t) \equiv u^r$ in $x > f'(u^r)t$ (just as in the no memory case). The problem is how to fill in the region $\Omega \equiv \{0 \le x \le f'(u^r)t\}$. Here, as in the no memory case, we will have to distinguish cases: Case (1) $u^{\ell} < u^r$, Case (2) $u^{\ell} > u^r$.

Our first clue as to how to fill in Ω comes from the proposition of Section 3 which tells us that $x = 0$ cannot be a contact discontinuity; that is, u must be continuous across $x = 0$. Hence we have

$$\begin{aligned} &u_t + f(u)_x + \alpha u = \alpha u^r, \quad x > 0 \\ &u(0,t) = u^{\ell}. \end{aligned} \tag{4.3}$$

The problem (4.3) can be solved by characteristics. Define $X(t,\bar{t})$, $U(t,\bar{t})$ by

$$\frac{dX}{dt} = f'(U), \qquad X(\bar{t},\bar{t}) = 0$$

$$\frac{dU}{dt} + \alpha U = \alpha u^r, \qquad U(\bar{t},\bar{t}) = u^{\ell}.$$

This yields

$$U(t,\bar{t}) = u^r + (u^{\ell}-u^r)e^{-\alpha(t-\bar{t})} \tag{4.4}$$

$$X(t,\bar{t}) = \int_{\bar{t}}^{t} f'(u^r + (u^{\ell}-u^r)e^{-\alpha(\tau-\bar{t})})d\tau.$$

Then to obtain u one solves $x = X(t,\bar{t})$ for $\bar{t} = T(x,t)$ and $u(x,t) = U(t,T(x,t))$.

It is not too difficult to verify that the above solution has the form

$$u(x,t) = \mathcal{U}(x;u^{\ell},u^r), \quad 0 < x < \chi(t;u^{\ell},u^r), \tag{4.5}$$

where

$$\chi(t;u^{\ell},u^r) = X(t,0) = \int_0^t f'(u^r + (u^{\ell}-u^r)e^{-\alpha\tau})d\tau \tag{4.6}$$

and $u(x;u^{\ell},u^{r})$ is the solution of

$$\frac{d}{dx} f(u) + \alpha u = \alpha u^{r}, \quad u(0) = u^{\ell}. \tag{4.7}$$

The solution of (4.7) is determined from

$$\begin{aligned} &\int_{u_\ell}^{u} \frac{f'(\xi)}{u_r-\xi}\, d\xi = \alpha x \quad \text{if} \quad u_\ell < u_r \\ &\int_{u}^{u_\ell} \frac{f'(\zeta)}{u_r-\zeta}\, d\zeta = -\alpha x \quad \text{if} \quad u_\ell > u_r. \end{aligned} \tag{4.8}$$

We need a second ingredient when $u^{\ell} < u^{r}$. This is the analog of the centered simple wave in the no memory case. We cannot of course get a solution depending only on x/t. In order to obtain the solution we want, it is helpful to introduce new variables $\xi = x/t$ and $\tau = t$, $u(x,t) = v(\xi,\tau)$. Then

$$v_\tau + \left(\frac{f'(v)}{\tau} - \frac{\xi}{\tau} \right) v_\xi + \alpha v = \alpha u^{r}. \tag{4.9}$$

Let us write the characteristic equations for (4.9). These are

$$\frac{dE}{d\tau} = \frac{f'(V)}{\tau} - \frac{E}{\tau}, \quad \frac{dV}{d\tau} + \alpha V = \alpha u^{r}, \tag{4.10}$$

$E = E(\tau,\bar{\xi})$, $V = V(\tau,\bar{\xi})$. We require that $V(0,\bar{\xi}) = \bar{\xi}$ and that $E(\tau,\bar{\xi})$ remain bounded as $\tau \to 0$. This yields

$$\begin{aligned} V(\tau,\bar{\xi}) &= u^{r} + (\bar{\xi}-u^{r})e^{-\alpha\tau}, \\ E(\tau,\bar{\xi}) &= \frac{1}{\tau}\int_0^{\tau} f'(u^{r} + (\bar{\xi}-u^{r})e^{-\alpha\eta})\,d\eta. \end{aligned} \tag{4.11}$$

Once again we determine our solution by eliminating the parameter. (Recall that E is to represent $x/t = x/\tau$.) We determine $c(x,t)$ by the formula

$$x = \int_0^t f'(u^r + (c-u^r)e^{-\alpha\eta})d\eta. \tag{4.12}$$

Observe that if $c = u^r$, (4.12) yields $x = f'(u^r)t$ while if $c = u^\ell$, (4.12) gives $x = \chi(t;u^\ell,u^r)$ as defined in (4.6). It can be verified that for any (x,t), $\chi(t;u^\ell,u^r) \le x \le f'(u^r)t$ and equation (4.12) has exactly one solution $c(x,t)$.

Now we define $W(x,t;u^\ell,u^r)$ by the formula (see (4.11))

$$W(x,t;u^\ell,u^r) = V(t,c(x,t)).$$

W will solve the equation in the region $\chi(t;u^r,u^\ell) < x < f'(u^r)t$ and, from the preceding paragraph, (4.11) and (4.5) we have

$$W(f'(u^r)t,t;u^\ell,u^r) \equiv u^r,$$

$$W(\chi(t;u^r,u^\ell),t;u^\ell,u^r) = u^r + (u^\ell-u^r)e^{-\alpha t} \equiv \mathcal{U}(x;u^\ell,u^r).$$

We have, then, a solution for Case (1) $u^\ell < u^r$. The solution is

$$\begin{aligned} &u \equiv u^\ell, \quad x \le 0; \quad u = \mathcal{U}, \quad 0 \le x \le \chi \\ &u = W, \quad \chi < x < f'(u^r)t; \quad u \equiv u^r, \quad x \ge f'(u^r)t. \end{aligned} \tag{4.13}$$

Observe that (4.13), as in Case (1) for no memory, consists only of acceleration waves. It is different however in that there is an additional acceleration wave at $x = 0$ and one at $x = \chi$. In loose language one may say that u "remembers" that it had a singularity initially at $x = 0$.

We have still to deal with Case (2), $u^\ell > u^r$. Here we do not have the simple wave available and we must introduce a shock. Let $x = s(t)$, $s(0) = 0$, be the shock. We propose to have $u = u^r$ for $x > s(t)$ and $u = \mathcal{U}(x;u^\ell,u^r)$ for $x < s(t)$. One checks that the Rankine - Hugoniot relation (3.1) yields then

$$\frac{ds}{dt} = \frac{f(u^r)-f(\mathcal{U}(s(t);u^\ell,u^r))}{u^r-\mathcal{U}(s(t);u^\ell,u^r)}, \quad s(0) = 0. \tag{4.14}$$

Observe that at t = 0, (4.14), (4.6) and the convexity of f give

$$\left.\frac{ds}{dt}\right|_{t=0} = \frac{f(u^r) - f(u^\ell)}{u^r - u^\ell} < f'(u^r) = \left.\frac{d\chi}{dt}\right|_{t=0}.$$

Hence, initially s(t) lies to the left of χ where u is defined. A little analysis shows that s continues to be in this region and is defined for all t. Moreover, again by the convexity of f,

$$f'(u(s(t);u^\ell,u^r)) > \frac{ds}{dt} > f'(u^r),$$

so the shock is admissible.

It is of interest to look at two limit cases. When α tends to zero, the equation tends to the no memory case and it can be verified that our solution tends to that at the beginning of the section. One can also check that when α tends to infinity our equation tends to $u_t = 0$. Correspondingly our solution tends to $u = u^\ell$ for $x < 0$, $u = u^r$ for $x > 0$.

It is quite instructive to consider a still more special case. Let us assume that $u^r = 0$ and that $f(u) = \frac{1}{2}u^2$*. One sees from (4.5) - (4.7) that

$$\begin{aligned} u(x;u^\ell,0) &= u^\ell - \alpha x \\ \chi(t;u^\ell,0) &= \frac{u^\ell}{\alpha}(1 - e^{-\alpha t}). \end{aligned} \tag{4.15}$$

With $u^r = 0$ we are always in Case (2) and (4.14) gives

$$\frac{ds}{dt} = \frac{1}{2}u(s(t);u^\ell,0) = \frac{1}{2}(u^\ell - \alpha s), \quad s(0) = 0.$$

Hence $s(t) = \frac{u^\ell}{\alpha}(1 - e^{-\frac{\alpha}{2}t})$. On this curve (4.15) yields

$$u(s(t)^-,t) = u(s(t);u^\ell,0) = u^\ell - \alpha s(t) = u^\ell e^{-\frac{\alpha}{2}t}. \tag{4.16}$$

Equations (4.15) and (4.16) illustrate two interesting memory effects. First, (4.15) shows that if the region on the right is "at rest" (that is,

*For this case (P_0) becomes the well-known Burgers equation. Note, however, that this f does not satisfy (F_1).

$u^r = 0$) then the shock is of finite extent in x. Second, (4.16) shows that the shock strength decays to zero exponentially. The latter fact is consistent with the damping effect of the memory term as discussed in [7].

We remark that if $f'(0) > 0$, then the first phenomenon does not occur. That is, $s(t) \to \infty$ as $t \to \infty$.

EXAMPLE II. $a(t) = \mu + (1-\mu)e^{-t}$, $0 < \mu < 1$; (C).

This is a prototype viscoelastic situation. It is considerably more complicated in that the resolvent kernel k is no longer constant, hence $(\tilde{P})$ is not so simple. We will describe our results for this case. The proofs are lengthy and will be presented elsewhere. These proofs consist mainly of a priori estimates for solutions and are obtained there by exploiting the fact that in the present case our problem can be rewritten as a hyperbolic system. It is not difficult to verify that the equivalent system is

$$\begin{aligned}
&u_t + \sigma_x = 0 \\
&u(x,0) = \phi^\ell \quad \text{on} \quad x < 0, \quad \phi^r \quad \text{on} \quad x > 0 \\
&\sigma_t + \sigma = f(u)_t + \mu f(u) \\
&\sigma(x,0) = \mu f(\phi^\ell) \quad \text{on} \quad x < 0, \quad \mu f(\phi^r) \quad \text{on} \quad x > 0.
\end{aligned} \tag{4.17}$$

We observe that one of the characteristic fields for this two by two system is degenerate, that is, lines x = constant.

We have completed the analysis only for special data, namely that in which $\phi^\ell > 0$ and $\phi^r = 0$. Thus the initial data in (4.17) are

$$(u(x,0),\sigma(x,0)) = \begin{cases} (\phi^\ell, \mu f(\phi^\ell)) & x < 0 \\ (0,0) & x > 0. \end{cases} \tag{4.18}$$

Our first observation about this problem comes from the proposition in section three. Since the initial history has a jump at $x = 0$ we must expect the solution to have a contact discontinuity at $x = 0$.

The results we have obtained depend on having the value ϕ^ℓ sufficiently small. Under this assumption we can show that there is an admissible shock solution which will be unique by Theorem 3. This solution has the following structure:

(i) $(u,\sigma) \equiv (\phi^{\ell}, \mu f(\phi^{\ell}))$ for $x < 0$.

(ii) There is a contact discontinuity at $x = 0$ with $u(0^+,t) = u_0(t)$ where $u_0(t)$ is defined by

$$f(u_0(t)) = \mu f(\phi^{\ell})e^{-\mu t} + f(\phi^{\ell})(1-e^{-\mu t}).$$

(iii) $(u,\sigma) \equiv (0,0)$ for $x > s(t)$ where $x = s(t)$ is an admissible shock curve.

(iv) For $0 < x < s(t)$, u and σ are differentiable with

$$u > 0, \quad u_t > 0, \quad u_x < 0, \quad \sigma > 0, \quad \sigma_t > 0, \quad \sigma_x < 0.$$

(v) $U(t) \equiv u(s^-(t),t)$ is decreasing and tends to zero as f tends to plus infinity.

Property (v) shows that the shock strength decays as t tends to infinity just as in Example (I).

The asymptotic behavior of the solution is very intriguing, particularly when compared to that in Example (I). Recall that in Example (I) a central role was played by the function $\mathcal{U}$ of (4.7). This is a steady solution, that is, independent of time. An analysis of the solution for Example (I) in the case $u^{\ell} > 0$, $u^r = 0$ and $f'(0) > 0$ shows that

$$u(x,t) \to u^{\ell}, \quad x < 0$$

$$u(x,t) \to \mathcal{U}(x;u^{\ell},0), \quad x > 0$$

as $t \to \infty$; that is, u tends to a steady state.

For Example (II) such steady solutions do not exist. What is possible are traveling waves (something not present in Example (I)). The equations (4.17) support solutions of the form $\hat{u}(\zeta)$, $\hat{\sigma}(\zeta)$, where $\zeta = x - \hat{c}(\phi^{\ell})t$, satisfying the limit relations

$$\lim_{\zeta\to-\infty} \hat{u}(\zeta) = \phi^{\ell}, \quad \lim_{\zeta\to+\infty} \hat{u}(\zeta) = 0. \tag{4.19}$$

These traveling waves are determined as follows:

$$\hat{c}(\phi^{\ell}) = \frac{\mu f(\phi^{\ell})}{\phi^{\ell}}$$

$$\hat{\sigma}(\zeta) = \hat{c}(\phi^{\ell})\hat{u}(\zeta) \tag{4.20}$$

$$\hat{c}(\phi^{\ell}) \frac{d}{d\zeta} (f(u)-\hat{c}(\phi^{\ell})u) = (\mu f(u)-\hat{c}(\phi^{\ell})u).$$

We have to distinguish two cases:

CASE (a). When $\hat{c}(\phi^{\ell}) < f'(0)$ there is a unique (to within a translation) smooth, strictly decreasing solution of (4.20) satisfying (4.19).

CASE (b). When $\hat{c}(\phi^{\ell}) > f'(0)$ there is a nonincreasing weak solution of (4.20), satisfying (4.19). This particular solution is unique to within a translation and satisfies

(i) $\hat{u}(\zeta) \equiv 0, \quad \zeta > 0$

(ii) $\hat{u}$ is continuous except for a jump at $\zeta = 0$ with $\hat{u}(0^-) = u_*$, $f(u_*) = \hat{c}(\phi^{\ell})u_*$.

Equation (4.20) and the convexity of f show that when ϕ^{ℓ} is small enough we will always be in Case (a).

We believe that as $t \to \infty$ the solution of our problem tends to some translate of the traveling wave, but we have not been able to prove this result. We do have a weak result in this direction which we describe now.

We need some preliminary remarks. Recall the function $u_0(t) \equiv u(0^+,t)$ in (ii). From the formula in (ii) we see that $u_0(0) < \phi^{\ell}$ while $u_0(t)$ is increasing and tends to ϕ^{ℓ} as $t \to \infty$. Next consider the function $U(t) = u(s(t)^-,t)$. We have $U(0) = u_0(0)$ and, as stated in (v), $\dot{U} < 0$ and $U(t) \to 0$ as $t \to \infty$. These facts yield $0 < u_0(t) - U(t)$ for all t and $u_0(t) - U(t) \to \phi^{\ell}$ as $t \to \infty$.

We now consider the level lines of u in the region $0 < x < s(t)$. These are curves $x(t,\alpha)$ such that $u(x(t,\alpha),\alpha) \equiv \alpha$. From the preceding paragraph $x(t,\alpha)$ is defined for $\alpha \in (0,\phi^{\ell})$ and $t \geq \tau(\alpha)$ where $\tau(\alpha)$ is defined by $U(\tau(\alpha)) = \alpha$ for $\alpha \in (0,u_0(0))$ and $u_0(\tau(\alpha)) = \alpha$ for $\alpha \in (u_0(0),\phi^{\ell})$. The quantity

$$C_{AV}(t) = \frac{1}{u_0(t)-U(t)} \int_{U(t)}^{u_0(t)} \frac{\partial x}{\partial t}(\alpha,t)d\alpha$$

is the average speed of propagation of the level lines. What we can show is that in the limit this quantity tends to the speed of propagation of the traveling wave; that is

$$C_{AV}(t) \to \hat{c}(\phi^{\ell}) \quad \text{as} \quad t \to \infty.$$

ACKNOWLEGDEMENTS

Professor Greenberg's research was partially supported by the National Science Foundation under grant MCS 80-18531.

Professor MacCamy's research was partially supported by the National Science Foundation under grant MCS 80-01944. Part of the work was initiated while he was visiting the Mathematics Research Center in Madison, Wisconsin.

REFERENCES

1. Dafermos, C. and L. Hsiao, private communication.

2. Glimm, J. and P. D. Lax, Decay of solutions of nonlinear hyperbolic conservation laws, *Mem. Amer. Math. Soc.*, No. 101 (1970.

3. Greenberg, J. M., The existence and qualitative properties of solutions of

$$\frac{\partial u}{\partial t} + \frac{1}{2}\frac{\partial}{\partial x}[u^2 + \int_0^t c(s)u^2(x,t-s)ds] = 0$$

Jour. Math. Anal. and Appl., Vol. 42, No. 1 (1973), 205-220.

4. Johnson, J. L. and J. A. Smoller, Global solutions for an extended class of hyperbolic systems of conservation laws, *Arch. Rat. Mech. and Anal.*, Vol. 32 (1969), 169-189.

5. Lax, P. D., Development of singularities of solutions of nonlinear hyperbolic partial differential equations, *Jour. Math. Phys.*, Vol. 5 (1964), 611-613.

6. Lax, P. D. Hyperbolic systems of conservation laws and the mathematical theory of shock waves, CBMS Reg. Conf. 1973.

7. MacCamy, R. C., An integro-differential equation with applications in heat flow, and a model for one-dimensional, nonlinear viscoelasticity, *Quart. of Appl. Math.*, Vol. 35, No. 1 (1977), 1-33

8. MacCamy, R. C. and V. J. Mizel, Existence and non-existence in the large of solutions of quasilinear wave equations, *Arch. Rat. Mech. and Anal.*, Vol. 25 (1967), 299-320.

9. MacCamy, R. C. and J. S. W. Wong, Stability theorems for some functional differential equations, *Trans. Amer. Math. Soc.*, Vo. 164 (1972), 1-37.

10. Slemrod, M., Instability of steady shearing flows in a nonlinear visco-elastic fluid, *Arch. Rat. Mech. and Anal.*, Vol. 68 (1978), 221-225.

AN EXAMPLE OF BOUNDARY LAYER IN DELAY EQUATIONS

Jack K. Hale
Luis Magalhães

Lefschetz Center for Dynamical Systems
Division of Applied Mathematics
Brown University
Providence, Rhode Island

Singular perturbation problems for functional differential equations have been studied by a number of authors [1-5]. These investigations were primarily concerned with the nature of convergence of the solutions to the degenerate problem for positive time. Very little is known about the boundary layer for the general case. The results of Halanay [4] and Klimushev [5] lead to a partial discussion of the boundary layer for a very special class of equations. In this note, we give a result for a special equation concerning necessary and sufficient conditions for the existence of an invariant subspace of finite codimension.

Let $r > 0$ be a given constant, $L^2 = L^2([-r,0],\mathbb{R})$, a_0, d_0 be real constants, $d_0 \neq 0$, a, b, c, d given functions in L^2. Consider the pair of equations

$$\begin{aligned} \dot{x}(t) &= a_0 x(t) + \int_{-r}^{0} a(\theta)x(t+\theta)d\theta + \int_{-r}^{0} b(\theta)y(t+\theta)d\theta \\ \mu\dot{y}(t) &= d_0 y(t) + \int_{-r}^{0} c(\theta)x(t+\theta)d\theta + \int_{-r}^{0} d(\theta)y(t+\theta)d\theta, \end{aligned} \tag{1}$$

where $\mu > 0$ is a small number. One could also consider a term $b_0 y(t)$ in the first equation in (1) and $c_0x(t)$ in the second. This will not affect the results because one can make a change of basis in $\mathbb{R}^2$ to reduce the discussion to (1) with the functions a,b,c,d being smooth functions of μ near $\mu = 0$. Let $x_t(\theta) = x(t+\theta)$, $y_t(\theta) = y(t+\theta)$, $-r \le \theta \le 0$, and $X = L^2 \times L^2 \times \mathbb{R} \times \mathbb{R}$.

For any $(\phi,\psi,\alpha,\beta) \in X$, equation (1) has a unique solution $(x(t),y(t))$, $t \ge -r$, satisfying the conditions $(x_0,y_0,x(0),y(0)) = (\phi,\psi,\alpha,\beta)$. If $T_\mu(t)(\phi,\psi,\alpha,\beta) = (x_t,y_t,x(t),y(t))$, $t \ge 0$, then $T_\mu(t)$, $t \ge 0$, is a C_0-semigroup on X.

If

$$\Sigma_0 = \{(\phi,\psi,\alpha,\beta) \in X:\ d_0\beta + \int_{-r}^{0} [c(\theta)\phi(\theta)+d(\theta)\psi(\theta)]d\theta = 0\},$$

then Σ_0 is a subspace of X of codimension one since $d_0 \ne 0$. For any $(\phi,\psi,\alpha,\beta) \in \Sigma_0$, the equation

$$\begin{aligned} \dot{x}(t) &= a_0x(t) + \int_{-r}^{0} [a(\theta)x(t+\theta)+b(\theta)y(t+\theta)]d\theta \\ 0 &= d_0y(t) + \int_{-r}^{0} [c(\theta)x(t+\theta)+d(\theta)y(t+\theta)]d\theta \end{aligned} \tag{2}$$

has a unique solution with initial data $(\phi,\psi,\alpha,\beta) \in \Sigma_0$.

In analogy with ordinary differential equations, one might expect that equation (1) has the following *convergence property*: there is a subspace Σ_μ of X which is invariant under $T_\mu(t)$, $\Sigma_\mu \to \Sigma_0$ as $\mu \to 0$ with this convergence being in the sense that the unit normals converge.

THEOREM. *If* $d_0 > 0$, *the convergence property holds for equation (1). If* $d_0 < 0$, *the convergence property holds for equation (1) if and only if* $c = 0$ $d = 0$; *that is, equation (1) has the form*

$$x = a_0x(t) + \int_{-r}^{0} a(\theta)x(t+\theta)d\theta + \int_{-r}^{0} b(\theta)y(t+\theta)d\theta$$

$$\mu\dot{y} = d_0y(t).$$

Proof. Let the inner product in X be denoted by

$$((\phi,\psi,\alpha,\beta),\ (g,h,\delta,\gamma)) = \alpha\delta + \beta\gamma + \int_{-r}^{0} [\phi(\theta)g(\theta) + \psi(\theta)h(\theta)]d\theta.$$

A few computations show that a normal to $\sum_0$ can be chosen as

$$(\bar{g},\bar{h},0,1),\quad \bar{g} = c/d_0,\quad \bar{h} = d/d_0. \tag{3}$$

If $(g,h,\delta,1)$ is a normal to $\sum_\mu$ and $\sum_\mu$ is invariant under the solutions of equation (1), then

$$x(t)\delta + y(t) + \int_{-r}^{0} [x(t+\theta)g(\theta) + y(t+\theta)h(\theta)]d\theta = 0,\quad t \geq 0 \tag{4}$$

for all initial data $(\phi,\psi,\alpha,\beta) \in X$. In particular, this must be true for (ϕ,ψ,α,β) in the domain $D(A_\mu)$ of the infinitesimal generator A_μ of $T_\mu(t)$. It is known that

$$D(A_\mu) = \{(\phi,\psi,\alpha,\beta):\ \phi,\psi \text{ are absolutely continuous,}$$

$$(\dot{\phi},\dot{\psi}) \in L^2 \times L^2,\ \phi(0) = \alpha,\ \psi(0) = \beta\}.$$

Also, if $(\phi,\psi,\alpha,\beta) \in D(A_\mu)$, then $(x(t),y(t))$ is locally absolutely continuous for $t \geq -r$, $(\dot{x}_t,\dot{y}_t) \in L^2 \times L^2$. Thus, we may differentiate (4) for these initial data to obtain

$$\dot{x}(t)\delta + \dot{y}(t) + \int_{-r}^{0} [\dot{x}(t+\theta)g(\theta) + \dot{y}(t+\theta)h(\theta)]d\theta = 0$$

for all $t \geq 0$. Substituting for $\dot{x}(t)$, $\dot{y}(t)$ from (1) and using the fact that (4) is satisfied, one obtains, for $t = 0$,

$$\left[a_0\delta - \frac{d_0}{\mu}\delta\right]\alpha + \int_{-r}^{0} \left[(a\delta + \frac{c}{\mu} - \frac{d_0}{\mu}g)\phi + \dot{\phi}g \right.$$
$$\left. + (b\delta + \frac{d}{\mu} - \frac{d_0}{\mu}h)\psi + \dot{\psi}h\right] d\theta = 0$$

for all $(\phi,\psi,\alpha,\beta) \in D(A_\mu)$. But this implies g, h are absolutely continuous with $\dot{g}, \dot{h} \in L^2$. Integrating by parts, one obtains the following relations:

$$\begin{aligned} &(a_0 - \frac{d_0}{\mu} - h(0))\delta + g(0) = 0 \\ &\dot{g} + (\frac{d_0}{\mu} + h(0))g = a\delta + \frac{c}{\mu} \\ &\dot{h} + (\frac{d_0}{\mu} + h(0))h = b\delta + \frac{d}{\mu} \\ &g(-r) = 0, \quad h(-r) = 0. \end{aligned} \tag{5}$$

To solve these equations, let $h(0) = h_0$, $g(0) = g_0$, $\lambda = h_0 + d_0/\mu$. The above equations imply that

$$\delta = -(a_0 - \lambda)^{-1} g_0$$

$$g_0 = \int_{-r}^{0} e^{\lambda\theta} [a(\theta)\delta + c(\theta)/\mu] d\theta$$

$$h_0 = \int_{-r}^{0} e^{\lambda\theta} [b(\theta)\delta + d(\theta)/\mu] d\theta$$

or

$$\delta = -(a_0 - \lambda)^{-1} g_0$$

$$g_0 = \frac{a_0 - \lambda}{\mu} \left[a_0 - \lambda + \int_{-r}^{0} e^{\lambda\theta} a(\theta) d\theta\right]^{-1} \int_{-r}^{0} e^{\lambda\theta} c(\theta) d\theta \tag{6}$$

$$\mu\lambda - d_0 = \int_{-r}^{0} e^{\lambda\theta} d(\theta) d\theta - \left[a_0 - \lambda + \int_{-r}^{0} e^{\lambda\theta} a(\theta) d\theta\right]^{-1} \left(\int_{-r}^{0} e^{\lambda\theta} b(\theta) d\theta\right) \int_{-r}^{0} e^{\lambda\theta} c(\theta) d\theta.$$

For $\sum_\mu$ to converge to $\sum_0$ as $\mu \to 0$, we must have $\delta \to 0$, $h \to d/d_0$, $g \to c/d_0$ as $\mu \to 0$. The latter two relations imply that $\lambda = d_0/\mu + \mathcal{O}(1)$ as $\mu \to 0$. Using this fact in the above expressions with $d_0 < 0$, one obtains $c = d = 0$. For the converse, one can use the above equations to construct $\sum_\mu$.

If $d_0 > 0$, then an application of the implicit function theorem shows equation (6) has a solution of the form $\lambda = d_0/\mu + \mathcal{O}(1)$ for any functions a,b,c, and d. This determines δ, g_0, h_0 which, in turn, define the functions g, h by (5). The formulas for δ, g and h define $\sum_\mu$, which converges to $\sum_0$. This completes the proof of the theorem.

REMARK 1. If $r = \infty$, the same type of proof applies when a, b, c, $d \in L^2 \equiv L^2(-r,0,\nu,\mathbb{R})$ with the Radon-Nikodym derivative of $\nu = e^{\eta\theta}$, $\eta > 0$, and $X = L_\nu^2 \times L_\nu^2 \times \mathbb{R} \times \mathbb{R}$.

REMARK 2. The proof can be modified to include also terms with discrete delays. The same conclusion as in the theorem is valid.

REMARK 3. Generalizations to the case where x and y are vectors are under investigation.

ACKNOWLEDGEMENTS

Professor Hale's research was supported in part by the National Science Foundation under contract MCS-7905774-05, by Air Force contract AFOSR-76-3092D, and by Army contract ARO-DAAG-29-76-C-0161.

Dr. Magalhães' research was supported in part by National Science Foundation contract MCS-7905774-05.

REFERENCES

1. Cooke K. L., The condition of regular degeneration for singularly perturbed linear differential-difference equations, *J. Differential Equations*, 1 (1965), 39-94.

2. Cooke, K. L. and K. R. Meyer, The condition of regular degeneration of singular perturbed systems of linear differential-difference equations, *J. Math. Anal. Appl.*, 14 (1966), 83-106.

3. Habets, P., Singular perturbations of functional differential equations, VII Internationale Konferenz über Nichtlineare Schuringungen 8-13 Sept. 1975. *Abhandlung Akad. Wissenschaften DDR*, no. 3, 1977, 307-313.

4. Halaney, A., Singular perturbations of systems with retarded arguments, *Rev. Math. Pures Appl.* 7 (1962), 301-308.

5. Klimushev, A. I., The dependence of solutions of a system of delay equations on a small parameter before the derivatives, *Ural Politsch. Inst. Sb.* 139 (1964), 5-11.

SOME RESULTS ON THE LIAPUNOV STABILITY OF FUNCTIONAL EQUATIONS

E. F. Infante

Division of Mathematical and Computer Sciences
National Science Foundation
Washington, D.C.
and
Division of Applied Mathematics
Brown University
Providence, Rhode Island

1. INTRODUCTION

It is well known that it is possible to obtain a wealth of asymptotic results for dynamical systems if one has available an appropriate Liapunov functional. Indeed, a large number of abstract results are available, results which yield specific stability properties for the dynamical system under consideration if a Liapunov functional with certain characteristics is available. Conversely, it is known that if a dynamical system has certain stability properties, then a Liapunov functional with certain characteristics does exist. Unfortunately such results are not very useful for the practical construction of Liapunov functionals, a process that remains somewhat of an art, especially for infinite dimensional dynamical systems such as functional and partial differential equations, be they linear or not.

The construction of quadratic Liapunov functionals for linear ordinary differential equations is well understood. In this brief paper we attempt to show that this method of construction can be imitated with profit in the case of functional differential equations. A number of results [1,2,3,4,5, 6,7,8,9] have recently appeared on this precise topic, Liapunov functionals of differing degrees of sophistication having been constructed for linear difference equations, linear difference-differential equations, integro-differential equations and difference-differential equations of the neutral type. This paper attempts to outline a procedure that appears to be common and to underlie all of these methods of construction and that naturally suggests the general type of appropriate quadratic form useful for construction of Liapunov functionals.

2. LINEAR DIFFERENTIAL EQUATIONS

Let us recall, for the purpose of inspiration, the well known method for the construction of quadratic Liapunov functionals for linear ordinary differential equations.

For this purpose, given the equation

$$\dot{x}(t) = Ax(t), \quad x(0) = x_0, \quad x \in \mathbb{R}^n, \tag{2.1}$$

let us suppose that this system is asymptotically stable, i.e., $\mathrm{Re}\lambda(A) < 0$, and consider the familiar Liapunov equation

$$A^T P + PA = -W, \tag{2.2}$$

where W is a symmetric positive definite n×n matrix. It is well known that a unique, positive definite n×n matrix solution P of this algebraic problem exists; moreover, a useful representation of this solution is given by

$$P = \int_0^\infty e^{A^T \xi} W e^{A\xi} d\xi. \tag{2.3}$$

This matrix generates a quadratic Liapunov functional

$$V(x) = x^T Px, \tag{2.4}$$

whose derivative along the solutions of equation (2.1) is given by

$$\overset{\circ}{V}(x) = x^T [A^T P + PA] x = -x^T Wx. \tag{2.5}$$

This transparently simple result is at the heart of a number of stability arguments. It should be noted that the choice of a Liapunov functional of the form $V(x) = x^T Px$ is natural for this problem, although a second thought on this point leads one to question the reason for this.

Of course, a simple explanation is easily provided. The essence of the method is to construct a function V(x) whose time derivative along the solutions of (2.1) is negative definite. A natural form of this derivative is

$$\overset{\circ}{V}(x) = -x^T W x, \quad W > 0, \tag{2.6}$$

which immediately leads to the function

$$V(x_0) = \int_0^\infty x^T(\xi) W x(\xi) d\xi. \tag{2.7}$$

This function, unfortunately, is in an undesirable representation. However, if we denote by $\hat{x}(s)$ the Laplace transform of the function $x(\xi)$, with $x(0) = x_0$, we have that

$$\hat{x}(s) = (sI-A)^{-1} x_0, \tag{2.8}$$

and application of Parseval's equality to equation (2.7) yields that

$$V(x_0) = x_0^T \{\int_{-\infty}^{\infty} (i\omega I-A)^{-1^*} W(i\omega I-A)^{-1} d\omega\} x_0, \tag{2.9}$$

which shows that the desired function $V(x_0)$ is of the form $V(x_0) = x_0^T P x_0$ with P a symmetric positive definite matrix.

This simple idea, applied to functional equations, will yield our desired forms.

3. LINEAR DIFFERENCE-DIFFERENTIAL EQUATIONS

Consider the problem of construction of a quadratic Liapunov functional for the retarded difference-differential equation

$$\dot{x}(t) = Ax(t) + Bx(t-\tau), \quad t \geq 0, \quad \tau > 0, \tag{3.1}$$

with initial conditions $x(0) = \xi \in \mathbb{R}^n$, $x_0 = \phi \in L_2$, $\phi:[-\tau,0] \to \mathbb{R}^n$, and where A and B are constant matrices. It is assumed that the condition $\det[\lambda I-A-Be^{-\lambda\tau}] \neq 0$, $\text{Re}\lambda \geq 0$, is satisfied, thus insuring asymptotic stability. Liapunov functionals for such an equation have been constructed in [5,6,8];

in particular, in [5,6] a general and useful functional was obtained through a somewhat laborious method of approximating (3.1) through a class of difference equations of increasing dimension. In the sequel, it is shown that the precise same form can be obtained, following the simple method shown in the previous section, in a straightforward manner.

Indeed, let W be a positive definite matrix and consider the functional whose derivative is

$$\overset{\circ}{V}(\xi,\phi) = -\xi^T W\xi \qquad (3.2)$$

along the solutions of equation (3.1). A representation of this functional is obviously given by

$$V(\xi,\phi) = \int_0^\infty x^T(\eta)Wx(\eta)d\eta, \qquad (3.3)$$

where x(t) is the solution of (3.1) with initial conditions $(\xi,\phi) \in \mathbb{R}^n \times L_2([-\tau,0],\mathbb{R}^n)$. This representation is not particularly desirable; however, denoting by $\hat{x}(s)$ the Laplace transform of the solution of equation (3.1) with initial conditions (ξ,ϕ) yields

$$[sI-A-Be^{-s\tau}]\hat{x}(s) = \xi + \int_{-\tau}^0 e^{-s(u+\tau)}B\phi(u)du,$$

or, with $M(s) = [sI-A-Be^{-s\tau}]^{-1}$,

$$\hat{x}(s) = M(s)\xi + \int_{-\tau}^0 M(s)e^{-s(u+\tau)}B\phi(u)du. \qquad (3.4)$$

Since Parseval's equality applied to equation (3.3) gives

$$V(\xi,\phi) = \int_{-\infty}^\infty \hat{x}^*(i\omega)W\hat{x}(i\omega)d\omega,$$

introduction of statement (3.4) into this equation yields that

$$V(\xi,\phi) = \xi^T[\int_{-\infty}^{\infty} M^*(i\omega)WM(i\omega)d\omega]\xi +$$
$$2\xi^T\int_{-\tau}^{0} [\int_{-\infty}^{\infty} M^*(i\omega)WM(i\omega)e^{-i\omega(u+\tau)}d\omega]B\phi(u)du + \quad (3.5)$$
$$\int_{-\tau}^{0}\int_{-\tau}^{0} \phi^T(u)B^T[\int_{-\infty}^{\infty} M^*(i\omega)WM(i\omega)e^{-i\omega(v-u)}d\omega]B\phi(v)dvdu.$$

Denoting by $Q(\alpha)$ the matrix function

$$Q(\alpha) = \int_{-\infty}^{\infty} M^*(i\omega)WM(i\omega)e^{-i\omega\alpha}d\omega, \quad -\tau \le \alpha \le \tau, \quad (3.6)$$

it follows that the functional (3.3) has the form

$$V(\xi,\phi) = \xi^TQ(0)\xi + 2\xi^T\int_{-\tau}^{0} Q(u+\tau)B\phi(u)du +$$
$$\int_{-\tau}^{0}\int_{-\tau}^{0} \phi^T(u)B^TQ(v-u)B\phi(v)dvdu. \quad (3.7)$$

This is precisely the principal part of the Liapunov functional obtained, in one case through a method of approximation, in the other case by intuition, in [5,6].

The functional given by (3.7) is of a natural quadratic form, which is particularly simple in that it is completely determined once the matrix function $Q(\alpha)$, $-\tau \le \alpha \le \tau$ is known. This functional is of this simplicity as a result of the particularly simple choice of $\overset{o}{V}$, as prescribed by equation (3.2). The representation of $Q(\alpha)$ given by (3.6) is not particularly desirable; it is a representation of a nature similar to that given by equations (2.3) and (2.9) for the linear ordinary differential case. It would be desirable to obtain $Q(\alpha)$ as a solution of an appropriate generalization of the familiar Liapunov equation (2.2). This is indeed possible, for a straightforward computation shows that the continuous function $Q(\alpha)$, $-\tau \le \alpha \le \tau$, is characterized by the property that $Q(-\alpha) = Q^T(\alpha)$, and by the functional equation

$$Q'(\alpha) = A^TQ(\alpha) + B^TQ^T(\tau-\alpha), \quad 0 < \alpha \le \tau, \tag{3.8}$$

with boundary conditions

$$A^TQ(0) + Q(0)A + B^TQ^T(\tau) + Q(\tau)B = -W. \tag{3.9}$$

That this functional equation represents a generalization of equation (2.2) is transparent. Less obvious is the fact that this equation is equivalent to a system of ordinary differential equations of order $2n^2$, hence a finite dimensional problem; the analysis of this functional equation has been presented in [7].

The key to the stability analysis presented in [5,6,8] lies in the computability of functional (3.7); however, it is difficult to show, in general, that this functional is positive definite on the natural space $\mathbb{R}^n \times L_2([-\tau,0],\mathbb{R}^n)$. For this reason in [5,6,8] the Liapunov functional

$$\tilde{V}(\xi,\phi) = V(\xi,\phi) + \xi^TM\xi + \int_{-\tau}^{0} \phi^T(\eta)R\phi(\eta)d\eta$$

is used in the final analysis, with $M > 0$, $R > 0$ arbitrary small positive definite matrices.

In an entirely analogous manner to the one presented above, it is possible to obtain in a straightforward manner appropriate quadratic functionals that lead to natural Liapunov functionals for neutral difference-differential equations [8] of the type

$$\frac{d}{dt}[x(t) + Cx(t-\tau)] = Ax(t) + Bx(t-\tau), \quad t > 0,$$

and for retarded difference-differential equations with multiple delays such as

$$\dot{x}(t) = Ax(t) + B_1x(t-\tau_1) + B_2x(t-\tau_2), \quad t > 0.$$

In this latter case, such results have been obtained through approximation methods [4] of a complicated and computationally complex nature which the present viewpoint completely avoids.

4. A VOLTERRA INTEGRODIFFERENTIAL EQUATION

The approach of the previous section is also applicable to Volterra integrodifferential equations, as exemplified by the equation

$$\dot{x}(t) = Ax(t) + \int_0^t B(t-\tau)x(\tau)d\tau, \quad t \geq t_0 > 0, \quad x \in \mathbb{R}^n, \tag{4.1}$$

where A and B are n×n matrices, A constant and B locally integrable. Liapunov functionals for such equations have recently been obtained [1,2,9]; the purpose here is to show how to obtain the basic quadratic forms that underlie these Liapunov functionals.

We consider the pair $(x_t(0),x_t) \in \mathbb{R}^n \times L_2[-t,0]$ where $x_t(\theta) = x(t+\theta)$ for $-t \leq \theta \leq 0$; note that equation (4.1) applies only for $t \geq t_0 > 0$. We also assume that (4.1) is asymptotically stable, i.e., that $\det[sI-A-\hat{B}(s)] \neq 0$ for every Re $s \geq 0$.

For equation (4.1) it is desired to find the functional $V(x_t(0),x_t)$ with the property that for an arbitrary positive definite matrix W one has that

$$\overset{\circ}{V}(x_t(0),x_t) = -x_t^T(0)Wx_t(0) \tag{4.2}$$

along the solutions of (4.1). Such a functional is obviously given by

$$V(x_t(0),x_t) = \int_0^\infty x_T^T(\eta)Wx_t(\eta)d\eta. \tag{4.3}$$

If one denotes by $\hat{x}(s)$ the Laplace transform of the solution of (4.1) with initial condition pair $(x_t(0),x_t)$, then one easily obtains that

$$[sI-A-\hat{B}(s)]\hat{x}(s) = x_t(0) + \int_{-t}^0 [\int_0^\infty e^{-s\xi}B(\xi-u)d\xi]x_t(u)du$$

or, in a more convenient notation,

$$\hat{x}(s) = M(s)x_t(0) + \int_{-t}^0 M(s)[\int_0^\infty e^{-s\xi}B(\xi-u)d\xi]x_t(u)du. \tag{4.4}$$

Application of Parseval's equality to equation (4.3) yields that

$$V(x_t(0),x_t) = \int_{-\infty}^{\infty} \hat{x}(i\omega)^* W\hat{x}(i\omega)d\omega,$$

and introduction of relation (4.4) into this expression, just as was done in the previous section, yields three terms. Rearrangement of the integrals in these terms, and the definition of the matrix function $Q(\alpha)$, $\alpha \in \mathbb{R}$, by

$$Q(\alpha) = \int_{-\infty}^{\infty} M^*(i\omega)WM(i\omega)e^{-i\omega\alpha}d\omega, \tag{4.5}$$

allows the functional $V(x_t(0),x_t)$ to be expressed in the form

$$\begin{aligned} V(x_t(0),x_t) = {} & x_t^T(0)Q(0)x_t(0) + \\ & 2x_t^T(0)\int_{-t}^{0}\left[\int_{-\theta}^{\infty} Q(v+\theta)B(v)dv\right]x_t(\theta)d\theta + \\ & \int_{-t}^{0}\int_{-t}^{0} x_t^T(\theta)P(\theta,\eta)x_t(\eta)d\eta d\theta, \end{aligned} \tag{4.6}$$

where

$$P(\theta,\eta) = \int_{-\theta}^{\infty}\int_{-\eta}^{\infty} B^T(v)Q(\eta-\theta + u-v)B(u)dudv.$$

The functional V, in a appropriate representation, is thus obtained. This is precisely the principal component of the Liapunov functional obtained in [1]. Again, the representation (4.5) for the matrix function $Q(\cdot)$ is not particularly a desirable one; it can be shown, [1], that this matrix function can be characterized as the unique exponentially decaying solution of the functional equation

$$\begin{aligned} & Q'(\alpha) = A^TQ(\alpha) + \int_0^{\infty} B^T(\beta)Q(\alpha-\beta)d\beta, \quad \alpha > 0, \\ & A^TQ(0) + Q(0)A + \int_0^{\infty} [B^T(\beta)Q^T(\beta) + Q(\beta)B(\beta)]d\beta = -W, \end{aligned} \tag{4.7}$$

with $Q(\alpha) = Q^T(-\alpha)$ for $\alpha \in \mathbb{R}$. Again, this functional equation represents a generalization of the algebraic Liapunov equation (2.2).

The relative simplicity of the functional (4.6), in its dependence on only one matrix function $Q(\alpha)$, is a consequence of the extremely simple choice for the derivative $\overset{o}{V}$ as expressed in (4.2). Again, the functional (4.6) does not have strong positivity properties. In general, it is desirable to add to this functional terms of the form $x_t^T(0)Mx_t(0)$ and $\int_{-t}^{0} x_t^T(\theta)Rx_t(\theta)d\theta$ where M and R are small positive definite matrices. It is precisely functionals of this type that have been used in [1,2,7] to study the asymptotic stability of Volterra integrodifferential equations of the form (4.1), as well as nonlinear equations and equations with nonconvolution kernels.

The construction of a Liapunov functional for the equation

$$\dot{x}(t) = Ax(t) + \int_{t-\tau}^{t} B(t-s)x(s)ds, \quad t > 0, \quad \tau \geq 0, \tag{4.8}$$

is carried out in exactly the same manner as used above, resulting in a functional on the Hilbert space $\mathbb{R}^n \times L_2([-\tau,0],\mathbb{R}^n)$.

The simple technique illustrated here is applicable to a broad class of problems, yielding a profitable method of construction of Liapunov functionals.

ACKNOWLEDGEMENTS

This research was partially supported by the National Science Foundation (Grant MCS-8041638), the Air Force Office of Scientific Research (Grant AFOSR-76-3092D), and the Army Research Office (Grant DAAG 79-C-0161).

REFERENCES

1. Abrahamson, D. L. and E. F. Infante, A Liapunov functional for linear Volterra integrodifferential equations, to appear.

2. Burton, T. A., Uniform stabilities for Volterra equations, *J. Diff. Eq.* 36 (1980), 40-53.

3. Carvalho, L. A. V., E. F. Infante and J. A. Walker, On the existence of simple Liapunov functions for linear retarded difference-differential equations, *Tôhoku Math. J.* 32 (1980), 283-297.

4. Castelan, W. B., A Liapunov functional for a matrix retarded difference-differential equation with several delays, *Lecture Notes in Mathematics, 799*, Springer-Verlag, 1980, 82-118.

5. Hale, J. K., *Theory of Functional Differential Equations*, Springer Verlag, 1977.

6. Infante, E. F. and W. B. Castelan, A Liapunov functional for a matrix difference-differential equation, *J. Diff. Eq.* 29 (1978), 439-451.

7. Infante, E. F. and W. B. Castelan, On a functional equation arising in the stability theory of difference-differential equations, *Q. Appl. Math.* 35 (1977), 311-319.

8. Infante, E. F. and W. B. Castelan, A Liapunov functional for a matrix neutral difference-differential equation with one delay, *J. Math. Anal. Appl.* 71 (1979), 105-130.

9. Miller, R. K., *Nonlinear Volterra Integral Equations*, W. A. Benjamin, 1971.

SOME SYNTHESIS PROBLEMS FOR INTEGRAL EQUATIONS

J. J. Levin

Department of Mathematics
University of Wisconsin
Madison, Wisconsin

We consider the equations

$$x'(t) + \int_{-\infty}^{\infty} g(x(t-\xi))dA(\xi) = f(t), \quad -\infty < t < \infty \tag{1f}$$

$$x(t) + \int_{-\infty}^{\infty} g(x(t-\xi))dA(\xi) = f(t), \quad -\infty < t < \infty, \tag{2f}$$

where $' = \frac{d}{dt}$,

$$g \in C(\mathbb{C}^N, \mathbb{C}^N), \quad A = (A_{ij}), \quad A_{ij} \in BV(\mathbb{R}^1, \mathbb{C}^1) \quad (i,j = 1,\dots,N) \tag{3}$$

$$f \in L^\infty(\mathbb{R}^1, \mathbb{C}^N), \quad \lim_{t\to\infty} f(t) = f(\infty) \text{ exists and, in (2f), } f \in \mathcal{B}(\mathbb{R}^1, \mathbb{C}^N). \tag{4}$$

$\mathcal{B}$ denotes Borel measurable. The limit equations associated with (1f) and (2f) are defined, respectively, by

$$y'(t) + \int_{-\infty}^{\infty} g(y(t-\xi))dA(\xi) = f(\infty), \quad -\infty < t < \infty \qquad (1^* f(\infty))$$

$$y(t) + \int_{-\infty}^{\infty} g(y(t-\xi))dA(\xi) = f(\infty), \quad -\infty < t < \infty. \qquad (2^* f(\infty))$$

For each $\mu \in \mathbb{C}^N$, with (3) holding, it is convenient to also consider the equations

$$y'(t) + \int_{-\infty}^{\infty} g(y(t-\xi))dA(\xi) = \mu, \quad -\infty < t < \infty \qquad (1^* \mu)$$

$$y(t) + \int_{-\infty}^{\infty} g(y(t-\xi))dA(\xi) = \mu, \quad -\infty < t < \infty. \qquad (2^* \mu)$$

Clearly, $(1^* f(\infty))$ is the special case of $(1^* \mu)$ in which $\mu = f(\infty)$; also, $(1^* \mu)$ is the special case of (1f) in which $f(t) \equiv \mu$. Similar remarks apply to (2f), $(2^* f(\infty))$, and $(2^* \mu)$.

Important problems for (1f) and (2f) are:

PROBLEM 1a. Analyze the asymptotic behavior of a bounded solution of (1f) as $t \to \infty$.

PROBLEM 1b. Analyze the asymptotic behavior of a bounded solution of (2f) as $t \to \infty$.

Of course, really detailed solutions to these problems require more hypothesis than (3), (4); however, some statements can be made without having any additional assumptions. It is well known that solutions to Problems 1a, 1b often involve the limit equations.

Our primary interest here is not in the preceding "analysis problems". Rather, roughly speaking, we study the following questions:

PROBLEM 2a. Synthesize, from a sequence of bounded solutions of $(1^* \mu)$ which satisfy a "closeness condition", an f which satisfies (4) and a corresponding solution, x, of (1f) such that $f(\infty) = \mu$ and such that the asymptotic behavior of x is related to (geometrical) properties (inherent in the closeness condition) of the given sequence.

PROBLEM 2b. Replace $(1^*\mu)$, (1f) by $(2^*\mu)$, (2f) in Problem 2a.

It will be seen, after "closeness condition" and other key words have been defined, that Problems 2a, 2b can be made precise and studied without reference to the analysis problems. However, the eventual purpose in studying these synthesis problems is, of course, to apply the results concerning them to the analysis problems. This will be discussed below in Theorems 7a, 7b.

There is, of course, an enormous literature pertaining to Problems 1a, 1b. Here we only state the principal definitions and results that are needed in the present study. We follow the references given below, which may be consulted for other references. Theorems I, Ia, Ib, II, IIa, IIb below are concerned with Problems 1 (below), 1a, 1b and appear in [3] (and, with some extensions, in [4]). Some discussion of Problem 2a appears in [2].

Part of our motivation is to develop tools for investigating the asymptotic behavior of bounded solutions of Volterra equations. The elementary lemmas detailing how such investigations are subsumed under investigations of (1f), (2f) may be found, for example, in [1]. Applications to Volterra equations motivate hypothesis (10) below. Theorems 7a, 7b may be used to obtain known results on scalar Volterra equations.

Let

$$\mathcal{Y} = \{y\colon \mathbb{R}^1 \to \mathbb{C}^N\}.$$

For brevity each element of $\mathcal{Y}$ will be called a *curve*, whether or not it is continuous (or even measurable). Clearly, every solution of each of the equations already mentioned is an element of $\mathcal{Y}$. We'll use the notation

$$C = C(\mathbb{R}^1, \mathbb{C}^N), \quad AC = AC(\mathbb{R}^1, \mathbb{C}^N), \ldots$$

for subsets of $\mathcal{Y}$. Also let

$$R(y) = \text{Range of } y \ (y \in \mathcal{Y}), \quad R(Y) = \bigcup_{y\in Y} R(y) \quad (Y \subset \mathcal{Y}),$$

$$y_\tau\colon t \to y(t+\tau) \quad (y \in \mathcal{Y},\ \tau \in \mathbb{R}^1),$$

$$x_n \to x \quad \text{c.o.} \quad (x_n, x \in \mathcal{Y}) \quad \text{if and only if}$$

$$x_n(t) \to x(t) \quad \text{uniformly on every compact subset of } \mathbb{R}^1.$$

Of course, c.o. denotes the compact open topology on $\mathcal{Y}$. For each $x \in \mathcal{Y}$, let

$$\Omega(x) = \{\omega \in \mathbb{C}^N | \ x(t_n) \to \omega \text{ for some } t_n \to \infty\},$$

$$\Gamma(x) = \{y \in \mathcal{Y} | \ x_{t_n} \to y \text{ c.o. for some } t_n \to \infty\},$$

$$A(x) = \{\alpha \in \mathbb{C}^N | \ x(t_n) \to \alpha \text{ for some } t_n \to -\infty\},$$

$$\Gamma^*(x) = \{y \in \mathcal{Y} | \ x_{t_n} \to y \text{ c.o. for some } t_n \to -\infty\}.$$

It turns out that although Problem 1a is a reasonable one, Problem 1b is not. This is because it is possible to have wildly behaved bounded solutions of (2f) under assumptions (3), (4) (see [1]). For this reason we consider the set, $T = T(\mathbb{R}^1, \mathbb{C}^N)$, of Tauberian curves in $\mathcal{Y}$ and its subset $\tilde{T}$ defined by

$$T = \{x \in \mathcal{Y} | \lim_{t\to\infty, \tau\to 0} |x(t+\tau)-x(t)| = 0\}$$

$$\tilde{T} = \{x \in T | \limsup_{t\to\infty} |x(t)| < \infty\},$$

where $|x| = \sum_{i=1}^{N} |x_i| \ (x \in \mathbb{C}^N)$. Later we denote $B_r = \{x \in \mathbb{C}^N | |x| < r\}$.

For each $\mu \in \mathbb{C}^N$ and $Q \subset \mathbb{C}^N$ the following sets of solutions of $(1^*\mu)$, $(2^*\mu)$ play important roles:

$$S_Q(1^*\mu) = \{y \in L^\infty \cap C_u^1 | \ y \text{ satisfies } (1^*\mu),\ R(y) \subset Q\}$$

$$S_Q(2^*\mu) = \{y \in L^\infty \cap C_u | \ y \text{ satisfies } (2^*\mu),\ R(y) \subset Q\},$$

where C_u denotes the uniformly continuous elements of $\mathcal{Y}$ and C_u^1 those with a uniformly continuous first derivative.

DEFINITION. We say that $\{t_m\} \subset \mathbb{R}^1$ is an expanding t-sequence if $t_m < t_{m+1}$ $(m = 1,2,\ldots)$ and if $t_m - t_{m-1} \to \infty$.

DEFINITION. A sequence $\psi_m \in C^\infty(\mathbb{R}^1, [0,1])$ $(m = 1,2,\ldots)$ is a ψ-sequence associated with an expanding t-sequence $\{t_m\}$ if: $\sum_{m=1}^{\infty} \psi_m(t) \equiv 1$ $(-\infty < t < \infty)$, $\lim_{m\to\infty} \|\psi_m^{(j)}\|_\infty = 0$ $(j = 1,2,\ldots)$, $\psi_1(t) \equiv 1$ $(t \le t_1)$, $\psi_1(t) \equiv 0$ $(t_2 \le t)$, $\psi_1'(t) \le 0$ $(t_1 \le t \le t_2)$, $\psi_m(t) \equiv 0$ $(t \le t_{m-1}, t \ge t_{m+1})$, $\psi_m(t_m) = 1$, $\psi_m'(t) \ge 0$ $(t_{m-1} \le t \le t_m)$, $\psi_m'(t) \le 0$ $(t_m \le t \le t_{m+1})$ $(m \ge 2)$.

Thus, a ψ-sequence, $\{\psi_m\}$, is a partition of unity for $\mathbb{R}^1$ in which the successive ψ_m vary slower and slower. It is easy to show that with any expanding t-sequence, there exists an associated ψ-sequence (in fact, infinitely many of them). It is sometimes convenient to assume that a ψ-sequence also satisfies the auxiliary condition (which can always be arranged for a given expanding t-sequence)

$$\sup_m \int_{t_m}^{t_{m+1}} |\psi_m''(t)|\,dt < \infty. \tag{5}$$

Problems 1a, 1b are special cases of

PROBLEM 1. Analyze the asymptotic behavior of a curve $x \in \tilde{T}$ as $t \to \infty$.

Part of the answer to Problem 1 (given in [3]) is contained in Theorem I, which describes how x approaches some sequence $\{y_m\} \subset \Gamma(x)$ as $t \to \infty$.

THEOREM I. If $x \in \tilde{T}$, then there exists a sequence $\{y_m\} \subset \Gamma(x)$ and an expanding t-sequence, $\{t_m\}$, such that

$$\lim_{m\to\infty} \{ \sup_{t_{m-2}\le t\le t_{m+2}} |x(t)-y_m(t)| \} = 0.$$

Moreover,

$$R(\Gamma(x)) = \Omega(x) \subset \overline{B}_{\limsup_{t\to\infty} |x(t)|},$$

and for any ψ-sequence, $\{\psi_m\}$, associated with $\{t_m\}$

$$x = \sum_{m=1}^{\infty} \psi_m y_m + \eta, \quad \lim_{t\to\infty} \eta(t) = 0.$$

The following consequences of Theorem I for (1f), (2f) (see [3]) are answers to Problems 1a, 1b.

THEOREM Ia. If (3), (4) hold and x is a bounded solution of (1f), then $x \in \tilde{T}$. Moreover, in addition to Theorem I,

$$\Gamma(x) \subset S_{\Omega(x)}(1^* f(\infty)), \quad \Omega(x) = R(S_{\Omega(x)}(1^* f(\infty))),$$

$$\lim_{m\to\infty} \{ \operatorname*{ess\,sup}_{t_{m-1} \le t \le t_{m+1}} |x'(t) - y'_m(t)| \} = 0,$$

$$\lim_{t\to\infty} \{ \operatorname*{ess\,sup}_{t \le \tau < \infty} |\eta'(\tau)| \} = 0.$$

THEOREM Ib. If (3), (4) hold and $x \in T$ is a bounded solution of (2f), then $x \in \tilde{T}$. Moreover, in addition to Theorem I,

$$\Gamma(x) \subset S_{\Omega(x)}(2^* f(\infty)), \quad \Omega(x) = R(S_{\Omega(x)}(2^* f(\infty))).$$

The numbering of subsequent theorems will be similar to that of Theorems I, Ia, Ib. Namely, Theorems n, na, nb refer, respectively, to some situation in $\mathcal{Y}$ (without specific reference to integral equations), its consequence for (1f), and its consequence for (2f). The lemmas will generally not be tied to a specific integral equation.

If $x \in \tilde{T}$, then from Theorem I and the triangle inequality it follows readily that

$$\lim_{m\to\infty} \{ \sup_{t_{m-1} \le t \le t_{m+2}} |y_m(t) - y_{m+1}(t)| \} = 0. \tag{6}$$

Formula (6) states that on longer and longer overlapping intervals tending toward ∞, successive y_m get closer and closer.

The point of departure for this paper is (6), which we call the "closeness condition". Now, however, it is an hypothesis (not a conclusion) on a sequence of curves in $\mathcal{Y}$. The basic construction (synthesis) of a curve x, from a sequence of curves, $\{y_m\}$, in $\mathcal{Y}$, which satisfy the closeness condition, is given in

LEMMA 1. Let $\{t_m\}$ be an expanding t-sequence, $\{\psi_m\}$ an associated ψ-sequence, and let $\{y_m\} \subset \mathcal{Y}$ satisfy (6). Define $x \in \mathcal{Y}$ by $x = \sum_{m=1}^{\infty} \psi_m y_m$. Then

$$\lim_{m\to\infty} \{ \sup_{t_{m-2} \le t \le t_{m+2}} |x(t) - y_m(t)| \} = 0.$$

In Lemma 1 through Theorem 4b below, the basic synthesis results, (6) is an hypothesis. However, in Theorems 5 and 6, which are applications of Lemma 1 - Theorem 4b, (6) is not an hypothesis but a conclusion. It is a consequence of some geometric structure that is the hypothesis of Theorems 5 and 6.

We now add various hypotheses to those of Lemma 1 and define curves p and q which will play a key role in the integral equations setting.

LEMMA 2. (i) Let (the hypothesis of) Lemma 1 and

$$y_m \in \mathcal{B} \ (m \ge 1), \quad \sup_m \{ \sup_{-\infty<t<\infty} |y_m(t)| \} \le K_1 < \infty$$

hold. Define $p \in \mathcal{Y}$ by $p = \sum_{m=1}^{\infty} \psi'_m y_m$. Then

$$x \in \mathcal{B}, \quad \sup_{-\infty<t<\infty} |x(t)| \le K_1,$$

$$p \in \mathcal{B}, \quad \sup_{-\infty<t<\infty} |p(t)| < \infty, \quad p(-\infty) = p(\infty) = 0.$$

(ii) Let (i) and $y_m \in C$ $(m \ge 1)$ hold. Then $x \in C$, $p \in C_u$.

(iii) Let (i) and $\{y_m\}$ is equicontinuous on $\mathbb{R}^1$ hold. Then $x \in C_u$.

(iv) Let (i) and (3) hold. Define $q \in \mathcal{Y}$ by

$$q(t) = \int_{-\infty}^{\infty} g(x(t-\xi))dA(\xi) - \sum_{m=1}^{\infty} \psi_m(t) \int_{-\infty}^{\infty} g(y_m(t-\xi))dA(\xi).$$

Then $q \in \mathcal{B}$, $\sup_{-\infty<t<\infty} |q(t)| < \infty$, $q(-\infty) = q(\infty) = 0$.

(v) Let (ii) and (iv) hold. Then $q \in C_u$.

The interpretation of the preceding abstract (or general) synthesis procedure for integral equations is given by the following results - which are partial answers to Problems 2a, 2b.

THEOREM 1a. Let (the hypothesis of) Lemma 2(iv), $\mu \in ¢^N$, and $y_m \in S_{\bar{B}_{K_1}}$ $(1^*\mu)$ $(m \geq 1)$ be satisfied. Then Lemma 2 (i) - (v) hold; moreover, x', $p' \in L^\infty \cap C_u$. Define $f \in \mathcal{Y}$ by

$$f(t) = x'(t) + \int_{-\infty} g(x(t-\xi))dA(\xi).$$

Then

$$f = \mu + p + q, \quad f \in L^\infty \cap C_u, \quad f(-\infty) = f(\infty) = \mu.$$

THEOREM 1b. Let (the hypothesis of) Lemma 2(iv), $\mu \in ¢^N$, and $y_m \in S_{\bar{B}_{K_1}}$ $(2^*\mu)$ $(m \geq 1)$ be satisfied. Then Lemma 2(v) holds. Define $f \in \mathcal{Y}$ by

$$f(t) = x(t) + \int_{-\infty}^{\infty} g(x(t-\xi))dA(\xi).$$

Then

$$f = \mu + q, \quad f \in L^\infty \cap C_u, \quad f(-\infty) = f(\infty) = \mu.$$

Note that the x, p, q of Theorems 1a, 1b are defined in Lemmas 1, 2. For brevity this cumulative use of notation and hypotheses is employed throughout.

For some applications the closeness condition (6), employed in Lemma 1, is not immediately useful. Rather, a sequence $\{\phi_k\} \subset \mathcal{Y}$ and a sequence of intervals, $\{I_k\}$, are available in these applications such that the I_k are disjoint, tend to ∞, length $(I_k) \to \infty$, and $\phi_k - \phi_{k+1}$ is increasingly small on I_k as $k \to \infty$. In Lemma 3 sequences $\{t_m\}$, $\{y_m\}$ satisfying Lemma 1 are constructed from such $\{\phi_k\}$, $\{I_k\}$. It should be noted that the hypothesis of Lemma 3 is quite minimal -- the details handle the transition between the sequences (so that no geometrical information is lost) and set up the machinery for later applications. It is through Lemma 3 that the hypotheses of Theorems 5 and 6 are seen to imply (6).6).

LEMMA 3. Let $\phi_k \in \mathcal{Y}$, $I_k = [\tau_k, \tau_k+\zeta_k]$, $\nu_k \ge 0$ $(k \ge 1)$ satisfy

$$\tau_k < \tau_k + \zeta_k < \tau_{k+1}, \quad \zeta_k \ge 7(k+1),$$

$$\sup_{t\in I_k} |\phi_k(t) - \phi_{k+1}(t)| \le \nu_k, \quad \nu_k \to 0.$$

Define $\{t_m\}$, $\{n_k\}$, $\{m_k\}$, $\{y_m\}$ by: t_1 is the smallest integer $\ge \tau_1$, $n_0 = 0$, n_k is the smallest integer such that

$$\tau_k + 3k \le t_1 + \sum_{j=0}^{k} jn_j \quad (k \ge 1),$$

and

$$m_k = 1 + \sum_{j=0}^{k} n_j \quad (k \ge 0),$$

$$t_m = t_1 + \sum_{j=0}^{k} jn_j + (k+1)(m-m_k) \quad (k \ge 0, \quad m_k \le m \le m_{k+1}),$$

$$y_1 = \phi_1, \quad y_m = \phi_{k+1} \quad (k \ge 0, \quad m_k + 1 \le m \le m_{k+1}).$$

Then $n_1 = 3$, $n_k \ge 1$ $(k \ge 1)$,

$$t_1 + \sum_{j=0}^{k} jn_j < \tau_k + \zeta_k - 3(k+1) \quad (k \ge 1),$$

$m_0 = 1$, $m_1 = 4$, $m_{k+1} \ge m_k + 7$ $(k \ge 1)$, $\{t_m\}$ is an expanding t-sequence, and

$$t_{m_k} = t_1 + \sum_{j=0}^{k} jn_j \quad (k \ge 0),$$

$$t_m = t_{m_k} + (k+1)(m-m_k) \quad (k \geq 0, \quad m_k \leq m \leq m_{k+1}),$$

$$t_m - t_{m-1} = k+1 \quad (k \geq 0, \quad m_k + 1 \leq m_{k+1})$$

$$[t_{m_k-3}, t_{m_k+3}] \subset I_k \quad (k \geq 1),$$

$$y_m = y_{m+1} \quad (m \neq m_k), \quad y_{m_k} - y_{m_k+1} = \phi_k - \phi_{k+1},$$

$$\sup_{t \varepsilon I_k} |y_{m_k}(t) - y_{m_k+1}(t)| \leq \nu_k \quad (k \geq 1).$$

If $\{\psi_m\}$ is a ψ-sequence associated with $\{t_m\}$ and if $x = \sum_{m=1}^{\infty} \psi_m y_m$, then Lemma 1 holds.

Lemma 4 and Theorems 2a, 2b are reformulations in the context of Lemma 3 of, respectively, Lemma 2 and Theorems 1a, 1b.

LEMMA 4. (i) Lemma 3 and $\phi_k \varepsilon \mathcal{B}$ $(k \geq 1)$, $\sup_k \{\sup_{-\infty<t<\infty} |\phi_k(t)|\} \leq K_1 < \infty$ imply that Lemma 2 (i) holds.

(ii) (i) and $\phi_k \varepsilon C$ $(k \geq 1)$ imply that Lemma 2 (ii) holds.

(iii) (i) and $\{\phi_k\}$ equicontinuous on $\mathbb{R}^1$ imply that Lemma 2 (iii) holds.

(iv) (i) and (3) imply that Lemma 2 (iv) holds.

(v) (ii) and (iv) imply that Lemma 2 (v) holds.

THEOREM 2a. Lemma 4 (iv), $\mu \varepsilon ¢^N$, $\phi_k \varepsilon S_{\bar{B}_{K_1}}$ $(1^*\mu)$ $(k \geq 1)$ imply that Lemma 4(i) - (v) and Theorem 1a hold.

THEOREM 2b. Lemma 4 (iv), $\mu \varepsilon ¢^N$, $\phi_k \varepsilon S_{\bar{B}_{K_1}}$ $(2^*\mu)$ $(k \geq 1)$ imply that Lemma 4 (v) and Theorem 1b hold.

The synthesized functions f of Theorems 2a, 2b (and, therefore, of Theorems 1a, 1b) are only guaranteed to satisfy $f(\pm\infty) = \mu$ at $t = \pm\infty$. Something stronger than this is often required. In Lemmas 5, 6 the machinery is set up for also obtaining $f-\mu \in L^1$ -- as stated in Theorems 3a and 3b. Of course, the functions f are different in Theorems 3a and 3b.

In connection with (3) we use the notation

$$\omega(\delta) = \omega_{K_1}(\delta) = \sup_{\substack{|\xi-\eta|\le\delta \\ |\xi|,|\eta|\le K_1}} |g(\xi) - g(\eta)|,$$

$V(t)$ = total variation of A on $(-\infty,t]$.

LEMMA 5. Let Lemma 4 (i) and $\sum_{k=1}^{\infty} \nu_k < \infty$ be satisfied. Then $p \in L^1$.

LEMMA 6. Let Lemma 4 (iv), $\sum_{k=1}^{\infty} k\omega(\nu_k) < \infty$, and

$$\int_{-\infty}^{0} |t|V(t)dt < \infty, \quad \int_{0}^{\infty} t[V(\infty) - V(t)]dt < \infty \tag{7}$$

be satisfied. Then $q \in L^1$.

THEOREM 3a. Let the hypotheses of Lemmas 5, 6 and Theorem 2a be satisfied. Then $f-\mu \in L^1$.

THEOREM 3b. Let the hypotheses of Lemma 6 and Theorem 2b be satisfied. Then $f-\mu \in L^1$.

Still more assumptions guarantee that p, q and the synthesized f's are of bounded variation on $\mathbb{R}^1$. Here g' denotes the matrix $g' = (\partial g_j/\partial x_i)$ and

$$\tilde{\omega}(\delta) = \tilde{\omega}_{K_1}(\delta) = \sup_{\substack{|\xi-\eta|\le\delta \\ |\xi|,|\eta|\le K_1}} |g'(\xi) - g'(\eta)|.$$

LEMMA 7. Let Lemma 5, (5), and

$$\begin{cases} \phi_k \in C^1, & \sup_k\{\sup_{-\infty<t<\infty} |\phi_k'(t)|\} < \infty, \\ \sup_{t\in I_k} |\phi_k'(t) - \phi_{k+1}'(t)| \le \tilde{\nu}_k, & \sum_{k=1}^{\infty} \tilde{\nu}_k < \infty \end{cases} \tag{8}$$

be satisfied. Then $x' \in L^\infty \cap C^1$, $p' \in L^\infty \cap C \cap L^1$ (hence, $p \in BV$).

LEMMA 8. Let Lemma 6, (8), $g \in C^1$, and

$$\sum_{k=1}^{\infty} k\nu_k < \infty, \quad \sum_{k=1}^{\infty} k\tilde{\nu}_k < \infty, \quad \sum_{k=1}^{\infty} k\tilde{\omega}(\nu_k) < \infty$$

be satisfied. Then $q' \in L^\infty \cap C \cap L^1$ (hence, $q \in BV$).

LEMMA 9. Let Lemma 6 and

$$\begin{cases} A(t) = \int_{-\infty}^{t} a(\xi)d\xi, & -\infty < t < \infty \\ a(t),\ t^2a(t),\ a'(t),\ t^2a'(t) \in L^1(\mathbb{R}^1, \mathbb{C}^{N^2}) \end{cases} \tag{9}$$

be satisfied. Then $q \in AC_{loc}$, $q' \in L^\infty \cap L^1$ (hence, $q \in BV$).

LEMMA 10. Let Lemma 6 and

$$\begin{cases} A(t) = 0, \quad -\infty < t \le 0; \quad A(t) = \int_0^t a(\xi)d\xi, \quad 0 \le t < \infty \\ a(t),\ t^2a(t),\ a'(t),\ t^2a'(t) \in L^1(\mathbb{R}^+, \mathbb{C}^{N^2}) \end{cases} \tag{10}$$

be satisfied. Then $q \in AC_{loc}$, $q' \in L^\infty \cap L^1$ (hence, $q \in BV$).

THEOREM 4a. Let the hypotheses of Theorem 3a, (5), and one of $g \in C^1$, (9), or (10) be satisfied. Then for sufficiently small ν_k and sufficiently large ζ_k Theorem 3a holds (with I_k replaced by suitable $\tilde{I}_k \subset I_k$) together with $f \in AC_{loc}$, $f' \in L^\infty \cap L^1$ (hence, $f \in BV$).

THEOREM 4b. Let the hypotheses of Theorem 3b and either (9) or (10) be satisfied. Then $f \in AC_{loc}$, $f' \in L^{\infty} \cap L^{1}$ (hence, $f \in BV$).

In the first application of the preceding synthesis results, it is assumed that a particular $y \in \mathcal{Y}$ acts at $t = -\infty$ in the same manner as a curve in $\tilde{T}$ does at $t = \infty$. It is shown how appropriate translates of y can be synthesized into an $x \in \tilde{T}$ such that $\Omega(x) = A(y)$. This is then interpreted for integral equations. In the latter setting this result emphasizes how the behavior of the solutions of the limit equation on all of $\mathbb{R}^1$ is relevant to the study of the asymptotic behavior as $t \to \infty$ of a particular bounded solution of either (1f) or (2f).

THEOREM 5. Let $y \in \mathcal{Y}$ satisfy $y(-t) \in \tilde{T}$. Then there exist ϕ_k, I_k, ν_k $(k \geq 1)$ satisfying the hypothesis of Lemma 3 and $\lambda_k \downarrow -\infty$ such that $\phi_k = y_{\lambda_k}$ $(k \geq 1)$, $A(y) = \Omega(x)$, and $x \in \tilde{T}$.

THEOREM 5a. (i) Let (3), $\mu \in ¢^N$, and $y \in S_{¢^N}(1^*\mu)$ be satisfied. Then Theorems 5 and 2a hold.

(ii) Let (i) and (7) be satisfied. Then Theorem 3a also holds.

(iii) Let (ii), (5), and one of $g \in C^1$, (9), or (10) be satisfied. Then Theorem 4a also holds.

THEOREM 5b. (i) Let (3), $\mu \in ¢^N$, and $y \in S_{¢^N}(2^*\mu)$ be satisfied. Then Theorem 5 and 2b hold.

(ii) Let (i) and (7) be satisfied. Then Theorem 3b also holds.

(iii) Let (ii) and either (9) or (10) be satisfied. Then Theorem 4b also holds.

In the next application of the synthesis results we need the

DEFINITION. $Y \subset \mathcal{Y}$ is of type 2 if either

$$\begin{cases} \text{there exist } c \in ¢^N \text{ and } y \in Y \text{ such that} \\ y(-\infty) = y(\infty) = c, \quad R(y) \neq c \end{cases} \tag{11}$$

or

$$\begin{cases} \text{there exist distinct } c_1,\ldots,c_n \in \mathbb{C}^N \text{ and} \\ y^{(1)},\ldots,y^{(n)} \in Y \quad (n \geq 2) \quad \text{such that} \\ y^{(j)}(-\infty) = c_j \quad (1 \leq j \leq n), \\ y^{(j)}(\infty) = c_{j+1} \quad (1 \leq j \leq n-1), \quad y^{(n)}(\infty) = c_1. \end{cases} \tag{12}$$

Thus, Y is of type 2 if there is a finite set of points in $\mathbb{C}^N$ which are "joined" by a closed "loop" of nonconstant elements of Y -- where each element of the loop is traversed in the direction of increasing t. Note that if $Y_1 \subset Y_2$ and Y_1 is of type 2, then Y_2 is of type 2.

If $k \geq 0$ and $n \geq 2$ are integers, then $[k]$ denotes the remainder on dividing k by n; thus

$$k = qn + [k], \quad q \in \{0,1,\ldots\}, \quad [k] \in \{0,\ldots,n-1\}.$$

THEOREM 6. Let Y be of type 2. Then there exist ϕ_k, I_k, ν_k $(k \geq 1)$ satisfying the hypothesis of Lemma 3 and $\lambda_k \downarrow -\infty$ such that

$$\phi_k = \begin{cases} y_{\lambda_{k-1}} & \text{(if (11) holds)} \\ y^{([k-1]+1)}_{\lambda_{k-1}} & \text{(if (12) holds)} \end{cases} \qquad (k \geq 1)$$

$$\Omega(x) = \begin{cases} \overline{R(y)} & \text{(if (11) holds)} \\ \bigcup_{j=1}^{n} \overline{R(y^{(j)})} & \text{(if (12) holds).} \end{cases}$$

In particular, $x(\infty)$ does not exist.

THEOREM 6a. (i) Let (3), $\mu \in \mathbb{C}^N$, and $Y \subset S_{\mathbb{C}^N}(1^*\mu)$ of type 2 be satisfied. Then Theorems 6 and 2a hold.

(ii) Let (i) and (7) be satisfied. Then Theorem 3a also holds.

(iii) Let (ii), (5), and one of $g \in C^1$, (9), or (10) be satisfied. Then Theorem 4a also holds.

THEOREM 6b. (i) Let (3), $\mu \in \mathbb{C}^N$, and $Y \subset S_{\mathbb{C}^N}(2^*\mu)$ of type 2 be satisfied. Then Theorems 6 and 2b hold.

(ii) Let (i) and (7) be satisfied. Then Theorem 3b also holds.

(iii) Let (ii) and either (9) or (10) be satisfied. Then Theorem 4b also holds.

The final topic discussed here is an application of Theorems 6a, 6b to the question of existence of the limit $x(\infty)$ of a bounded solution, x, of either (1f) or (2f). First we need some preliminary material from [3]. The critical points, CP, of $(1^*\mu)$ and $(2^*\mu)$ are defined by

$$CP(1^*\mu) = \{c \in ¢^N | g(c)A(\infty) = \mu\}$$

$$CP(2^*\mu) = \{c \in ¢^N | c + g(c)A(\infty) = \mu\},$$

where in view of (3) we have without loss of generality taken $A(-\infty) = 0$. It is easily shown that

LEMMA 11a. Let (3), $\mu \in ¢^N$, and $y \in S_{¢^N}(1^*\mu)$ be satisfied. Then $y(\infty) \in CP(1^*\mu)$ if $y(\infty)$ exists and $y(-\infty) \in CP(1^*\mu)$ if $y(-\infty)$ exists.

LEMMA 11b. Let (3), $\mu \in ¢^N$, and $y \in S_{¢^N}(2^*\mu)$ be satisfied. Then $y(\infty) \in CP(2^*\mu)$ if $y(\infty)$ exists and $y(-\infty) \in CP(2^*\mu)$ if $y(-\infty)$ exists.

For $Y \subset \mathcal{Y}$ and $\alpha,\beta \in ¢^N$ denote $Y^{(\alpha,\beta)} = Y \cap \{y(-\infty) = \alpha, y(\infty) = \beta\}$

DEFINITION. $Y \subset \mathcal{Y}$ is of type 1 if either $Y = \emptyset$ or there exist distinct $c_1,\ldots,c_n \in ¢^N$ $(n \geq 1)$ and $E \subset \{(i,j) | 1 \leq i, j \leq n\}$ such that

$$Y = \bigcup_{i=1}^{n} [Y \cap \{y(t) \equiv c_i\}] \cup [\bigcup_{(i,j)\in E} Y^{(c_i,c_j)}],$$

where either $E = \emptyset$ or satisfies:

$$\begin{cases} i \neq j \text{ and } Y^{(c_i,c_j)} \neq \emptyset \text{ for each } (i,j) \in E, \\ \text{there does not exist a sequence} \\ \{(i_k,j_k)\}_{k=1}^{p} \subset E \text{ for any } p \geq 2 \text{ such that} \\ i_{k+1} = j_k \ (1 \leq k \leq p-1) \text{ and } i_1 = j_p. \end{cases}$$

Thus, loosely speaking, Y is of type 1 if it contains no loops. Clearly,

LEMMA 12. If $Y_1 \subset Y_2$ and Y_2 is of type 1, then Y_1 is of type 1.

From [3] we have:

THEOREM II. Let $x \in \tilde{T}$. Then $x(\infty)$ exists if and only if $\Gamma(x)$ is of type 1.

THEOREM IIa. Let (3), (4) hold and let x be a bounded solution of (1f). Then $x(\infty)$ exists if and only if $S_{\Omega(x)}(1^*f(\infty))$ is of type 1.

THEOREM IIb. Let (3), (4) hold and let $x \in T$ be a bounded solution of (2f). Then $x(\infty)$ exists if and only if $S_{\Omega(x)}(2^*f(\infty))$ is of type 1.

Theorems IIa, IIb, and Lemma 12 immediately yield:

COROLLARY IIa. Let (3), (4), x is a bounded solution of (1f), and $S_{\bar{B}_r}(1^*f(\infty))$ is of type 1 for all $r \geq 0$ all hold. Then $x(\infty)$ exists.

COROLLARY IIb. Let (3), (4), $x \in T$ is a bounded solution of (2f) and $S_{\bar{B}_r}(2^*f(\infty))$ is of type 1 for all $r \geq 0$ all hold. Then $x(\infty)$ exists.

The following results are direct consequences of Theorem 6a, Corollary IIa and of Theorem 6b, Corollary IIb. The point of these results is, loosely speaking, that if the existence of $x(\infty)$ of a bounded solution, x, of an integral equation is guaranteed under certain conditions which include fairly stringent hypotheses on f, then $x(\infty)$ also exists under very weak hypothesis on f.

THEOREM 7a. Let (3), $\mu \in \mathbb{C}^N$, and

$$\begin{cases} y(\infty) \text{ and } y(-\infty) \text{ exist for all } y \in S_{\mathbb{C}^N}(1^*\mu), \\ CP(1^*\mu) \cap V \text{ is a finite set for all compact } V \subset \mathbb{C}^N \end{cases} \tag{13}$$

hold. Furthermore, let $x(\infty)$ exist for any bounded solution, x, of (1f), whenever one of (i), (ii), or (iii) holds, where

(i) $f \in L^{\infty} \cap C_u$, $f(\infty) = \mu$ are satisfied.

(ii) (i), $f-\mu \in L^1$, and (7) are satisfied.

(iii) (ii), $f \in AC_{loc}$, $f' \in L^{\infty} \cap L^1$, and one of $g \in C^1$, (9), or (10) are satisfied.

Then $S_{\bar{B}_r}(1^*\mu)$ is of type 1 for all $r \geq 0$. (In particular, $x(\infty)$ exists for any bounded solution, x, of (1f), whenever f merely satisfies (4) and $f(\infty) = \mu$ -- and g, A satisfy (3), (13), and the appropriate additional restrictions of (ii) or (iii).)

THEOREM 7b. Let (3), $\mu \in \mathbb{C}^N$,

$$\begin{cases} y(\infty) \text{ and } y(-\infty) \text{ exist for all } y \in S_{\mathbb{C}^N}(2^*\mu), \\ CP(2^*\mu) \cap V \text{ is a finite set for all compact } V \subset \mathbb{C}^N, \end{cases} \tag{14}$$

and

$$S_{\bar{B}_r}(2^*\mu) \quad \text{is equicontinuous for each } r \geq 0 \tag{15}$$

hold. Furthermore, let $x(\infty)$ exist for any $x \in T$ bounded solution of (2f), whenever one of (i), (ii), or (iii) holds, where

(i) $f \in L^{\infty} \cap C_u$, $f(\infty) = \mu$ are satisfied

(ii) (i), $f-\mu \in L^1$, and (7) are satisfied

(iii) (ii), $f \in AC_{loc}$, $f' \in L^{\infty} \cap L^1$, and either (9) or (10) are satisfied (note, (9) or (10) implies (15) holds).

Then $S_{\bar{B}_r}(2^*\mu)$ is of type 1 for all $r \geq 0$. (In particular, $x(\infty)$ exists for any bounded solution $x \in T$ of (2f) provided only that f satisfies (4) and $f(\infty) = \mu$ -- and g, A satisfy (3), (14), (15), and the appropriate additional restrictions of (ii) or (iii).)

ACKNOWLEDGEMENT

This research was supported by National Science Foundation Grant No. MCS-8001524.

REFERENCES

1. J. J. Levin and D. F. Shea, On the asymptotic behavior of the bounded solutions of some integral equations, I, II, III, *J. Math. Anal. Appl.* 37 (1972), 42-82, 288-326, 537-575.

2. J. J. Levin, On some geometric structures for integrodifferential equations, *Adv. Math.* 22 (1976), 146-186.

3. J. J. Levin, Tauberian curves and integral equations, *J. Integral Eqs.* 2 (1980), 57-91.

4. J. J. Levin, Remarks on Tauberian curves, *Amer. J. Math.*, Supplement (1981), 185-190.

ON VOLTERRA EQUATIONS WITH LOCALLY FINITE MEASURES AND L^∞-PERTURBATIONS

Stig-Olof Londen

Mathematics Research Center
University of Wisconsin-Madison
Madison, Wisconsin
and
Institute of Mathematics
Helsinki University of Technology
Otaniemi, Finland

In this lecture we discuss the asymptotics of the scalar, real, nonlinear Volterra differential equation of convolution type

$$\begin{aligned} &x'(t) + \int_{[0,t]} g(x(t-s))d\mu(s) = f(t), \quad t \in R^+ = [0,\infty), \\ &x(0) = x_0. \end{aligned} \tag{1}$$

Here g and f are given functions, μ is a given real Borel measure on R^+ and x(t) is the unknown solution. In particular, we concentrate on the case when both μ and f are large; that is, on the case when μ is only locally finite and f vanishes at infinity, but $f \in L^p(R^+)$, for some $p < \infty$, does not necessarily hold.

The following basic assumptions will be made throughout this lecture:

$$\begin{cases} g \in C(R) \\ \mu \text{ is a real, locally finite, positive definite measure on } R^+ \\ f \in L^1_{loc}(R^+) \\ x \in (L^\infty \cap LAC)(R^+), \ x \text{ satisfies (1) a.e. on } R^+. \end{cases} \tag{2}$$

In particular we note that we do assume the existence of a uniformly bounded, locally absolutely continuous solution of (1).

For u,v functions, ν a measure, all supported on R^+, we define the convolutions $(u*v)(t)$, $(u*\nu)(t)$ for $t \in R^+$ by

$$(u*v)(t) = \int_0^t u(t-\tau)v(\tau)d\tau, \quad (u*\nu)(t) = \int_{[0,t]} u(t-\tau)d\nu(\tau).$$

Also let

$$\tilde{u}(z) = \int_{R^+} e^{-zt}u(t)dt, \quad \tilde{\nu}(z) = \int_{R^+} e^{-zt}d\nu(t),$$

and write $\hat{\nu}(\omega) = \tilde{\nu}(i\omega)$, $\hat{u}(\omega) = \tilde{u}(i\omega)$; $\omega \in R$.

The classical approach to obtain information on the asymptotic behavior of solutions of (1) is to define the quadratic form Q for $\phi \in L^2_{loc}(R^+)$, $T > 0$, by

$$Q(\phi,\mu,T) = \int_0^T \phi(t)(\phi*\mu)(t)dt$$

and then try to arrange things so that

$$\sup_{T>0} Q(g(x(t)),\mu,T) < \infty. \tag{3}$$

As is well-known [7,8], a detailed knowledge of the asymptotics of $g(x(t))$ can be obtained from (3). The straightforward way to get (3) is to take f small; that is, $f \in L^1(R^+)$, which together with $x \in L^\infty(R^+)$ yields (3).

In case f only satisfies $\lim_{t\to\infty} f(t) = 0$ then (3) is, in general, out of reach. But by instead taking μ small enough, specifically, by assuming

$$\int_{R^+} t\,d|\mu|(t) < \infty,$$

and by working with the limit equation corresponding to (1) which is

$$y'(t) + \int_{R^+} g(y(t-s))d\mu(s) = 0, \quad t \in R, \tag{4}$$

one may even now obtain asymptotic results on (1), [2,9].

If the continuity assumption on $g(x)$ is strengthened to $g(x)$ locally Lipschitzian, then no moment condition on μ is needed; μ finite suffices. In fact, one has the following

THEOREM 1, [3]. Let μ be a finite real positive definite Borel measure on R^+ and assume $\hat{\mu}(\omega) = 0$ for $\omega \in Z \equiv \{\omega | \mathrm{Re}\ \hat{\mu}(\omega) = 0\}$. Suppose $0 \notin Z$ and let $g(x)$ be locally Lipschitzian. For any positive constant K define

$$Y_K = \{y(\tau) | y \in LAC(R),\ y \text{ satisfies (4) a.e. on } R,\ |y(\tau)| \le K,\ \tau \in R\}.$$

Then

$$\sup_{y \in Y_K} \|g(y(\tau))\|_{L^2(R)} < \infty, \quad \sup_{y \in Y_K} \|y'(t)\|_{L^2(R)} < \infty. \tag{5}$$

Recall that $\mathrm{Re}\ \hat{\mu}(\omega) \ge 0$, $\omega \in R$, provided μ is a (finite) positive definite measure.

From (5) one may, under the condition $\lim_{t\to\infty} f(t) = 0$, deduce that $\lim_{t\to\infty} g(x(t)) = \lim_{t\to\infty} x'(t) = 0$.

The obvious next question to pose is what results one may obtain in the case when f satisfies only $\lim_{t\to\infty} f(t) = 0$ and μ is merely locally finite. A difficulty that one now encounters is, of course, that the limit equation (4) becomes meaningless.

A result providing a partial answer to this question is the following Theorem 2 which essentially is due to Gripenberg [1], although he formulates it somewhat differently; in particular, he takes μ finite.

THEOREM 2. Let z be the solution of

$$z'(t) + (z*\mu)(t) = f(t), \quad z(0) = x_0, \quad t \ge 0. \tag{6}$$

Assume that

$$r' \in (L^1 \cap NBV)(R^+), \quad \int_{R^+} t d|r'|(t) < \infty,$$

$$|\hat{\mu}(\omega)| < \infty, \quad \omega \neq 0,$$

$$\text{Im } \hat{\mu}(\omega) = 0 \quad \text{if} \quad \omega \in Z, \quad \omega \neq 0.$$

Then, if $\lim_{t\to\infty} z(t)$ exists and is finite and $\lim_{t\to\infty} z'(t) = 0$, one has

$$\lim_{t\to\infty} x'(t) = 0$$

and

$$\lim_{t\to\infty} [r(\infty)x(t) + [1 - r(\infty)]g(x(t))] = z(\infty).$$

The functions g,f and the measure μ are, of course, the same in (1) and in (6).

Thus, if the solution z(t) of the linear equation (6) tends to a limit, then so do, under certain hypotheses, bounded solutions of the nonlinear equation (1).

In Theorem 2, r(t) denotes the differential resolvent of μ and is defined as the solution of

$$r'(t) + (r*\mu)(t) = 0, \quad r(0) = 1. \tag{7}$$

Note that $z(t) = x_0r(t) + (r*f)(t)$. In applications one frequently has $r(\infty) = 0$ in which case the second conclusion of Theorem 2 reduces to $\lim_{t\to\infty} g(x(t)) = z(\infty)$.

The set Z of Theorem 2 is defined as in Theorem 1, but by $\hat{\mu}(\omega)$, $\omega \neq 0$, we now mean $\lim_{\substack{s\to i\omega \\ \text{Re } s>0}} \tilde{\mu}(s)$. To see that this is well-defined, one notes that as μ is a positive definite measure then it is a tempered distribution [7], and so $\tilde{\mu}(s)$ exists for Re s > 0. Next observe that

$$-\int_{R^+} e^{-st} dr'(t) = s\tilde{\mu}(s)[s + \tilde{\mu}(s)]^{-1}, \quad \text{Re } s > 0.$$

But by assumption the left side is continuous for Re s ≥ 0. Hence the limit of the right side as $s \to i\omega$ exists and so $\lim_{\substack{s\to i\omega \\ \text{Re } s>0}} \tilde{\mu}(s)$ is well-defined for $\omega \neq 0$, although possibly infinite. This last possibility is, however, excluded by assumption.

There are two serious drawbacks in Theorem 2. The first is the moment condition on the variation of $r'(t)$ which, in general, is a very difficult hypothesis to check. It is satisfied if $d\mu(t) = a(t)dt$ with $(-1)^k a^{(k)}(t) \geq 0$ for $k = 0,1,2,3$; but apart from this particular example it is hard to find classes of locally finite measures which do satisfy this assumption. The second drawback is the requirement that $|\hat{\mu}(\omega)|$ be finite for $\omega \neq 0$ which excludes some typical positive definite measures like $d\mu(t) = t^{-\alpha}\cos t\, dt$, $0 < \alpha < 1$.

Note that in applications the moment condition frequently implies that $\hat{\mu}$ is finite for $\omega \neq 0$. This is seen as follows. Let $-\nu$ denote the finite measure generated by the variation of $r'(t)$. Thus $\hat{\nu}(\omega) = -\int_{R^+} e^{-i\omega t} dr'(t)$. From the moment condition it follows that $\hat{\nu}(\omega) \in C^1(R)$ and, in particular, that

$$\text{Re } \hat{\nu}(\omega) = |i\omega + \hat{\mu}(\omega)|^{-2}\omega^2 \text{Re } \hat{\mu}(\omega) \in C^1(R). \tag{8}$$

Let $\omega_0 \neq 0$ be a point such that $|\hat{\mu}(\omega_0)| = \infty$. Then $\text{Re } \hat{\nu}(\omega_0) = 0$ and

$$[\omega-\omega_0]^{-1}[\text{Re } \hat{\nu}(\omega) - \text{Re } \hat{\nu}(\omega_0)] \begin{cases} \geq 0 & \text{if } \omega > \omega_0 \\ \leq 0 & \text{if } \omega < \omega_0. \end{cases}$$

Consequently, if ((9) does hold in most applications)

$$\limsup_{|\omega-\omega_0|\to 0} |\hat{\mu}(\omega)|^{-2}|\omega-\omega_0|^{-1}\text{Re } \hat{\mu}(\omega) > 0, \tag{9}$$

then a contradiction follows and $|\hat{\mu}(\omega_0)| < \infty$.

To realize why $|\hat{\mu}(\omega)| < \infty$, $\omega \neq 0$, and the moment condition are crucial to the proof, it suffices to at first observe that the proof of Theorem 1 basically consists of at first transforming (1) to an equation of the same type as (1) but with a finite measure, namely to

$$x'(t) + \int_{[0,t]} h(x(t-s))d\nu(s) = z'(t), \tag{10}$$

and then of applying previously known results [2,9]. In (10), $h(x) \equiv g(x) - x$ and, by (8), ν is a positive definite measure. The existing asymptotic results on (10) (assuming ν positive definite, finite; $h \in C(R)$, $z'(\infty) = 0$) do, however, require the imaginary part of the transform to vanish on the set where the real part vanishes. To check this requirement one notices at first that

$$\{\omega | \text{Re } \hat{\nu}(\omega) = 0\} = \{0\} \cup Z_1 \cup Z_2 \tag{11}$$

where

$$Z_1 \equiv \{\omega \neq 0 | \text{Re } \hat{\mu}(\omega) = 0\}, \quad Z_2 \equiv \{\omega \neq 0 | \ |\hat{\mu}(\omega)| = \infty\}. \tag{12}$$

Obviously $\hat{\nu}(\omega) = 0$ for $\omega \in Z_1$. But for $\omega \in Z_2$ we have $\hat{\nu}(\omega) = i\omega \neq 0$ and consequently we are led to the assumption that Z_2 be empty.

Secondly, one observes that the existing asymptotic results on (10) do, in case $0 \in \{\omega | \text{Re } \hat{\nu}(\omega) = 0\}$ which now holds, require the measure to be not only finite but also to satisfy a first moment condition.

So far it has not been possible to remove the assumption $|\hat{\mu}(\omega)| < \infty$, $\omega \neq 0$, and the moment condition without adding some other hypotheses. What we do have, however, is the following:

THEOREM 3, [4,5]. Let $g(x)$ be locally Lipschitzian and assume

$$xg(x) > 0, \quad x \neq 0; \quad \liminf_{|x|\to 0} x^{-1}g(x) > 0,$$

and

$$\text{Im } \hat{\mu}(\omega) = 0, \quad \omega \in Z_1. \tag{13}$$

Furthermore, suppose that the solution $r(t)$ of (7) satisfies

$$r' \in (L^1 \cap \text{NBV})(R^+)$$

and that

$$\lim_{t\to\infty} z'(t) = 0, \quad \lim_{t\to\infty} z(t) \text{ exists and is finite.}$$

Finally, let $\gamma \equiv \lim_{|\omega|\to 0} i\omega[\hat{\mu}(\omega)]^{-1}$ be finite. Then $\lim_{t\to\infty} x'(t) = 0$ and, in addition,

$$\lim_{t\to\infty} \{x(t) + \gamma^{-1}g(x(t))\} = x_0 + \lim_{\substack{s\to 0+ \\ s \text{ real}}} \tilde{f}(s) \quad \text{if} \quad \gamma > 0 \tag{14}$$

$$\lim_{t\to\infty} g(x(t)) = \lim_{\substack{s\to 0+ \\ s \text{ real}}} \{s[\tilde{\mu}(s)]^{-1}\tilde{f}(s)\} \quad \text{if} \quad \gamma = 0. \tag{15}$$

Let us make a few comments on Theorem 3. Note at first that the assumption $r' \in L^1(R^+)$ and the positive definiteness of μ yield that γ exists and is nonnegative. The assumption that γ is finite can be shown to be equivalent to $r(\infty) < 1$; in fact, one has $\gamma = r(\infty)[1-r(\infty)]^{-1}$ for $r(\infty) < 1$, $\gamma = \infty$ for $r(\infty) = 1$. (If we let $b(t) \equiv \mu([0,t])$, then $\gamma^{-1} = \int_{R^+} b(s)ds$.)

Secondly, observe that the existence of all the limits in (14), (15) is part of the conclusion. We also wish to point out that the conclusions of Theorems 2 and 3 are the same (except that Theorem 3 does not allow the case $r(\infty) = 1$). However, in Theorem 3 we have preferred to formulate the results in terms of the a priori given g, μ and f.

Finally, and most important, note that no moment condition is imposed on the variation of r'; neither is $|\hat{\mu}(\omega)| < \infty$ for $\omega \neq 0$ required. The key added condition is that g be locally Lipschitzian whereas (13), although essential to the proof, is of a more technical nature. These assumptions on g have the advantage of being easily checked, and they are not overly restrictive.

The proof of Theorem 3 is rather lengthy and involves some fairly complicated estimates. The key ingredient is the following auxiliary lemma concerning solutions of the limit equation

$$y(t) + \int_{R^+} g_1(y(t-s))\alpha([0,s])ds = 0, \quad t \in R. \tag{16}$$

LEMMA, [4]. Let g_1 be locally Lipschitzian and satisfy

$$xg_1(x) \geq 0, \quad x \in R. \tag{17}$$

Suppose α is a real, finite, positive definite Borel measure on R^+ and that

$\alpha([0,t]) \in L^1(R^+)$. Define Z_α by

$$Z_\alpha = \{\omega \mid \operatorname{Re} \hat{\alpha}(\omega) = \operatorname{Re} \int_{R^+} e^{-i\omega t} d\alpha(t) = 0\}$$

and suppose that Z_α can be written as the union of three pairwise disjoint sets Z_1, Z_2, $\{0\}$ such that

$$\operatorname{Im} \hat{\alpha}(\omega) = 0, \quad \omega \in Z_1,$$

$$\inf_{\omega \in Z_2 \cup \{0\}} \operatorname{Re} \int_{R^+} e^{-i\omega t} \alpha([0,t])dt > 0.$$

Define, for any positive constant K,

$$Y_K = \{y \mid y \in LAC(R), \ y \text{ satisfies (16) on } R, \ \|y\|_{L^\infty(R)} \leq K\}.$$

Then

$$\sup_{y \in Y_K} \|g_1(y(t))\|_{L^2(R)} < \infty, \quad \sup_{y \in Y_K} \|y'(t)\|_{L^2(R)} < \infty.$$

Considering the comments made early in this lecture and the transformation of (1) to (10), a very natural question to ask is when does

$$\sup_{T>0} Q(h(x(t)),\nu,T) < \infty \tag{18}$$

hold and, in case it does, what information on the asymptotics of $x(t)$ can be extracted from this relation. Below we give some preliminary results in this direction.

Note at first that $z' \in L^1(R^+)$ together with $x \in L^\infty(R^+)$ gives (18). As

$$z'(t) = x_0 r'(t) + f(0)r(t) + (f' * r)(t),$$

it is clear that sufficient conditions for $z' \in L^1(R^+)$ to hold are

$$r, r', f' \in L^1(R^+). \tag{19}$$

To simplify the statements below we also assume

$$x \text{ is uniformly continuous on } R^+. \tag{20}$$

Let $\Gamma(h(x(t)))$ denote the positive limit set of $h(x(t))$ and let $\sigma(\Gamma(h(x(t))))$ denote the spectrum of this limit set. One has (see [7,8] and use (11)) that (18) implies

$$\sigma(\Gamma(h(x(t)))) \subset \{0\} \cup Z_1 \cup Z_2. \tag{21}$$

The obvious question now is what points, if any, can be removed from the right side of (21). Some answers to this are given in Theorems 4-6 below. To shorten these statements we let (22) be defined as follows

$$\begin{cases} (2),\ (19),\ (20) \text{ hold} \\ \hat{\mu}(\omega) = 0, \quad \omega \in Z_1, \\ Z_2 \text{ is (at most) denumerable.} \end{cases} \tag{22}$$

Also observe that as $r' \in L^1(R^+)$, $\operatorname{Re} \hat{r}'(Z_1) = 0$, $\operatorname{Re} \hat{r}'(Z_2) = -1$ then one has that Z_2 is a bounded set and $\operatorname{dist}(Z_1,Z_3) > 0$. ($Z_3 \equiv Z_2 \cup \{0\}$.)

We may now formulate

THEOREM 4, [5]. Let (22) hold and assume that Z_1 is empty. Then $\lim_{t\to\infty} g(x(t)) = z(\infty)$.

Thus the points of Z_2, if alone (except for $\{0\}$) on the right side of (21), do not create any oscillatory behavior at infinity.

In case neither Z_1 nor Z_2 is empty we obtain the following somewhat surprising result:

THEOREM 5. Let (22) hold and assume that

$$\hat{r}'(\omega) \text{ is Hölder continuous with exponent } 1/2 \text{ on } \{\omega \mid \operatorname{dist}(\omega,Z_3) \geq 1/2 \operatorname{dist}(Z_1,Z_3)\}. \tag{23}$$

Then

$$\sigma(\Gamma(x(t))) \subset \{0\} \cup Z_2, \quad \sigma(\Gamma(g(x(t)))) \subset \{0\} \cup Z_1. \tag{24}$$

The condition (23) is needed as the proof makes use of a result by Pollard [6]. Requiring Hölder continuity on a neighbornood of Z_3 would, of course, be unrealistic and (23) does indeed exclude a vicinity of Z_3 from the set where Hölder continuity is to hold.

How to proceed beyond (24) without making additional assumptions on Z_2 is at this moment an open problem. It appears that with appropriately chosen Z_1, Z_2 and g one may have an oscillatory behavior at infinity. Note, however, that if the points of Z_2 are isolated and the technical condition (25) is satisfied, then the conclusion (24) may be strengthened by using [8, p. 252]. In fact, we obtain

THEOREM 6. Let (22) hold and suppose that Z_2 contains only isolated points. Furthermore, for each $\omega_2 \in Z_2$, let

$$\lim_{\varepsilon \downarrow 0} \left[\int_{\omega_2+\varepsilon}^{\omega_2+1} + \int_{\omega_2-1}^{\omega_2-\varepsilon}\right] \frac{\mathrm{Re}\ \hat{\mu}(\omega)}{|\hat{\mu}(\omega)|^2(\omega-\omega_2)^2}\, d\omega = \infty. \tag{25}$$

Then

$$\sigma(\Gamma(h(x(t)))) \subset \{0\} \cup Z_1.$$

If in addition (23) is satisfied, then $\lim_{t\to\infty} g(x(t)) = z(\infty)$.

Note that (25) holds in most cases arising in applications.

ACKNOWLEDGEMENT

This research was sponsored by the United States Army under Contract No. DAAG29-80-C-0041.

REFERENCES

1. G. Gripenberg, On nonlinear Volterra equations with nonintegrable kernels, *SIAM J. Math. Anal.* 11 (1980), 668-682.

2. S.-O. Londen, On a Volterra integrodifferential equation with L^∞-perturbation and noncountable zero-set of the transformed kernel, *J. Integral Eqs.* 1 (1979), 275-280.

3. S.-O. Londen, On an integral equation with L^∞-perturbation, *J. Integral Eqs.*, to appear.

4. S.-O. Londen, On some integral equations with locally finite measures and L^∞-perturbations, MRC Technical Summary Report #2224, Mathematics Research Center, University of Wisconsin-Madison, Madison, Wisconsin, 1981.

5. S.-O. Londen, manuscript in preparation.

6. H. Pollard, The harmonic analysis of bounded functions, *Duke Math. J.* 20 (1953), 499-512.

7. O. J. Staffans, Positive definite measures with applications to a Volterra equation, *Trans. Amer. Math. Soc.* 218 (1976), 219-237.

8. O. J. Staffans, Tauberian theorems for a positive definite form, with applications to a Volterra equation, *Trans. Amer. Math. Soc.* 218 (1976), 239-259.

9. O. J. Staffans, On a nonlinear integral equation with a nonintegrable perturbation, *J. Integral Eqs.* 1 (1979), 291-307.

A NONLINEAR CONSERVATION LAW WITH MEMORY

J. A. Nohel

Mathematics Research Center
University of Wisconsin - Madison
Madison, Wisconsin

1. INTRODUCTION

In this paper we study the model nonlinear Volterra functional differential equation (with infinite memory)

$$u_t + \phi(u)_x + \int_{-\infty}^{t} a'(t-\tau)\psi(u(\tau,x))_x d\tau = f(t,x), \quad -\infty<t<\infty, \quad 0\le x\le 1, \tag{1.1}$$

where $\phi,\psi\colon \mathbb{R} \to \mathbb{R}$ are given smooth constitutive functions, $a\colon [0,\infty) \to \mathbb{R}$ is a given memory kernel, and $f\colon \mathbb{R} \times [0,1] \to \mathbb{R}$ is a given function representing an external force; subscripts denote partial derivatives and $' = d/dt$. The motivation and the assumptions under which (1.1) is studied are provided by the more complex physical problem of the extension of a finite, homogeneous, elastoviscous body moving under the action of an assigned body force. The viscoelastic problem, formulated in Section 2, was recently studied by Dafermos and Nohel [5]; references to earlier literature are given in Section 2.

The model problem (1.1), which is of independent interest, is simpler in that it is of first order, while the equation of motion (2.8) below, is of second order; otherwise (1.1) incorporates the interesting features of (2.8). The most important of these is the following. If $\psi \equiv 0$, $f \equiv 0$, (1.1) reduces to Burgers' equation (conservation law of gas dynamics):

$$u_t + \phi(u)_x = 0; \tag{1.2}$$

note that (1.2) is quasilinear and hyperbolic. It is classical, see Lax [10], that the Cauchy problem consisting of Burgers' equation and the initial condition $u(0,x) = u_0(x)$, $x \in \mathbb{R}$, does not, in general, possess a classical smooth solution, no matter how smooth (and "small") one takes the initial datum u_0; if ϕ is convex Lax [10] has shown that the solution develops a singularity (shock) in finite time due to the crossing of characteristics. More precisely, if $u_0'(x) > 0$, u_0 smooth, and ϕ is convex, (1.2) has a smooth solution for all $t > 0$, while if $u_0'(x_0) < 0$ for some x_0, the characteristics of (1.2) will cross and shocks will develop in finite time; a similar result also holds for systems of conservation laws in one space dimension. If ϕ is not convex, e.g., in nonlinear elasticity, similar results have been established by MacCamy and Mizel [12]; for recent closely related literature see also general results by John [8], Kleinerman and Majda [9], Malek-Madani [13].

The purpose of this paper is: First, in Section 2 we formulate the problem of motion of a nonlinear viscoelastic body as analyzed recently by Dafermos and Nohel [5] using a combination of energy methods and properties of strongly positive kernels. Second, under assumptions motivated by the viscoelastic problem, we show in Sections 3 and 4 that equation (1.1), together with an assigned periodic boundary condition and an assigned smooth history

$$u(t,x) = v(t,x), \quad -\infty < t \le 0, \quad 0 \le x \le 1,$$

which satisfies (1.1) for $t \le 0$, possesses a unique classical solution for all $t > 0$, provided the history function v and the forcing term f are sufficiently smooth and "small" in suitable norms. This result, in which the same strategy as in [5] is used, exhibits the dissipative character of the integral term in (1.1); its proof serves to illustrate a general energy technique for hyperbolic, nonlinear Volterra problems. Finally, in Section 5 we formulate the problem of development of singularities in finite time and we present a recent result, analogous to the result for Burgers' equation, in an important special case, [15]; the proof will appear elsewhere. The fact that the question of development of singularities for the physically interesting viscoelastic problem remains open, provides the principal motivation for studying the simpler model problem (1.1). It should be noted that the problem discussed in Section 5 is different from the study of weak solutions for the Riemann problem for (1.1) (Greenberg and Ling Hsiao [6], Dafermos and Ling Hsiao [3]; see also the paper of Greenberg, Ling Hsiao and MacCamy in these proceedings).

2. A HYPERBOLIC, NONLINEAR VOLTERRA EQUATION IN VISCOELASTICITY

Problems arising in continuum mechanics can often be modeled by quasilinear hyperbolic systems in which the characteristic speeds are not necessarily constant. Such systems have the property that waves may be amplified and solutions that were initially smooth may develop discontinuities ("shocks") in finite time. Of particular interest are situations in which the destabilizing mechanism arising from nonlinear effects can coexist and compete with dissipative effects.

In certain cases dissipation is so powerful that waves cannot break and solutions remain smooth for all time. A more interesting situation arises when the amplification and decay mechanisms are nearly balanced so that the outcome of their confrontation cannot be predicted at the outset. A dimensional analysis indicates that the breaking of waves develops on a time scale inversely proportional to wave amplitude, whereas dissipation proceeds at a rate roughly independent of amplitude. Therefore, when the initial data are small it might be expected that the dissipation effects would prevail and waves would not break. Results of this type have been obtained by Nishida [17] for a model problem concerning a quasilinear second order wave equation in one space dimension. Nishida's analysis uses the method of Riemann invariants and is therefore restricted to one space dimension. Using energy methods, Matsumura [16] has studied the case of more than one space dimension, and was able to prove the existence of smooth solutions for all time for quasilinear hyperbolic systems with frictional damping. Burgers' equation (1.2) shows the necessity of the presence of some form of frictional damping to avoid the formation of "shocks". The rather delicate situation of thermal damping in thermoelasticity is discussed by Slemrod [18].

A different and subtler dissipative mechanism is induced by memory effects of elastico-viscous materials. Dafermos and Nohel [5] have recently developed and analyzed a one-dimensional nonlinear model for the homogeneous extension of an elastico-viscous rod whose ends are free of traction. Their simple, one-dimensional, model corresponds to the following constitutive relation, suggested by the theory developed by Coleman and Gurtin [1],

$$\sigma(t,x) = \phi(e(t,x)) + \int_{-\infty}^{t} a'(t-\tau)\psi(e(\tau,x))d\tau, \tag{2.1}$$

where σ is the stress, e the strain, a the relaxation function with $' = d/dt$, and ϕ,ψ assigned constitutive functions. The relaxation function is normalized so that $a(\infty) = 0$. When the reference configuration is a natural state, $\phi(0) = \psi(0) = 0$. Experience indicates that $\phi(e)$, $\psi(e)$, as well as the equilibrium stress

$$\chi(e) \equiv \phi(e) - a(0)\psi(e) \tag{2.2}$$

are increasing functions of e, at least near equilibrium (i.e., $|e|$ small). Moreover, the effect of viscosity is dissipative. To express mathematically the above physical requirements, we impose upon $a(t)$, $\phi(e)$, $\psi(e)$ and $\chi(e)$ the following assumptions:

$$a(t) \in W^{2,1}(0,\infty), \quad a(t) \text{ is strongly positive definite on } [0,\infty); \tag{2.3}$$

$$\phi(e) \in C^3(-\infty,\infty), \quad \phi(0) = 0, \quad \phi'(0) > 0; \tag{2.4}$$

$$\psi(e) \in C^2(-\infty,\infty), \quad \psi(0) = 0, \quad \psi'(0) > 0; \tag{2.5}$$

$$\chi'(0) = \phi'(0) - a(0)\psi'(0) > 0. \tag{2.6}$$

Assumption (2.3), which requires that $a(t) - \alpha \exp(-t)$ be a positive definite kernel on $[0,\infty)$ for some $\alpha > 0$, expresses the dissipative character of viscosity. Smooth, integrable, nonincreasing, convex relaxation functions, e.g.,

$$a(t) = \sum_{k=1}^{K} \nu_k \exp(-\mu_k t), \quad \nu_k > 0, \quad \mu_k > 0, \tag{2.7}$$

which are commonly employed in the applications of the theory of viscoelasticity, satisfy (2.3).

We now consider a homogeneous, one-dimensional body (string or bar) with reference configuration [0,1] of density $\rho = 1$ (for simplicity) and constitutive relation (2.1), which is moving under the action of an assigned body force $g(t,x)$, $-\infty < t < \infty$, $0 \le x \le 1$, with the ends of the rod free of traction. We let u(t,x) denote the displacement of particle x at time t, in which case the strain is $e(t,x) = u_x(t,x)$. Thus the equation of motion $\rho u_{tt} = \sigma_x + \rho g$ here takes the form of the nonlinear (hyperbolic) Volterra functional differential equation

$$u_{tt} = \phi(u_x)_x + \int_{-\infty}^{t} a'(t-\tau)\psi(u_x)_x d\tau + g, \quad -\infty < t < \infty, \quad 0 \le x \le 1. \tag{2.8}$$

The physical problem of the motion of a vicscoelastic body suggests that the history of the motion of the body up to time t = 0 is assumed known, i.e.,

$$u(t,x) = v(t,x), \quad -\infty < t \le 0, \quad 0 \le x \le 1, \tag{2.9}$$

where v(t,x) is a given sufficiently smooth function which satisfies equation (2.8) for $t \le 0$, together with appropriate boundary conditions. In order to show that the motion of the viscoelastic bar remains smooth for all $t > 0$, the mathematical task is to determine a smooth extension u(t,x) of v(t,x) on $(-\infty,\infty) \times [0,1]$ which satisfies (2.8) together with assigned boundary conditions, for $-\infty < t < \infty$.

Dafermos and Nohel [5, Theorem 1.1] establish such a global result for the problem (2.8), (2.9) together with homogeneous Neumann boundary conditions

$$u_x(t,0) = u_x(t,1) = 0, \quad -\infty < t < \infty; \tag{2.10}$$

these are shown to be equivalent to the statement: the boundary of the body is free of traction ($\sigma(t,0) = \sigma(t,1) = 0$, $-\infty < t < \infty$, σ given by (2.1)). Their global result is valid for sufficiently smooth and "small" external body forces g and history functions v. Other boundary conditions and various generalizations are also considered.

The general strategy used in [5] is as follows. First one establishes the existence of a unique smooth local solution u defined on a maximal interval $(-\infty,T_0) \times [0,1]$, with the property that when $T_0 < \infty$ a certain norm of the solution becomes infinite as $t \to T_0^-$; this is done by a fixed point argument (combined with a standard energy method for linear problems) on a suitably chosen abstract space of functions. Second, energy methods are combined with properties of strongly positive kernels to show that due to the viscous dissipation of the integral term in (2.8), the aforementioned norm remains uniformly bounded on the maximal interval, provided the data g and v are sufficiently smooth and small. By standard theory for nonlinear problems this means that $T_0 = +\infty$ and the smooth solution exists globally in t. This part of the analysis involves obtaining a priori estimates of certain norms of the derivatives (in one space dimension, up to and including order 3) directly from the equation (2.8). It is here that it becomes convenient to use the equivalent form

$$u_{tt} = \chi(u_x)_x + \int_{-\infty}^{t} a(t-\tau)\psi(u_x)_{\tau x}d\tau + g, \quad -\infty < t < \infty, \quad 0 \leq x \leq 1 \quad (2.11)$$

of equation (2.8) (equation (2.11) is obtained from (2.8) by integrating by parts with respect to τ and by using the definition of the equilibrium stress χ); assumption (2.6) plays a crucial role.

For the special case $\psi(e) \equiv \phi(e)$ various global existence results for (2.8) were established by MacCamy [11], Dafermos and Nohel [4] and Staffans [20]. The assumption $\psi \equiv \phi$ allows one to invert the linear Volterra integral operator on the right-hand side of (2.8) and thus express $\phi(u_x)_x$ in terms of $u_{tt} - g$ through an inverse Volterra integral operator using the resolvent kernel associated with a'. One may then transfer time derivatives from u_{tt} to the resolvent kernel via integration by parts. This procedure, which was also discussed by Dafermos [2] in a recent expository paper in a number of special cases, reveals the instantaneous character of dissipation and, at the same time, renders the memory term linear and milder, thus simplifying the analysis considerably. On the other hand, the above approach is somewhat artificial: By inverting the right-hand side of (2.8), one loses sight of the original equation and of the physical interpretation of the derived a priori estimates. More importantly, the physical appropriateness of the restriction $\psi = \phi$ is by no means clear.

REMARK. The present normalization of the kernel a with $a(\infty) = 0$ is different from that in the existing literature (see [4], [11], [20]). The reader should note a', not a, enters the constitutive relation (2.1) as well as the equation of motion (2.8). In the earlier literature in which only the special case $\psi \equiv \phi$ was studied, the normalization $a(t) = a_\infty + A(t)$ for $0 \leq t < \infty$, $a(0) = 1$, $a_\infty > 0$, $A \in W^{2,1}(0,\infty)$, A strongly positive was used. The normalization used here is crucial for generating the a priori estimates directly from equation (2.11) (equivalent to (2.8)). The reader should note that the present normalization and (2.6) imply that $0 < a(0) < 1$, if $\phi \equiv \psi$.

The question of the development of singularities of solutions of (2.8) in finite time for sufficiently "large", smooth data (which have been observed for viscoelastic bodies [1]) is under active study. Some partial results with $\psi \equiv \phi$ in (2.8) have been obtained by Hattori [7]; for a viscoelastic fluid see Slemrod [19]. However, in the general case of (2.8) the problem is far from settled. For this reason we believe that the approach in Section 5 for the considerable simpler problem (1.1) is particularly useful and suggestive.

3. A CONSERVATION LAW WITH FADING MEMORY

We study the model nonlinear, history-boundary value problem

$$u_t + \phi(u)_x + \int_{-\infty}^{t} a'(t-\tau)\psi(u(\tau,x))_x d\tau = f(t,x), \quad -\infty<t<\infty, \quad 0\le x\le 1 \tag{3.1}$$

subject to the periodic boundary condition

$$u(t,0) = u(t,1); \tag{3.2}$$

the history of the solution u is assumed to be known up to time $t = 0$, i.e.,

$$u(t,x) = v(t,x), \quad -\infty < t \le 0, \quad 0 \le x \le 1, \tag{3.3}$$

where v is a given smooth $(C^1((-\infty,0] \times [0,1]))$ function which satisfies (3.1), (3.2) for $t \le 0$. In (3.1) $\phi,\psi\colon \mathbb{R} \to \mathbb{R}$, $f\colon \mathbb{R} \times [0,1] \to \mathbb{R}$, and $a\colon [0,\infty) \to \mathbb{R}$ are given functions satisfying assumptions analogous to those for the viscoelastic problem outlined in Section 2. Our task in this section is to determine a smooth extension $u(t,x)$ of $v(t,x)$ on $(-\infty,\infty) \times [0,1]$ which satisfies (3.1) and (3.2), and to study the asymptotic properties of u as $t \to \infty$. In order to do this, the history v and the forcing function f will have to be taken sufficiently smooth and "small" in suitable norms.

The requirement that the history function v should satisfy (3.1) and (3.2) for $t < 0$ is motivated by the viscoelastic problem described in Section 2. The history value problem in which the function v is sufficiently smooth for $t \le 0$ (but need not satisfy (3.1)) is also of interest, and can be studied by the same methods (see further remarks below).

The basic assumptions for the global existence theory are as follows. Concerning the constitutive functions ϕ, ψ:

$$\phi, \psi \in C^2(\mathbb{R}), \quad \phi(0) = \psi(0) = 0, \quad \phi'(0) > 0, \quad \psi'(0) > 0; \tag{c}$$

concerning the kernel a:

$$a \in W^{2,1}[0,\infty) \text{ and } a \text{ is strongly positive on } [0,\infty); \tag{a}$$

concerning the forcing term f:

$$\begin{cases} f,\ f_t,\ f_x \in C[(-\infty,\infty);\ L^2(0,1)] \cap L^2[(-\infty,\infty);\ L^2(0,1)], \\ f(t,x) = f_1(t,x) + f_2(t,x), \\ f_{1tt},\ f_{2tx} \in L^2[(-\infty,\infty);\ L^2(0,1)], \\ \int_0^1 f(t,x)\,dx = 0, \quad -\infty < t < \infty, \end{cases} \tag{f}$$

and to measure the "size" of the forcing term we define

$$F = \sup_{(-\infty,\infty)} \int_0^1 \{f^2 + f_t^2 + f_x^2\}(t,x)\,dx + \int_{-\infty}^{\infty}\int_0^1 \{f^2 + f_t^2 + f_x^2 + f_{1tt}^2 + f_{2tx}^2\}\,dx\,dt;$$

concerning the history v:

$$\begin{cases} v,\ v_t,\ v_x,\ v_{tt},\ v_{tx},\ v_{xx} \in C[(-\infty,\infty);\ L^2(0,1)] \cap L^2[(-\infty,\infty);\ L^2(0,1)], \\ \int_0^1 v(t,x)\,dx = 0, \quad -\infty < t \quad 0, \\ \text{and } v \text{ satisfies (3.1), (3.2) for } t \le 0. \end{cases} \tag{H}$$

Analogous to the "equilibrium stress" for the viscoelastic body we define the constitutive function $\chi\colon \mathbb{R} \to \mathbb{R}$ by

$$\chi(\cdot) = \phi(\cdot) - a(0)\psi(\cdot),$$

and we assume that

$$\chi'(0) = \phi'(0) - a(0)\psi'(0) > 0. \tag{3.4}$$

The following equation, obtained from (3.1) by carrying out an integration by parts with respect to τ, will play a crucial role in the analysis:

$$u_t + \chi(u)_x + \int_{-\infty}^{t} a(t-\tau)\psi(u(\tau,x))_{x\tau}d\tau + f(t,x), \quad -\infty<t<\infty, \quad 0\le x\le 1. \tag{3.5}$$

It is clear that the problem consisting of (3.5), (3.2), (3.3) is equivalent to the original problem (3.1), (3.2), (3.3).

Upon setting

$$h(t,x) = -\int_{-\infty}^{0} a'(t-\tau)\psi(v(\tau,x))_x d\tau + f(t,x), \quad 0\le t<\infty, \quad 0\le x\le 1 \tag{3.6}$$

$$u_0(x) = v(0,x) \quad 0 \le x \le 1, \tag{3.7}$$

the history-boundary value problem (3.1) - (3.3) reduces to the initial-boundary value problem

$$\begin{cases} u_t + \phi(u)_x + \int_0^t a'(t-\tau)\psi(u(\tau,x))_x d\tau = h(t,x), & 0<t<\infty, \quad 0\le x\le 1 \\ u(t,0) = u(t,1), & 0\le t<\infty \\ u(0,x) = u_0(x), & 0\le x\le 1, \end{cases} \tag{3.8}$$

where $\int_0^1 h(t,x)dx = 0$ for $0 \le t < \infty$. (Use $v(t,0) = v(t,1)$ and assumptions (f).) Conversely, a solution of (3.8), where $\int_0^1 u_0(x)dx = 0$, can be shown to solve (3.1) - (3.3) by constructing a smooth function $v(t,x)$ on $(-\infty,0] \times [0,1]$, satisfying (3.2), $\int_0^1 v(t,x)dx = 0$ $(-\infty < t \le 0)$, and also requiring $v(0,x) = u_0(x)$, as well as

$$\begin{cases} v_t(0,x) = -\phi(u_0(x))_x + h(0,x) \equiv u_1(x) \\ v_{tt}(0,x) = -\phi''(u_0(x))u_0'(x)u_1(x) - \phi'(u_0(x))u_1'(x) \\ \qquad - a'(0)\psi(u_0(x))_x + h_t(0,x), \end{cases} \tag{3.9}$$

and defining

$$f(t,x) = \begin{cases} v_t + \phi(v)_x + \int_{-\infty}^{t} a'(t-\tau)\psi(v(\tau,x))_x d\tau, & -\infty<t\leq 0, \quad 0\leq x\leq 1 \\ h(t,x) + \int_{-\infty}^{0} a'(t-\tau)\psi(v(\tau,x))_x d\tau, & 0 < t < \infty, \quad 0 \leq x \leq 1; \end{cases} \tag{3.10}$$

the requirements (3.9) insure that f defined by (3.10) has the necessary smoothness properties across $t = 0$ required in the existence theory.

The main global result is

THEOREM 3.1. Let the assumptions (a), (c), and (3.4) be satisfied. There exists a constant $\mu > 0$ such that for every forcing term f satisfying assumptions (f) with $F \leq \mu^2$, and for any history function v on $(-\infty,0] \times [0,1]$ satisfying assumptions (H), there exists a unique function u on $(-\infty,\infty) \times [0,1]$ with $u, u_t, u_x, u_{tt}, u_{tx}, u_{xx} \in C[(-\infty,\infty); L^2(0,1)] \cap L^2[(-\infty,\infty); L^2(0,1)]$ satisfying (3.1) - (3.3), as well as $\int_0^1 u(t,x)dx = 0$, $-\infty < t < \infty$. Moreover,

$$u(t,x),\ u_t(t,x),\ u_x(t,x) \to 0 \quad \text{as} \quad t \to \infty \tag{3.11}$$

uniformly on [0,1].

The proof of Theorem 3.1 will be given in Section 4. It uses the general strategy developed in [5], although there are technical differences in details. One first establishes the existence of a local solution u on a maximal interval $(-\infty,T_0) \times [0,1]$ with the property that when $T_0 < \infty$ a certain norm of u becomes infinite as $t \to T_0^-$ (see Proposition 4.1 below). One then uses a combination of energy methods and properties of strongly positive kernels to show that the integral in (3.1) exerts a dissipative effect resulting in the aforementioned norm remaining uniformly bounded, independent of T_0, provided the constant μ in Theorem 3.1 is sufficiently small. Thus, in particular, $T_0 = +\infty$ and the smooth solution exists globally.

REMARK 3.2. It follows by standard regularity techniques that the solution u of (3.1), (3.2), (3.3) is C^1 smooth on $\mathbb{R} \times [0,1]$.

REMARK 3.3. A result similar to Theorem 3.1 (established by the same methods) holds for the boundary-initial value problem

$$\begin{cases} u_t + \phi(u)_x + \int_0^t a'(t-\tau)\psi(u(\tau,x))_x d\tau = h(t,x), & 0 < t < \infty, \quad 0 \leq x \leq 1 \\ u(t,0) = u(t,1), & 0 \leq t < \infty \\ u(0,x) = u_0(x), & 0 \leq x \leq 1, \quad u_0(0) = u_0(1); \end{cases} \tag{3.12}$$

here ϕ, ψ, and a satisfy the assumptions of Theorem 3.1, h is defined on $[0,\infty) \times [0,1]$ and satisfies assumptions (f) modified in the obvious way, and the initial datum $u_0 \in H^2(0,1)$, with $\int_0^1 u_0(x)dx = 0$. For the theorem to hold one must require that $\|u_0\|_{H^2}$, and a suitable norm of h are sufficiently small, as can be seen from a detailed examination of the corresponding estimates. (See Remark 4.3 below.)

REMARK 3.4. The requirement that the history v satisfies (3.1), (3.2) for $t \leq 0$, and the condition: the constant μ (where $F \leq \mu^2$) of Theorem 3.1 is "small", imply that the history function v, as well as the forcing term f in (3.1), are both small in suitable norms.

REMARK 3.5. If $\phi \equiv \psi$ in (3.1), Theorem 3.1 can be applied without any change, provided $0 < a(0) < 1$ (in order that (3.4) is satisfied). However, in this special case the somewhat different energy techniques of Dafermos and Nohel [4], or of Staffans [20], or the method of Riemann invariants of MacCamy [11] can also be used.

REMARK 3.6. It will follow from the proof of Theorem 3.1 in Section 4 that because the problem (3.1) - (3.3) is in one space dimension and on a finite space interval it is sufficient to obtain global estimates for the derivatives (in this case estimates of u_{tt}, u_{tx}, u_{xx}, because (3.1) is of first order) in the $L^\infty(L^2(0,1))$ and $L^2(L^2(0,1))$ norms, and then to make use of the Poincaré inequality to estimate lower order terms. However, we cannot apply this method to obtain global estimates for the pure Cauchy problem consisting of the Volterra equation

$$u_t + \phi(x)_x + \int_0^t a'(t-\tau)\psi(u(\tau,x))_x d\tau = h(t,x), \quad 0 < t < \infty, \quad x \in \mathbb{R} \quad (3.13)$$

together with the initial condition $u(0,x) = u_0(x)$, $x \in \mathbb{R}$, because $x \in \mathbb{R}$ and the Poincaré inequality does not apply. For the local existence result (analogue of Proposition 4.1) one can circumvent this difficulty. The same comments apply to the analogous pure Cauchy problem associated with the second-order nonlinear Volterra equation (2.8). If $\phi \equiv \psi$ in (3.13) or (2.8), the analogous pure Cauchy problems can be treated by either the methods of [4], [11] or of [20], because global estimates of u and of the derivatives of u in appropriate norms are obtained successively from the equations. However, if $\phi \not\equiv \psi$ in (3.13) or in the pure Cauchy problem associated with (2.8), these mathematically interesting problems remain to be tackled.

4. PROOF OF THEOREM 3.1

a. Local Theory

This is carried out for the boundary-initial value problem (which was shown to be equivalent to the history value problem (3.1) - (3.4) in Section 3 for a suitable choice of the forcing term f - see (3.10)):

$$\begin{cases} u_t + \phi(u)_x + \int_0^t a'(t-\tau)\psi(u(\tau,x))_x d\tau = h(t,x), & 0 < t < \infty, \quad 0 \le x \le 1 \\ u(t,0) = u(t,1), & 0 \le t < \infty \\ u(0,x) = u_0(x), & x \in [0,1]. \end{cases} \quad (4.1)$$

We make the following assumptions:

$$a', a'' \in L^1(0,\infty), \quad \phi, \psi \in C^2(\mathbb{R}), \quad \phi(0) = \psi(0) = 0, \quad \psi'(0) > 0,$$

there exists a constant $\kappa > 0$ such that $\phi'(\xi) \ge \kappa > 0$ $(\xi \in \mathbb{R})$,
$h: [0,\infty)\times[0,1] \to \mathbb{R}$, $h(t,x) = h_1(t,x) + h_2(t,x)$, h, h_t, $h_x \in C([0,\infty); L^2(0,1))$,
$\int_0^1 h(t,x)dx = 0$ for $t \in [0,\infty)$, $h_{1tt}, h_{2tx} \in L^2([0,\infty); L^2(0,1))$, and $u_0 \in H^2(0,1)$,

$\int_0^1 u_0(x)dx = 0$. The reader will note that the full strength of assumptions (a) and assumption (3.4) are not used in Proposition 4.1 below; on the other hand ϕ' is now required to be bounded away from zero (compare with assumptions (c)).

PROPOSITION 4.1. Under the above assumptions there exists a number T_0, $0 < T_0 \leq \infty$, and a unique function $u \in C^1([0,T_0) \times [0,1])$ with u_{tt}, u_{tx}, u_{xx} $\in C([0,T]; L^2(0,1))$ for every T, $0 < T < T_0$, such that u satisfies (4.1) on $[0,T_0) \times [0,1]$ and $\int_0^1 u(t,x)dx = 0$. Moreover, if $T_0 < \infty$

$$\limsup_{t\to T_0^-} \int_0^1 [u^2(t,x) + u_t^2(t,x) + \ldots + u_{xx}^2(t,x)]dx = +\infty. \tag{4.2}$$

It is clear that with h defined by (3.6), u_0 by (3.7), the problem (3.1) - (3.3) satisfies the assumptions of Proposition 4.1.

REMARK 4.2. A similar result holds for the pure Cauchy problem associated with (4.1), i.e., no boundary condition is specified and $u(0,x) = u_0(x)$, $u_0 \in H^2(\mathbb{R})$; however, the function space X(M,T) below must be specified differently.

The proof of Proposition 4.1 is very similar to that of Theorem 2.1 in [5], and will only be sketched. Let $M > 0$, $T > 0$ and let X(M,T) denote the set of functions w(t,x) on $[0,T] \times [0,1]$ with w, w_t, w_x, w_{tt}, w_{tx}, w_{xx} $\in$ $C([0,T];L^2(0,1))$, $w(t,0) = w(t,1)$, $w(0,x) = u_0(x)$, $\int_0^1 w(t,x)dx = 0$, $\int_0^1 w_t(t,x)dx = 0$, $0 \leq t \leq T$, and

$$\sup_{[0,T]} \int_0^1 [w_{tt}^2(t,x) + w_{tx}^2(t,x) + w_{xx}^2(t,x)]dx \leq M^2. \tag{4.3}$$

It follows from the Poincaré-type inequalities (see application of Lemmas A.1 and A.2 in Appendix) and from (4.3) that if $w \in X(M,T)$ then

$$w^2(t,x) + w_x^2(t,x) + w_t^2(t,x) \leq M^2, \quad 0 \leq t \leq T, \quad 0 \leq x \leq 1. \tag{4.4}$$

Let $S: X(M,T) \to C^1([0,T] \times [0,1])$ be the mapping which carries $w \in X(M,T)$ into the solution of the linear problem consisting of

$$u_t + \phi'(w)u_x = -\int_0^t a'(t-\tau)\psi(w(\tau,x))_x d\tau + h(t,x), \quad 0<t\leq T, \quad 0\leq x\leq 1 \qquad (4.5)$$

and of the boundary and initial conditions in (4.1). It is clear that a fixed point of S is a solution of (4.1) on $[0,T] \times [0,1]$. Also note $\phi'(w)$ is $W^{1,\infty}$ smooth and $\phi'(w)_t$ and $\phi'(w)_x$ are in $L^\infty([0,T]; L^2(0,1))$. Moreover, if $g(t,x)$ denotes the right-hand side of (4.5), $g(t,x) = g_1(t,x) + g_2(t,x)$, then g satisfies the same assumptions as h does, and by fairly standard theory for linear problems $u_{tt}, u_{tx}, u_{xx} \in C([0,T]; L^2(0,1))$; embedding type arguments then yield that $u \in C^1([0,T] \times [0,1])$. Thus it suffices to show that the map S has a unique fixed point u in $X(M,T)$. Once this has been demonstrated it follows from the assumption $\int_0^1 h(t,x)dx = 0$, $t \in [0,1]$, the equation in (4.1), and from the boundary condition that $\frac{\partial}{\partial t}\int_0^1 u(t,x)dx = 0$ for $t \in [0,T]$; since $\int_0^1 u_0(x)dx = 0$, one then also has $\int_0^1 u(t,x)dx = 0$, $t \in [0,T]$.

The remainder of the proof of Proposition 4.1 is completed in the following steps: (i) analogous to the proof of Lemma 2.1 in [5] (here the argument is shorter), use the standard energy method for linear problems to show that when M is sufficiently large and $T > 0$ is sufficiently small, S maps $X(M,T)$ into itself; (ii) define the metric

$$\rho(u,\bar{u}) = \max_{[0,T]} \{\int_0^1 [(u_t-\bar{u}_t)^2 + (u_x-\bar{u}_x)^2](t,x)dx\},$$

where $u,\bar{u} \in X(M,T)$; by the lower semicontinuity of norms in a Banach space, $X(M,T)$ is complete under ρ; analogous to Lemma 2.2 in [5] show that S is a strict contraction on $X(M,T)$ under the metric ρ; (iii) by Banach's fixed point theorem the map S has a unique fixed point $u \in C^1([0,T] \times [0,1])$ for M sufficiently large and T sufficiently small, which solves (4.1); the existence of the maximal interval of validity $[0,T_0) \times [0,1]$ of the solution u satisfying (4.2) is established in a standard manner (see [5]). This completes the sketch of the proof of Proposition 4.1.

b. Global Theory

By the constitutive assumptions (c) and (3.4) there exist $\delta > 0$, $\kappa > 0$ such that

$$\phi'(\xi) \geq \kappa, \quad \psi'(\xi) \geq \kappa, \quad \chi'(\xi) \geq \kappa \quad (|\xi| \leq \delta). \tag{4.6}$$

If necessary, modify ϕ outside $[-\delta,\delta]$ such that $\phi \in C^2(\mathbb{R})$ and $\phi'(\xi) \geq \kappa$ for all $\xi \in \mathbb{R}$. To prove the existence of the global solution u on $(-\infty,\infty) \times [0,1]$ of the history value problem (3.1) - (3.3) asserted in Theorem 3.1, let u be the unique solution on the maximal interval $(-\infty,T_0) \times [0,1]$ guaranteed by Proposition 4.1, and assume that $0 < T_0 < \infty$.

For $0 < T < T_0$ let

$$U(T) = \sup_{(-\infty,T]} \int_0^1 [u^2(t,x) + u_t^2(t,x) + u_x^2(t,x) + \dots + u_{xx}^2(t,x)]dx$$
$$+ \int_{-\infty}^{T} \int_0^1 [u^2 + u_t^2 + u_x^2 + \dots + u_{xx}^2]dxdt, \tag{4.7}$$

where ... stand for the terms u_{tt}^2 and u_{tx}^2. Recall that T_0 is characterized by (4.2) and thus the first integral in U(T) tends to infinity as $T \to T_0^-$; also recall that $u \equiv v$ for $t \leq 0$. The basic strategy of the proof is the same as in [5]; we wish to show that there exist constants $\nu > 0$ ($\nu \leq \delta$ in (4.6)) and $K > 0$, independent of T, such that if

$$|u(t,x)|^2 + |u_x(t,x)|^2 + |u_t(t,x)|^2 \leq \nu^2, \quad \nu \leq \delta, \tag{4.8}$$

on $(-\infty,T] \times [0,1]$ then

$$U(T) \leq KF, \tag{4.9}$$

where F is the constant defined under assumptions (f). The proof that (4.8) implies (4.9) will be outlined below using energy estimates. Once this claim is established the proof of Theorem 3.1 is completed as follows. First, by the assumptions on the history v(t,x), (4.8) holds as a strict inequality for $t > 0$ sufficiently small. Next, by the Poincaré inequality (see Lemma A.1 in Appendix) and the definition of U(T),

$$|u(t,x)|^2 + |u_x(t,x)|^2 + |u_t(t,x)|^2 \tag{4.10}$$

$$\leq \int_0^1 [u_x^2(t,x) + u_{xx}^2(t,x) + u_{tx}^2(t,x)]dx < U(T)$$

on $(-\infty,T] \times [0,1]$. Choosing the constant $\mu^2 < \frac{\nu^2}{K}$ and $F \leq \mu^2$, (4.10) shows that (4.9) implies (4.8) (as a strict inequality). Therefore, if $F \leq \mu^2 < \frac{\nu^2}{K}$, the estimates (4.8) and (4.9) both hold for every $T \in (-\infty,T_0)$, and, consequently, by Proposition 4.1 (see especially (4.2)) $T_0 = +\infty$. Moreover, (4.7) and (4.9) imply that

$$u,\ u_t,\ u_x,\ u_{tt},\ u_{tx},\ u_{xx} \in L^\infty((-\infty,\infty);\ L^2(0,1)) \cap L^2((-\infty,\infty);\ L^2(0,1)). \tag{4.11}$$

But $u,\ u_t,\ u_x,\ u_{tt},\ u_{tx},\ u_{xx} \in L^2((-\infty,\infty);\ L^2(0,1))$ also implies that

$$u,\ u_t,\ u_x \to 0 \quad \text{in} \quad L^2(0,1) \quad \text{as} \quad t \to \infty, \tag{4.12}$$

which in view of $u,\ u_t,\ \ldots,\ u_{xx} \in L^\infty((-\infty,\infty);\ L^2(0,1))$ yields (3.11) and completes the proof.

It remains to establish that (4.8) implies (4.9). For this purpose we will need the following properties of strongly positive kernels. Introduce the notation

$$(a*g)(t) = \int_{-\infty}^{t} a(t-\tau)g(\tau)d\tau$$

and

$$Q[a,w;s] = \int_{-\infty}^{s} w(t)(a*w)(t)dt.$$

Let assumptions (a) of Theorem 3.1 be satisfied. There exist constants β, $\gamma > 0$ such that for every $s \in \mathbb{R}$ and for every $w \in L^2(-\infty,s)$

$$\int_{-\infty}^{s} [(a*w)(t)]^2dt \leq \beta Q[a,w;s], \tag{4.13}$$

where

$$\beta = \frac{1}{\alpha} \|a\|^2_{L^1(0,\infty)} + \frac{4}{\alpha} \|a'\|^2_{L^1(0,\infty)},$$

and

$$\int_{-\infty}^{s} [(a' * w)(t)]^2 dt \le \gamma Q[a,w;s], \tag{4.14}$$

where

$$\gamma = \frac{1}{\alpha} \|a'\|^2_{L^1(0,\infty)} + \frac{4}{\alpha} \|a''\|^2_{L^1(0,\infty)}.$$

The estimates (4.13), (4.14), which have also been used in [5], are essentially contained in Staffans [20]. Another important property deals with the resolvent kernel k of the linear Volterra operator

$$\phi'(0)y + \psi'(0)a' * y$$

defined to be the solution of the linear Volterra equation

$$\phi'(0)k(t) + \psi'(0) \int_0^t a'(t-\tau)k(\tau)d\tau = -\psi'(0)a'(t), \quad 0 \le t < \infty. \tag{k}$$

LEMMA 4.2. If assumptions (a) are satisfied, and if $\phi'(0) > 0$, $\psi'(0) > 0$ and (3.4) holds, then $k \in L^1(0,\infty)$.

The proof of Lemma 4.2 is given in [5].

The first estimate needed for the proof that (4.8) implies (4.9) is obtained from equation (3.5). (Recall that problem (3.5), (3.2), (3.3) is equivalent to the original problem (3.1) - (3.3), and it is assumption (3.4) concerning χ which will play an important role.) Multiply (3.5) by $\psi(u)_{xt}$ and integrate over $(-\infty,s] \times [0,1]$, $s < T_0$. After several integrations by parts in which the boundary condition is invoked, we obtain

$$\frac{1}{2} \int_0^1 \chi'(u(s,x))\psi'(u(s,x))u_x^2(s,x)dx + \int_0^1 Q[a,\psi(u)_{xt};s]dx \tag{4.15}$$

$$= -\frac{1}{2} \int_{-\infty}^{s} \int_0^1 [\chi'(u)\psi''(u) + \chi''(u)\psi'(u)]u_t u_x^2 dxdt$$

$$- \frac{1}{2} \int_{-\infty}^{s} \int_{0}^{1} \psi''(u) u_x u_t^2 \, dx dt + \int_{0}^{1} f(s,x) \psi(u(s,x))_x dx$$

$$- \int_{-\infty}^{s} \int_{0}^{1} f_t \psi(u)_x dx dt.$$

In contrast to the analogous calculation in [5, see (3.21)], no useful information is extracted here from integration of the term $u_t \psi(u)_{xt}$ over $(-\infty, s] \times [0,1]$.

REMARK 4.3. When obtaining the analogous estimates for the boundary-initial value problem (3.12) (see Remark 3.3), the analogue of equation (3.5) from which the global estimates are calculated is

$$u_t + \chi(u)_x + \int_{0}^{t} a(t-\tau) \psi(u(\tau,x))_{x\tau} d\tau = h(t,x) - a(t) \psi(u_0)_x .$$

To simplify several of the estimates which follow, we make the additional assumption that ϕ, ψ (and hence also χ) $\varepsilon\ C^3(\mathbb{R})$; the alternative is to employ difference quotients and pass to limits as in [5]. Differentiate (3.5) with respect to t (use $(a*g)(t) = \int_0^\infty a(\xi) g(t-\xi) d\xi$, differentiate, and then change variables) to obtain

$$u_{tt} + \chi(u)_{xt} + a*\psi(u)_{xtt} = f_t, \quad -\infty < t < \infty, \quad 0 \le x \le 1. \tag{4.16}$$

Multiply (4.16) by $\psi(u)_{xtt}$ and integrate over $(-\infty, s] \times [0,1]$. After several integrations by parts the result of this tedious calculation is

$$\frac{1}{2} \int_{0}^{1} \chi'(u(s,x)) \psi'(u(s,x)) u_{xt}^2(s,x) dx + \int_{0}^{1} Q[a, \psi(u)_{xtt}; s] dx \tag{4.17}$$

$$= -(I_1 + \frac{1}{2} I_2 + I_3) - (J_1 + \ldots + J_7) + \int_{0}^{1} f_t(s,x) \psi(u(s,x))_{xs} dx$$

$$- \int_{-\infty}^{s} \int_{0}^{1} f_{tt} \psi(u)_{xt} dx dt + \frac{1}{2} \int_{-\infty}^{s} \int_{0}^{1} [\chi''(u) \psi'(u) + \chi'(u) \psi''(u)] u_{xt}^2 u_t \, dx dt;$$

the terms I_k in (4.17) come from

$$I = \int_{-\infty}^{s} \int_{0}^{1} u_{tt}\psi(u)_{xtt} dxdt = I_1 + I_2 + I_3 + I_4$$

where

$$I_1 = \int_{-\infty}^{s} \int_{0}^{1} \psi'''(u)u_{tt}u_t^2 u_x dxdt,$$

$$I_2 = \int_{-\infty}^{s} \int_{0}^{1} \psi''(u)u_{tt}^2 u_x dxdt,$$

$$I_3 = 2\int_{-\infty}^{s} \int_{0}^{1} \psi''(u)u_{tt}u_{xt}u_t dxdt,$$

$$I_4 = \int_{-\infty}^{s} \int_{0}^{1} \psi'(u)u_{tt}u_{xtt} dxdt = -\frac{1}{2} I_2;$$

the terms J_k in (4.17) come from

$$J = \int_{-\infty}^{s} \int_{0}^{1} \psi(u)_{xtt}\chi(u)_{xt} dxdt = J_1 + \dots + J_7 + J_8,$$

where

$$J_8 = \int_{-\infty}^{s} \int_{0}^{1} \chi'(u)\psi'(u)u_{xt}u_{xtt} dxdt$$

$$= \frac{1}{2} \int_{0}^{1} \chi'(u(s,x))\psi'(u(s,x))u_{xt}^2(s,x)dx$$

$$- \frac{1}{2} \int_{-\infty}^{s} \int_{0}^{1} [\chi''(u)\psi'(u) + \chi'(u)\psi''(u)]u_{xt}^2 u_t dxdt$$

which respectively give the first term on the left side and last term on the right side of (4.17),

$$J_1 = \int_{-\infty}^{s} \int_{0}^{1} \chi''(u)\psi'''(u)u_x^2 u_t^3 dxdt,$$

$$J_2 = 2\int_{-\infty}^{s} \int_{0}^{1} \chi''(u)\psi''(u)u_{xt}u_t^2 u_x dxdt,$$

$$J_3 = \int_{-\infty}^{s} \int_{0}^{1} \chi''(u)\psi''(u)u_{tt}u_x^2 u_t dxdt,$$

$$J_4 = \int_0^1 \chi''(u(s,x))\psi'(u(s,x))u_t(s,x)u_x(s,x)u_{xt}(s,x)dx$$

$$- \int_{-\infty}^{s}\int_0^1 \chi''(u)\psi'(u)u_{xt}^2 u_t dxdt - \int_{-\infty}^{s}\int_0^1 u_{xt}u_x[\chi''(u)\psi'(u)u_t]_t dxdt,$$

$$J_5 = \int_{-\infty}^{s}\int_0^1 \chi'(u)\psi'''(u)u_{xt}u_x u_t^2 dxdt,$$

$$J_6 = 2\int_{-\infty}^{s}\int_0^1 \chi'(u)\psi''(u)u_{xt}^2 u_t dxdt,$$

$$J_7 = \int_{-\infty}^{s}\int_0^1 \chi'(u)\psi''(u)u_{xt}u_{tt}u_x dxdt.$$

The next estimate follows from the identity

$$a(0)\psi(u)_{xt} = a'*\psi(u)_{xt} + a*\psi(u)_{xtt} \tag{4.18}$$

which is derived by integrating $a*\psi(u)_{xtt}$ by parts. Multiply (4.18) by u_{xt} and integrate over $(-\infty,s] \times [0,1]$; then in each term on the right hand side use the Cauchy Schwarz inequality and (4.13), (4.14). This leads to the estimate

$$a(0)\int_{-\infty}^{s}\int_0^1 u_{tx}\psi(u)_{xt}dxdt \tag{4.19}$$

$$\leq [\int_{-\infty}^{s}\int_0^1 u_{tx}^2 dxdt]^{1/2}\{(\gamma\int_0^1 Q[a,\psi(u)_{xt};s]dx)^{1/2} + (\beta\int_0^1 Q[a,\psi(u)_{xtt};s]dx)^{1/2}\}.$$

Using $\psi(u)_{xt} = \psi'(u)u_{xt} + \psi''(u)u_x u_t$ and (4.6) on the left side of (4.19) and $ab \leq \frac{a^2}{2\varepsilon} + \frac{\varepsilon}{2}b^2$, $\varepsilon = 2/\kappa a(0)$ on the right side of (4.19) gives the useful estimate

$$\frac{\kappa a(0)}{2}\int_{-\infty}^{s}\int_0^1 u_{tx}^2 dxdt - \frac{\gamma}{\kappa a(0)}\int_0^1 Q[a,\psi(u)_{xt};s]dx \tag{4.20}$$

$$- \frac{\beta}{\kappa a(0)}\int_0^1 Q[a,\psi(u)_{xtt};s]dx \leq a(0)\int_{-\infty}^{s}\int_0^1 \psi''(u)u_{xt}u_x u_t dxdt.$$

Next write (4.16) in the form

$$u_{tt} = -(\chi''(u)u_x u_t + \chi'(u)u_{xt} + a*\psi(u)_{xtt}) + f_t;$$

square both sides, integrate over $(-\infty,s] \times [0,1]$, use $(\sum_{i=1}^{n} a_i)^2 \le n \sum_{i=1}^{n} a_i^2$ with $n = 4$, and use (4.13) to obtain the estimate

$$\int_{-\infty}^{s}\int_0^1 u_{tt}^2 dxdt - 4\int_{-\infty}^{s}\int_0^1 [\chi'(u)]^2 u_{xt}^2 dxdt \tag{4.21}$$
$$- 4\beta\int_0^1 Q[a,\psi(u)_{xtt};s]dx \le 4\int_{-\infty}^{s}\int_0^1 [\chi''(u)]^2 u_x^2 u_t^2 dxdt + 4\int_{-\infty}^{s}\int_0^1 f_t^2 dxdt.$$

We now return to (3.1), differentiate with respect to t and obtain

$$u_{tt} = -(\phi'(u)u_{xt} + \phi''(u)u_x u_t + a'*\psi'(u)u_{xt} + a'*\psi''(u)u_x u_t) + f_t,$$

square both sides, integrate with respect to x over [0,1], and evaluate at $t = s$. This gives the estimate

$$\int_0^1 u_{tt}^2(s,x)dx - 5\int_0^1 [\phi'(u(s,x))]^2 u_{tx}^2(s,x)dx \tag{4.22}$$
$$- 5\|a'\|^2_{L^1(0,\infty)} \times \sup_{(-\infty,s]} \int_0^1 [\psi'(u(t,x))]^2 u_{tx}^2(t,x)dx$$
$$\le 5\int_0^1 [\phi''(u(s,x))]^2 u_x^2(s,x)u_t^2(s,x)dx + 5\int_0^1 f_t^2(s,x)dx$$
$$+ 5\|a'\|^2_{L^1(0,\infty)} \times \sup_{-\infty<t\le s} \int_0^1 [\psi''(u(t,x))]^2 u_x^2(t,x)u_t^2(t,x)dx.$$

Next, differentiate (3.1) with respect to x to obtain

$$u_{tx} + \phi(u)_{xx} + a'*\psi(u)_{xx} = f_x,$$

and write $\phi(u)_{xx} = \phi'(u)u_{xx} + \phi''(u)u_x^2 = \phi'(0)u_{xx} + [\phi'(u)-\phi'(0)]u_{xx} + \phi''(u)u_x^2$, and similarly for $\psi(u)_{xx}$. This gives the equation

$$\phi'(0)u_{xx} + \psi'(0)a'*u_{xx} = X(t,x), \tag{4.23}$$

where

$$X(t,x) = -u_{tx} + f_x - [\phi'(u) - \phi'(0)]u_{xx} - \phi''(u)u_x^2$$
$$- a'*[\psi'(u) - \psi'(0)]u_{xx} - a'*\psi''(u)u_x^2.$$

Letting k be the resolvent of the operator on the left side of (4.23), we have

$$\phi'(0)u_{xx}(t,x) = X(t,x) + (k*X)(t,x). \tag{4.24}$$

By Lemma 4.2 $k \in L^1(0,\infty)$, and this gives the estimates

$$[\phi'(0)]^2 \int_0^1 u_{xx}^2(s,x)dx \le 2\int_0^1 X^2(s,x)dx + 2\|k\|_{L^1}^2 \times \sup_{(-\infty,s]} \int_0^1 X^2(t,x)dx \tag{4.25}$$

and

$$[\phi'(0)]^2 \int_{-\infty}^s \int_0^1 u_{xx}^2 dxdt \le 2[1 + \|k\|_{L^1}] \int_{-\infty}^s \int_0^1 X^2 dxdt. \tag{4.26}$$

To complete the proof that (4.8) implies (4.9), we first use (4.8), (4.6), (f), and the fact that for $|u(t,x)| \le \nu$, the derivatives $|\phi'(u(t,x))|$, $|\phi''(u(t,x))|$, $|\phi'''(u(t,x))|$, $|\psi'(u(t,x))|$, $|\psi''(u(t,x))|$, $|\psi'''(u(t,x))|$, $|\chi'(u(t,x))|$, $|\chi''(u(t,x))|$, $|\chi'''(u(t,x))|$ are bounded by a constant $C > 0$ in order to simplify the basic estimates derived above; here we have used the additional simplifying assumption that ϕ,ψ and hence also $\chi \in C^3(\mathbb{R})$; this can be avoided as in [5].

Thus (4.15) becomes

$$\frac{\kappa^2}{4}\int_0^1 u_{xt}^2(s,x)dx + \int_0^1 Q[a,\psi(u)_{xt};s]dx \le \frac{3}{2}\nu C^2 \int_{-\infty}^s \int_0^1 u_x^2 dxdt$$
$$+ \frac{C\nu}{2}\int_{-\infty}^s \int_0^1 u_t^2 dxdt + FC\left[\frac{C}{\kappa^2} + \frac{1}{2C^2\sqrt{\nu}}\right]; \tag{4.27}$$

similarly (4.17) simplifies to

$$\frac{\kappa^2}{4}\int_0^1 u_{xt}^2(s,x)\,dx + \int_0^1 Q[a,\psi(u)_{xtt};s]\,dx \tag{4.28}$$

$$\le C^2\nu\int_0^1 u_{xt}^2(s,x)\,dx + C\nu\left[\frac{1}{2} + C\right]\int_0^1 u_x^2(s,x)\,dx$$

$$+ C\nu[F + C + C\nu + C\nu^2]\int_{-\infty}^{s}\int_0^1 u_x^2\,dxdt + C^2\nu^2\int_{-\infty}^{s}\int_0^1 u_t^2\,dxdt$$

$$+ C\nu\left[\frac{3}{2} + \left(\frac{9}{2} + \frac{7}{2}\nu\right)C\right]\int_{-\infty}^{s}\int_0^1 u_{tx}^2\,dxdt$$

$$+ C\nu\left[2 + \frac{5}{2}C + \left(\frac{1}{2} + C\right)\nu\right]\int_{-\infty}^{s}\int_0^1 u_{tt}^2\,dxdt + CF\left(\nu + \frac{1}{\kappa^2} + \frac{1}{2\nu}\right);$$

the estimate (4.20) becomes

$$\frac{\kappa a(0)}{2}\int_{-\infty}^{s}\int_0^1 u_{tx}^2\,dxdt - \frac{\gamma}{\kappa a(0)}\int_0^1 Q[a,\psi(u)_{xt};s]\,dx \tag{4.29}$$

$$- \frac{\beta}{\kappa a(0)}\int_0^1 Q[a,\psi(u)_{xtt};s]\,dx \le a(0)\,\frac{\nu C}{2}\left(\int_{-\infty}^{s}\int_0^1 u_{xt}^2\,dxdt + \int_{-\infty}^{s}\int_0^1 u_x^2\,dxdt\right);$$

estimate (4.21) simplifies to

$$\int_{-\infty}^{s}\int_0^1 u_{tt}^2\,dxdt - 4C^2\int_{-\infty}^{s}\int_0^1 u_{xt}^2\,dxdt \tag{4.30}$$

$$- 4\beta Q[a,\psi(u)_{xtt};s] \le 4C^2\nu^2\int_{-\infty}^{s}\int_0^1 u_t^2\,dxdt + 4F;$$

estimate (4.22) simplifies to

$$\int_0^1 u_{tt}^2(s,x)dx - 5C^2\int_0^1 u_{tx}^2(s,x)dx \tag{4.31}$$

$$- 5C^2\|a'\|_{L^1} \times \sup_{-\infty<t\le s} \int_0^1 u_{tx}^2(t,x)dx \le 5C^2\nu^2\int_0^1 u_x^2(s,x)dx$$

$$+ 5C^2\nu^2\|a'\|_{L^1} \times \sup_{-\infty<t\le s} \int_0^1 u_x^2(t,x)dx + 5F;$$

the estimate (4.25) becomes

$$\kappa^2\int_0^1 u_{xx}^2(s,x)dx \le 2\int_0^1 X^2(s,x)dx + 2\|k\|_{L^1}^2 \times \sup_{(-\infty,s]} \int_0^1 X^2(t,x)dx; \tag{4.32}$$

finally, (4.26) becomes

$$\kappa^2\int_{-\infty}^s \int_0^1 u_{xx}^2 dxdt \le 2[1 + \|k\|_{L^1}]\int_{-\infty}^s \int_0^1 X^2 dxdt, \tag{4.33}$$

where in the estimates (4.32) and (4.33) $X(t,x)$ is the function (depending on u, u_x, u_{xx}, u_{tx}, a, and f) given preceding formula (4.24).

We now focus our attention on the simplified estimates (4.27) through (4.33).

By the Poincaré inequality (see Appendix) $U(T)$, given by (4.7), can be majorized by

$$U(T) \le \sup_{(-\infty,T]} [\int_0^1 [2u_x^2(t,x) + u_{tt}^2(t,x) + 2u_{tx}^2(t,x) + u_{xx}(t,x)]dx]$$
$$+ \int_{-\infty}^T \int_0^1 [u_{tt}^2 + 2u_{tx}^2 + 3u_{xx}^2]dxdt. \tag{4.34}$$

Moreover, each term on the right hand side of (4.34) can be majorized by a suitable linear combination of the left hand sides of the estimates (4.27-4.33). On the other hand, each term on the right hand sides of the estimates (4.27-4.33) can be majorized by terms of the form $O(f)$, or $O(\nu)U(T)$, or $\varepsilon U(T) + c(\varepsilon)F(T)$ for any $\varepsilon > 0$; the last of these comes from estimating the right hand sides of (4.32) and (4.33). This combined with (4.34) yields the final estimate

$$U(T) \le \{\mathcal{O}(\nu) + \mathcal{O}(\varepsilon)\}U(T) + c(\varepsilon)F. \tag{4.35}$$

Therefore, fixing $\nu > 0$, $\varepsilon > 0$ sufficiently small, (4.35) yields (4.9), assuming that (4.8) holds. This completes the proof of Theorem 3.1.

5. DEVELOPMENT OF SINGULARITIES

We consider the pure Cauchy problem

$$\begin{cases} u_t + \phi(u)_x + a'*\psi(u)_x = 0 \\ u(0,x) = u_0(x), \quad x \in \mathbb{R}; \end{cases} \tag{5.1}$$

throughout this section $*$ will denote the convolution on $[0,t]$ (not $(-\infty,t]$). For a discussion of the existence of smooth solutions of (5.1) on $[0,\infty) \times \mathbb{R}$ for sufficiently smooth and "small" data we refer the reader to Remarks 3.3, 3.6. It is known (see Proposition 4.1 and Remark 4.2) that if a, ϕ, ψ satisfy the assumptions of Proposition 4.1 and $u_0 \in H^2(\mathbb{R})$, then there exists a unique smooth solution of (5.1) on a maximal interval $[0,T_0) \times \mathbb{R}$, $0 < T_0 \le \infty$.

Our objective is to study the problem of the development of singularities (shocks) of the solution in finite time such that a physically meaningful entropy condition will be satisfied (see [14]), assuming that a local smooth solution exists.

Let $\xi \in \mathbb{R}$ and let $u(t,x)$ be a smooth solution of (5.1). We define the characteristic through ξ of (5.1) to be the curve $x = x(t,\xi)$ in the t,x plane specified by the initial value problem

$$\begin{cases} \dfrac{dx}{dt} = \phi'(u(t,x)), \quad t > 0 \\ x(0,\xi) = \xi, \quad \xi \in \mathbb{R}. \end{cases} \tag{5.2}$$

It should be noted that the total derivative of $u(t,x)$ along the characteristic $x(t,\xi)$ is

$$\frac{d}{dt} u(t,x(t,\xi)) = u_t(t,x(t,\xi)) + u_x(t,x(t,\xi)) \frac{dx}{dt}$$

$$= u_t(t,x(t,\xi)) + \phi'(u(t,x(t,\xi)))u_x(t,x(t,\xi))$$

$$\equiv u_t(t,x(t,\xi)) + \phi(u(t,x(t,\xi)))_x.$$

Let $u(t,\xi) = u(t,x(t,\xi))$. If u is a solution of (5.1), its derivative along the characteristic $x(t,\xi)$ satisfies the integrodifferential equation

$$\begin{cases} \dfrac{du}{dt} = - \displaystyle\int_0^t a'(t-\tau)\psi(u(\tau,x(t,\xi)))_x d\tau, \quad t > 0 \\ \\ u(0,x(0,\xi)) = u_0(x(0,\xi)) = u_0(\xi), \quad \xi \in \mathbb{R} . \end{cases} \tag{5.3}$$

The reader should note that (5.3) is no longer of convolution type, because of the term $\psi(u(\tau,x(t,\xi)))_x = \psi'(u(\tau,x(t,\xi)))u_x(\tau,x(t,\xi))$ under the integral.

Let $x(t,\xi)$ be the characteristic of (5.1) through ξ and define $v(t,\xi) = \frac{\partial}{\partial \xi} x(t,\xi)$; note that $v(0,\xi) = 1$. The function v measures the growth of the characteristics with respect to ξ. Let $\phi''(\cdot) \neq 0$. According to Lemma 2.1 of [14], a singularity will develop in the solution u of (5.1) in finite time, if it can be shown that there exists a number $\bar{T}$, $0 < \bar{T} < \infty$ such that $v(\bar{T},\xi) < 0$. For, in this case there exists a $0 < T < \bar{T}$ and $\xi_1 \neq \xi_2 \in \mathbb{R}$ such that $x(T,\xi_1) = x(T,\xi_2)$ (i.e., the characteristics through ξ_1 and ξ_2 cross at T), and $u(\bar{T}^-,x(\bar{T},\xi_1)) \neq u(\bar{T}^-,x(\bar{T},\xi_2))$. This is the definition of the development of a "shock" at $\bar{T}$ in the solution u of (5.1). It is explained in [14] that this "shock" solution satisfies the physically meaningful entropy condition.

We shall therefore set up a differential equation for $v(t,\xi)$. Let $w(t,\xi) = \frac{\partial u}{\partial \xi}(t,x(t,\xi))$, and note that $w(0,\xi) = u_0'(\xi)$. Then using (5.2), we have

$$\frac{dv}{dt} = \frac{d}{dt} x_\xi(t,\xi) = \frac{\partial}{\partial \xi}\frac{dx}{dt} = \frac{\partial}{\partial \xi} \phi'(u(t,x(t,\xi)))$$

$$= \phi''(u(t,x(t,\xi)))w(t,\xi).$$

Thus v satisfies the initial value problem

$$\frac{dv}{dt} = \phi''(u(t,x(t,\xi)))w, \quad v(0,\xi) = 1. \tag{5.4}$$

The equation satisfied by w is found by differentiating (5.3) with respect to ξ, obtaining the initial value problem

$$\frac{dw}{dt} = -v(t,\xi)\int_0^t a'(t-\tau)[\psi''(u(\tau,x(t,\xi)))u_x^2(\tau,x(t,\xi)) \tag{5.5}$$

$$+ \psi'(u(\tau,x(t,\xi)))u_{xx}(\tau,x(t,\xi))]d\tau, \quad w(0) = u_0'(\xi).$$

Our objective is to use the system of four nonlinear equations (5.2) - (5.5) for the quantities $x(t,\xi)$, $u(t,\xi)$, $v(t,\xi)$, $w(t,\xi)$ satisfying the indicated conditions to establish the development of a shock in the sense described above. This problem, which is under active study, remains to be solved in this generality.

We restrict ourselves to the special case $\psi \equiv \phi$ in (5.1). By Remarks 3.5, 3.6 the Cauchy problem (5.1) has a unique smooth solution u on $[0,\infty) \times \mathbb{R}$ if a and ϕ satisfy the assumptions of Theorem 3.1, $0 < a(0) < 1$, and if $\|u_0\|_{H^2(\mathbb{R})}$ is sufficiently small. In the special case $\psi \equiv \phi$ we can use the method of MacCamy [11] and Dafermos and Nohel [4]; introduce the resolvent kernel k of a' defined by the equation

$$k(t) + (a'*k)(t) = -a'(t), \quad 0 \le t < \infty \tag{k}$$

and write (5.1) in the equivalent form

$$\begin{cases} u_t + \phi(u)_x + k(0)u + k'*u = k(t)u_0(x), & 0 < t < \infty, \quad x \in \mathbb{R} \\ u(0,x) = u_0(x), & x \in \mathbb{R}. \end{cases} \tag{5.6}$$

Note that since a satisfies assumptions (a), $k(0) = -a'(0) > 0$. The method of Lemma 4.2 applied to equation (k) shows that since $0 < a(0) < 1$, one has

$$k, k' \in L^1(0,\infty). \tag{5.7}$$

REMARK. (5.7) also holds if $a(t) = a_\infty + A(t)$, $a(0) = 1$, $a_\infty > 0$, and A satisfies assumptions (a).

To establish the development of singularities in a smooth solution u of (5.1) with $\psi \equiv \phi$ (equivalently, of (5.6)) we study the system of nonlinear equations corresponding to (5.2) - (5.5). In this case, it is easily seen that the quantities $x(t,\xi)$, $u(t,\xi) = u(t,x(t,\xi))$, $v(t,\xi) = \frac{\partial}{\partial\xi} x(t,\xi)$, $w(t,\xi) = \frac{\partial}{\partial\xi} u(t,x(t,\xi))$ satisfy the initial value problem

$$\frac{dx}{dt} = \phi'(u(t,x(t,\xi)))$$

$$\frac{du}{dt} + k(0)u + \int_0^t k'(t-\tau)u(\tau,x(t,\xi))d\tau = k(t)u_0(x)$$

$$\frac{dv}{dt} = \phi''(u(t,x(t,\xi)))w \tag{5.8}$$

$$\frac{dw}{dt} + k(0)w + (\int_0^t k'(t-\tau)u_x(\tau,x(t,\xi))d\tau - u_0'(t,x(t,\xi)))v = 0$$

$$x(0,\xi) = \xi, \quad u(0,\xi) = u_0(\xi), \quad v(0,\xi) = 1, \quad w(0,\xi) = u_0'(\xi).$$

In recent joing work with Malek-Madani [15] we have established the following result; its proof, which uses (5.7), (5.8) and the general strategy for the formation of shocks outlined above, will appear elsewhere. A similar result was stated by MacCamy in his lecture; see note in [6].

THEOREM 5.1. Let a $(0 < a(0) < 1)$ satisfy assumptions (a). In addition, let the resolvent kernel k of a' satisfy

$$k(t) \geq 0, \quad k'(t) \leq 0, \quad 0 \leq t < \infty. \tag{5.9}$$

Let $\phi \in C^2(\mathbb{R})$, $\phi(0) = 0$, $\phi'(\cdot) > 0$, $\phi''(\cdot) > 0$. Let $u_0 \in C^2(\mathbb{R})$. If $u_0'(\xi) < 0$ and $|u_0'(\xi)|$, $\xi \in \mathbb{R}$, is sufficiently large, every (necessarily) smooth solution $u(t,x)$ of the Cauchy problem (5.6) (equivalently, of (5.1) if $\psi \equiv \phi$) will develop a shock in finite time. If $\phi''(\cdot) \geq \beta > 0$, an upper bound for the time at which a shock develops is

$$\bar{T} = \frac{1}{k(0)} \log \frac{u_0'(\xi)\beta}{u_0'(\xi)\beta + k(0)}.$$

The following considerations provide examples of kernels a in (5.6) (equivalently, of (5.1) if $\psi \equiv \phi$) for which Theorem 5.1 can be applied.

REMARK 5.2. If $a(t) = \beta e^{-\alpha t}$ $(0 < \beta < 1)$, a simple calculation shows that $k(t) = \alpha\beta e^{-\alpha(1-\beta)t}$, and evidently the inequalities (5.9) are satisfied in the strict sense.

More generally, one has the following result established by elementary consideration from the resolvent equation (k).

LEMMA 5.3. Let $a \in C^2[0,\infty)$, $(-1)^j a^{(j)}(t) \geq 0$, $j = 0,1,2$; $0 \leq t < \infty$, and assume that

$$a'(t)a'(0) - a''(t) < 0, \quad 0 \leq t < \infty.$$

Then $k(t) > 0$, $k'(t) < 0$ for $0 \leq t < \infty$. If also $0 < a(0) < 1$, and $a \in W^{2,1}[0,\infty)$, then $k, k' \in L^1(0,\infty)$.

COROLLARY 5.4. Let $a(t) = \sum_{j=1}^{m} \beta_j e^{-\alpha_j t}$, $\beta_j > 0$, $\alpha_j > 0$; if

$$\alpha_i > \sum_{j=1}^{m} \alpha_j \beta_j, \quad i = 1,2,\ldots,m,$$

then $k(t) > 0$, $k'(t) < 0$ for $0 \quad t < \infty$; if also $\sum_{j=1}^{m} \beta_j < 1$ then $k, k' \in L^1(0,\infty)$.

Finally we remark that the general approach used to prove Theorem 5.1 can also be used to show that if by contrast, $u_0'(\xi) > 0$, and if the other assumptions of Theorem 5.1 hold, no singularities develop in the solution u in finite time. Thus Theorem 5.1, together with this remark, form the analogue for the conservation law with memory (5.1) of Lax's classical result for the conservation law (1.2).

APPENDIX

For the convenience of the reader we state and prove the following elementary inequalities which were used in the proof of Theorem 3.1 and which are generally referred to as Poincaré inequalities.

LEMMA A.1. Let $g, g' \in L^2(0,1)$, g real, and let $\bar{g} = \int_0^1 g(x)dx$. Then

$$\int_0^1 g^2(x)dx \le [\bar{g}]^2 + \int_0^1 [g'(x)]^2 dx;$$

in particular, if $\bar{g} = 0$, then

$$\int_0^1 g^2(x)dx \le \int_0^1 [g'(x)]^2 dx.$$

PROOF. Take $0 \le x_0 < x \le 1$. Then

$$g(x) - g(x_0) = \int_{x_0}^{x} g'(\xi)d\xi$$

and by Cauchy-Schwarz

$$(g(x) - g(x_0))^2 \le (x - x_0)\int_{x_0}^{x} (g'(\xi))^2 d\xi \le \int_0^1 (g'(\xi))^2 d\xi.$$

Thus

$$\int_0^1 (g(x) - g(x_0))^2 dx = \int_0^1 g^2(x)dx - 2g(x_0)\bar{g} + g^2(x_0)$$

$$\le \int_0^1 [g'(x)]^2 dx.$$

By the continuity of g choose x_0 such that $g(x_0) = \bar{g}$, and the first inequality is immediate.

LEMMA A.2. Let $g, g', g'' \in L^2(0,1)$, g real, $g(1) = g(0)$, $\bar{g} = \int_0^1 g(x)dx = 0$. Then

$$g^2(x) + [g'(x)]^2 \le \int_0^1 [g''(x)]^2 dx, \quad 0 \le x \le 1.$$

PROOF. Let $0 \le y < x \le 1$; we have

$$g'(x) - g'(y) = \int_y^x g''(\xi)d\xi.$$

Squaring both sides and using Cauchy-Schwarz gives

$$[g'(x)]^2 + [g'(y)]^2 - 2g'(x)g'(y) \le \int_0^1 [g''(x)]^2dx.$$

Integrating with respect to y over [0,1] and using $\int_0^1 g'(y)dy = g(1) - g(0) = 0$ we have

$$[g'(x)]^2 + \int_0^1 [g'(y)]^2dy \le \int_0^1 [g''(x)]^2dx.$$

Since $\bar{g} = 0$, the conclusion follows from the inequality

$$[g(x)]^2 \le \int_0^1 [g'(y)]^2dy,$$

the proof of which is contained in that of Lemma 1.

APPLICATION. If $w \in X(M,T)$ defined in the proof of Proposition 4.1, then

$$w^2(t,x) + w_x^2(t,x) + w_t^2(t,x) \le \int_0^1 [w_{xx}^2(t,x) + w_{tx}^2(t,x)]dx, \quad 0 \le x \le 1.$$

ACKNOWLEDGEMENT

This research was sponsored by the United States Army under Contract No. DAAG29-80-C-0041.

REFERENCES

1. B. D. Coleman and M. E. Gurtin, Waves in materials with memory II. On the growth and decay of one-dimensional acceleration waves, *Arch. Rational Mech. Anal.* 19 (1965), 239-265.

2. C. M. Dafermos, Can dissipation prevent the breaking of waves?, *Trans. of the 26th Conf. of Army Mathematicians*, ARO Rep. 81-1, pp. 187-198.

3. C. M. Dafermos and Ling Hsiao, Hyperbolic systems of balance laws with inhomogeneity and dissipation, preprint.

4. C. M. Dafermos and J. A. Nohel, Energy methods for nonlinear hyperbolic Volterra integrodifferential equations, *Comm. P. D. E.* 4 (1979), 219-278.

5. C. M. Dafermos and J. A. Nohel, A Nonlinear hyperbolic Volterra equation in viscoelasticity, *American J. Math.*, Supplement (1981), 87-115.

6. J. M. Greenberg and Ling Hsiao, The Riemann problem for the system

$$u_t + \sigma_x = 0$$

$$(\sigma - \hat{\sigma}(u))_t + \frac{1}{\varepsilon}(\sigma - \hat{\sigma}(u)) = 0,$$

preprint.

7. H. Hattori, Breakdown of smooth solutions in dissipative nonlinear hyperbolic equations, Ph.D. Thesis, Rensselaer Polytechnic Inst., 1981.

8. F. John, Formation of singularities in one-dimensional nonlinear wave propagation, *Comm. Pure Appl. Math.* 27 (1974), 377-405.

9. S. Kleinerman and A. Majda, Formation of singularities for wave equations including nonlinear vibrating strings, *Comm. Pure Appl. Math.* 33 (1980), 241-263.

10. P. Lax, Development of singularities of solutions of non-linear hyperbolic differential equations, *J. Math. Phys.* 5 (1964), 611-613.

11. R. C. MacCamy, A model for one-dimensional, nonlinear viscoelasticity, *Quart. Appl. Math.* 35 (1977), 21-33.

12. R. C. MacCamy and V. Mizel, Existence and nonexistence in the large of solutions of quasilinear wave equations, *Arch. Rational Mech. Anal.* 25 (1967), 299-320.

13. R. Malek-Madani, Energy criteria for finite hyperelasticity, *Arch. Rational Mech. Anal.*, 77 (1981), 177-188.

14. R. Malek-Madani, Formation of singularities for a conservation law with damping term, *Volterra and Functional Differential Equations*, Marcel Dekker, New York, 1982 (this volume).

15. R. Malek-Madani and J. A. Nohel, Development of singularities for conservation laws with memory, in preparation.

16. A. Matsumura, Global existence and asymptotics of the solutions of the second order quasilinear hyperbolic equations with first order dissipation, *Publ. Res. Inst. Math. Sci., Kyoto Univ., Ser A* 13 (1977), 349-379.

17. T. Nishida, Global smooth solutions for the second-order quasilinear wave equation with the first order dissipation, Nonlinear hyperbolic equations and related topics in fluid dynamics, Lecture Notes, Univ. de Paris-Sud (Orsay), 1978.

18. M. Slemrod, Global existence, uniqueness and asymptotic stability of classical smooth solutions in one-dimensional nonlinear thermoelasticity, *Arch. Rational Mech. Anal.* 76 (1981), 97-134.

19. M. Slemrod, Instability of steady shearing flows in a nonlinear visco-elastic fluid, *Arch. Rational Mech. Anal.* 68 (1978), 211-225.

20. O. J. Staffans, On a nonlinear hyperbolic Volterra equation, *SIAM J. Math. Anal.* 11 (1980), 793-812.

FUNCTIONAL EQUATIONS AS CONTROL CANONICAL FORMS FOR DISTRIBUTED PARAMETER CONTROL SYSTEMS AND A STATE SPACE THEORY FOR CERTAIN DIFFERENTIAL EQUATIONS OF INFINITE ORDER

David L. Russell

Department of Mathematics
University of Wisconsin
Madison, Wisconsin

1. FINITE DIMENSIONAL DISCRETE SYSTEMS

Let us begin by discussing the canonical form theory as it is manifested in the control of finite dimensional discrete linear control systems with scalar input:

$$x_{k+1} = Ax_k + bu_k, \quad k = 0,1,2,... \tag{1.1}$$

$$y_k = (x_k,h) \equiv h^* x_k, \tag{1.2}$$

wherein A is an n×n matrix, x, b and h are in E^n and u,y are scalar. The objective is to replace (1.1) by an equivalent system, called the control canonical form of (1.1) which, as we will see, has a number of useful features.

Let the characteristic polynomial of A be

$$\det(\lambda I-A) \equiv p(\lambda) = \lambda^n + a_1\lambda^{n-1} + ... + a_{n-1}\lambda + a_n. \tag{1.3}$$

Applying the "variation of parameters" formula to (1.1) we have

$$x_{k+r} = A^r x_k + \sum_{\ell=0}^{r-1} A^{r-\ell-1} bu_{k+\ell}, \qquad r = 0,1,2,...,$$

from which we see (cf. (1.3)) that the x_k, u_k satisfy a certain recursion relation, namely

$$x_{k+n} + a_1 x_{k+n-1} + \cdots + a_{n-1} x_{k+1} + a_n x_k$$
$$= p(A)x_k + \sum_{\ell=0}^{n-1} \left(\sum_{r=\ell+1}^{n} a_{n-r} A^{r-\ell-1} b \right) u_{k+\ell}.$$

Using the Cayley - Hamilton theorem, we have $p(A) = 0$. Defining

$$\beta_{n-\ell} = \sum_{r=\ell+1}^{n} a_{n-r} A^{r-\ell-1} b, \quad \ell = 0,1,\ldots,n-1,$$

we thus have

$$x_{k+n} + a_1 x_{k+n-1} + \cdots + a_{n-1} x_{k+1} + a_n x_k$$
$$= \beta_1 u_{k+n-1} + \cdots + \beta_{n-1} u_{k+1} + \beta_n u_k. \tag{1.4}$$

This is a vector equation in which the a_i, u_i are scalars and the β_i, x_i are n-vectors. Forming y_k as in (1.2) we have

$$y_{k+n} + a_1 y_{k+n-1} + \cdots + a_{n-1} y_{k+1} + a_n y_k = \gamma_1 u_{k+n-1} + \cdots + \gamma_n u_k \tag{1.5}$$

where

$$\gamma_j = h^* \beta_j, \quad j = 1,2,\ldots,n$$

are scalars.

The system represented by (1.5) has a certain significance for arbitrary observation vectors h. For example, suppose (1.1), (1.2) represents a scalar input-output process wherein A, b, h are unknown. It may be desired to determine these from the time histories of the y_k, u_k. The form (1.5) shows that in general this is not possible; an equivalent input-output relation can be realized with (1.5) which involves only 2n parameters rather than the $n^2 + 2n$ present in (1.1), (1.2). It is shown in [6], e.g., that the a_i, γ_i can be uniquely determined from the time histories of the y_k, u_k if (h,A) is observable and (A,b) is controllable. Thus, from the viewpoint of the number of parameters involved, (1.5) is minimal.

If (A,b) is controllable, the vectors b, Ab,..., $A^{n-1}b$ are linearly independent and hence β_1, $\beta_2,\ldots,\beta_n$ are also linearly independent. It follows then that there is exactly one observation vector h for which

$$h^*\beta_1 = 1, \; h^*\beta_j = 0, \quad j = 2,3,\ldots,n. \tag{1.6}$$

If the y_k are defined, as in (1.2), to be the observations corresponding to this particular observation vector h, then (1.5) becomes

$$y_{k+n} + a_1 y_{k+n-1} + \cdots + a_{n-1}y_{k+1} + a_n y_k = u_{k+n-1}, \tag{1.7}$$

which is the control canonical form for (1.1). It has a number of features which make it attractive. The problem of null controllability, steering from an arbitrary initial state to zero, is very easily solved. Assuming $y_{-(n-1)}$, $y_{-(n-2)},\ldots, y_0$ given, (1.7) can be used to determine u_0, $u_1,\ldots$, u_{n-1} successively so that y_1, $y_2,\ldots,y_n$ are all zero. Then with the remaining controls equal to zero, we will have $y_{n+1} = y_{n+2} = \cdots = 0$ thereafter. The problem of spectral assignment is also easily dealt with in this context. If the required exponents for the closed loop system are the zeros of the polynomial

$$g(\lambda) = \lambda^n + c_1\lambda^{n-1} + \cdots + c_{n-1}\lambda + c_n,$$

the causal feedback law

$$u_{k+n-1} = (a_1-c_1)y_{k+n-1} + \cdots + (a_{n-1}-c_{n-1})y_{k+1} + (a_n-c_n)y_k \tag{1.8}$$

carries (1.7) into

$$y_{k+n} + c_1 y_{k+n-1} + \cdots + c_{n-1}y_{k+1} + c_n y_k = 0, \tag{1.9}$$

which has the specified exponents. It has been shown (see, e.g., [11], [17]) that the vector state x_{k+n-1} of (1.1) is related to $y_{k+n-1},\ldots,y_k$ by

$$x_{k+n-1} = \beta \begin{pmatrix} y_{k+n-1} \\ \vdots \\ y_k \end{pmatrix} \equiv \beta\eta_k,$$

where β is a nonsingular $n\times n$ matrix independent of the applied controls u_j, provided only that (A,b) is controllable. Writing (1.8) as

$$u_{k+n-1} = f^* \eta_k,$$

the corresponding relation in the system

$$x_{k+n} = Ax_{k+n-1} + bu_{k+n-1}$$

(obviously equivalent to (1.1)) is

$$u_{k+n-1} = f^* \beta^{-1} x_{k+n-1} \equiv g^* x_{k+n-1},$$

and it is then easy to show that

$$x_{k+n} = (A+bg^*)x_{k+n-1}$$

(equivalently, $x_{k+1} = (A+bg^*)x_k$) is such that $A + bg^*$ has the zeros of $g(\lambda)$ as its spectrum, showing that any n complex numbers can be realized as eigenvalues of the closed loop system. The usefulness of the control canonical form (1.7) lies in the fact that the needed feedback relation is obvious in the context of that system.

If the observation vector h is chosen so that

$$h^* \beta_\nu = 1, \quad h^* \beta_j = 0, \quad j = 1,\ldots,n, \quad j \neq \nu, \tag{1.10}$$

then we obtain in place of (1.7)

$$y_{k+n} + a_1 y_{k+n-1} + \cdots + a_{n-1} y_{k+1} + a_n y_k = u_{k+n-\nu}, \tag{1.11}$$

which is just as useful for theoretical purposes but suffers from the drawback that the feedback law

$$u_{k+n-\nu} = (a_1-c_1)y_{k+n-1} + \cdots + (a_{n-1}-c_{n-1})y_{k+1} + (a_n-c_n)y_k$$

needed to pass from (1.11) to (1.9) is not causal. Similar (but worse) problems arise in connection with infinite dimensional systems, particularly those of hyperbolic type.

Of course, any of the requirements (1.6), (1.10) are very special and an observation (1.2) available from measurement of some component of an actual system is not likely to satisfy any such condition. It is therefore rather significant that canonical compensators exist, as indicated in the following

THEOREM. If in (1.1), (1.2) (h,A) is observable and (A,b) is controllable, then there is an integer n_1, $n_1 \leq n$, such that whenever B is an m×m matrix ($m \geq n_1$) whose minimal and characteristic polynomials coincide, then there exist m dimensional vectors j,r,d such that the augmented system (with $y_k = h^* x_k$ still)

$$x_{k+1} = Ax_k + bu_k \tag{1.12}$$

$$z_{k+1} = y_k r + Bz_k + du_k \tag{1.13}$$

with augmented observation

$$w_k = h^* x_k + j^* z_k \quad (= y_k + j^* z_k) \tag{1.14}$$

is canonical, that is, w_k satisfies

$$w_{k+n+m} + \alpha_1 w_{k+n+m-1} + \cdots + \alpha_{n+m} w_k = u_{k+n+m-1}$$

where

$$\det\left[\lambda I_{n+m} - \begin{pmatrix} A & 0 \\ rh^* & B \end{pmatrix}\right] = \lambda^{n+m} + \alpha_1 \lambda^{n+m-1} + \cdots + \alpha_{n+m}.$$

The proof is given in [13] and will not be repeated here. We remark, however, that the second system (1.13), the compensator, would ordinarily be realized electronically so that, no matter what j may be (in general, it is not unique), the observation increment $j^* z_k$ is, in principle at least, available for use.

2. A GENERAL FREQUENCY DOMAIN APPROACH

The canonical reduction process of Section 1 can be extended quite directly to some infinite dimensional systems, particularly those associated with certain classes of hyperbolic partial differential equations. This has been done, e.g., in [11], [17], [13]. Here, as in [13], we will use a frequency domain procedure which, as we will see, relies on a partial fractions decomposition of the system transfer function. In general this approach makes use of many connections with completeness and independence results for sets of complex exponentials, spaces of entire and meromorphic functions, interpolation theorems, etc.

We consider a system

$$\dot{x} = Ax + bu, \quad x \in H, \tag{2.1}$$

where H is a separable Hilbert space and A generates a strongly continuous semi-group of bounded operators, e^{At}, in H. The control input element b may belong to H or it may be (as is required, for example, to treat boundary value control in this context) an admissible control input element (see [11]). Along with (2.1) we assume an observation process

$$y = \langle x,h \rangle \tag{2.2}$$

where either $h \in H$ or else h is an admissible observation or output element. We further suppose that A has distinct (not essential, but convenient here) eigenvalues λ_k, $k = 1,2,\ldots$, and corresponding eigenvectors ϕ_k, $k = 1,2,\ldots$, which form a uniform (i.e., Riesz) basis for H. The Hille - Yosida theorem ([2]) implies that the λ_k must lie in some left half plane of the complex domain $\mathbb{C}$.

We may assume expansions

$$x(t) = \sum_{k=1}^{\infty} x_k(t)\phi_k, \tag{2.3}$$

$$b = \sum_{k=1}^{\infty} b_k\phi_k, \tag{2.4}$$

$$h = \sum_{k=1}^{\infty} h_k\psi_k \tag{2.5}$$

where $\{\psi_k\}$ is the dual, biorthogonal basis associated with $\{\phi_k\}$. The expansion (2.3) is convergent in H; indeed, the uniform basis property implies the existence of positive numbers c, C such that

$$c^{-2}\|x(t)\|_H^2 \le \sum_{k=1}^{\infty} |x_k(t)|^2 < C^2\|x(t)\|_H^2.$$

When b and/or h belong to H the coefficients b_k, h_k have similar properties. If b and/or h are admissible input and/or output elements, then the b_k and/or h_k need not be square summable but will satisfy boundedness or growth requirements as required for the particular system in hand (see [11] for details).

Formally, (2.3), (2.4), (2.5) imply that

$$y(t) = \sum_{k=1}^{\infty} h_k x_k(t) \tag{2.6}$$

and the $x_k(t)$ may be seen to satisfy

$$\dot{x}_k = \lambda_k x_k + b_k u, \quad k = 1,2,\ldots . \tag{2.7}$$

Thus the transfer function for the input output process relating u and y is

$$T(\lambda) = \sum_{k=1}^{\infty} \frac{h_k b_k}{\lambda - \lambda_k} . \tag{2.8}$$

The admissibility requirements on b, h imply that (2.8) is convergent in some appropriate sense. Sometimes this will involve a grouping of the terms.

Now let P be a linear functional or distribution which, at the very least, is defined on all exponential functions and polynomials. The characteristic function of P is

$$p(\lambda) = \langle e^{\lambda\cdot}, P\rangle.$$

In the significant examples $p(\lambda)$ is either entire or else holomorphic in some region which includes a right half plane of $\mathbb{C}$. We assume that P can be selected so that its zeros, all simple, are precisely the eigenvalues, λ_k, of the generator A in (2.1).

Defining

$$(P \circ f)(t) = \langle f(\cdot+t), P\rangle$$

for functions f such that $f(\cdot+t)$ lies in the domain of P for all $t \geq 0$ (at least), our candidate for the "office" of control canonical form of (2.1) is the functional equation

$$P \circ y = u. \tag{2.9}$$

Implicit in this choice is the assumption that this equation should constitute a well-defined system of some sort with a recognizable state space E so that one can meaningfully discuss the idea of a transformation $Q: H \to E$ which transforms the system (2.1) into its putative control canonical form (2.9).

Before we can get into any of this, a more immediate question presents itself. If (2.1), (2.2) and (2.9) are to yield the same input-output relationship, they must, clearly, have the same transfer function. The transfer function for (2.9) is (assuming identity output)

$$\mathcal{T}(\lambda) = \frac{1}{p(\lambda)} .$$

Comparing this with (2.8), $T(\lambda) = \mathcal{T}(\lambda)$ gives

$$\frac{1}{p(\lambda)} = \sum_{k=1}^{\infty} \frac{h_k b_k}{\lambda - \lambda_k} , \tag{2.10}$$

which should be valid for some form of convergence of the right hand side. Preferably it should be possible to relate that convergence to the topologies on the admissible spaces of input and output functions. But however that may turn out in individual cases, we can draw, tentatively, some very simple conclusions. For (2.10) to be true in any reasonable sense, the residues should match. Since the residue of $1/p(\lambda)$ at $\lambda = \lambda_k$ is $1/p'(\lambda_k)$ (and $p'(\lambda_k)$ is assumed non-zero since the λ_k are taken to be simple zeros), we conclude that

$$\frac{1}{p'(\lambda_k)} = h_k b_k. \tag{2.11}$$

If one assumes that the input element is given, so that the b_k are fixed (and different from zero if (2.7) is to be controllable in any general sense), then we see that the h_k are uniquely determined by

$$h_k = \frac{1}{p'(\lambda_k)b_k}, \tag{2.12}$$

which shows that the canonical observation vector is uniquely determined by the characteristic function $p(\lambda)$ and the input element b.

In some cases the form (2.6) of the observation process must be modified. This is discussed in [13] and we will have a little more to say about it later in this article.

3. SOME EXAMPLES

Let us first consider the wave equation

$$\frac{\partial^2 w}{\partial t^2} - \frac{\partial^2 w}{\partial x^2} = 0 \tag{3.1}$$

with the boundary conditions

$$w(0,t) = 0, \quad \frac{\partial w}{\partial x}(1,t) = u(t). \tag{3.2}$$

The eigenvalues of the operator

$$Lw = -w'' \tag{3.3}$$

with homogeneous boundary conditions conformable with (3.2) are

$$\mu_k = \left(\frac{2k-1}{2}\right)^2\pi^2, \quad k = 1,2,3,\ldots \tag{3.4}$$

and the corresponding orthonormal eigenfunctions are

$$\phi_k(x) = \sqrt{2}\sin\left(\frac{2k-1}{2}\pi x\right), \quad k = 1,2,3,\ldots . \tag{3.5}$$

Expanding

$$w(x,t) = \sum_{k=1}^{\infty} w_k(t)\phi_k(x),$$

it may be verified that (3.1), (3.2) give

$$w_k''(t) + \mu_k w_k(t) = (-1)^{k-1}\sqrt{2}\,u(t), \quad k = 1,2,3,\ldots . \tag{3.6}$$

Letting

$$\begin{pmatrix} w_k(t) \\ w_k'(t) \end{pmatrix} = \begin{pmatrix} \frac{1}{i\omega_k} & \frac{-1}{i\omega_k} \\ 1 & 1 \end{pmatrix} \begin{pmatrix} \eta_k(t) \\ \eta_{-k}(t) \end{pmatrix}, \tag{3.7}$$

with $\omega_k = (\frac{2k-1}{2})\pi$, we obtain

$$\begin{pmatrix} \eta_k'(t) \\ \eta_{-k}'(t) \end{pmatrix} = \begin{pmatrix} i\omega_k & 0 \\ 0 & -i\omega_k \end{pmatrix} \begin{pmatrix} \eta_k(t) \\ \eta_{-k}(t) \end{pmatrix} + \begin{pmatrix} b_k \\ b_{-k} \end{pmatrix} u(t), \tag{3.8}$$

an infinite collection of first order equations with

$$b_k = b_{-k} = \frac{(-1)^{k-1}}{\sqrt{2}}, \quad k = 1,2,3,\ldots . \tag{3.9}$$

This corresponds to a system (2.7) in ℓ^2 with A having eigenvalues

$$\lambda_k = i\omega_k, \quad \lambda_{-k} = -i\omega_k, \quad k = 1,2,3,\ldots , \tag{3.10}$$

and, following the development in [12], the indicated control input element (3.9) is admissible.

An entire function having the complex numbers (3.10) as its zeros is

$$p(\lambda) = e^{\lambda} + e^{-\lambda} = 2 \cosh \lambda, \tag{3.11}$$

which is the characteristic function of the distribution

$$\delta_{(1)} + \delta_{(-1)}.$$

Thus our candidate control canonical form is the neutral zero order delay equation

$$y(t+1) + y(t-1) = u(t). \tag{3.12}$$

It remains to identify the observation on (3.1), (3.2) which satisfies this equation. We have

$$p'(i\omega_k) = 2\sinh(i\omega_k) = e^{i\omega_k} - e^{-i\omega_k}$$
$$= 2i\sin\omega_k = (-1)^{k-1}2i, \quad k = 1,2,3,\ldots \tag{3.13}$$

and, since p is an even function, p' is odd and

$$p'(-i\omega_k) = (-1)^k 2i, \quad k = 1,2,3,\ldots . \tag{3.14}$$

Following (2.12), and using (3.13), (3.14), we see that the coefficients of the canonical output element are

$$h_k = \frac{1}{p'(i\omega_k)b_k} = \frac{1}{(-1)^{k-1}2i\left(\frac{(-1)^{k-1}}{\sqrt{2}}\right)} = \frac{-i}{\sqrt{2}}$$

$$h_{-k} = \frac{1}{p'(-i\omega_k)b_k} = \frac{i}{\sqrt{2}},$$

that is, the canonical observation is given by

$$y(t) = \sum_{k=1}^{\infty}\left(\frac{-i}{\sqrt{2}}\eta_k(t) + \frac{i}{\sqrt{2}}\eta_{-k}(t)\right).$$

Inverting (3.7) this becomes

$$y(t) = \sum_{k=1}^{\infty}\left(\frac{-1}{\sqrt{2}}\left[\frac{i\omega_k}{2}w_k(t) + \frac{1}{2}v_k(t)\right] + \frac{i}{\sqrt{2}}\left[\frac{-i\omega_k}{2}w_k(t) + \frac{1}{2}v_k(t)\right]\right)$$
$$= \sum_{k=1}^{\infty}\left(\frac{\omega_k}{\sqrt{2}}\right)w_k(t) \equiv \sum_{k=1}^{\infty} g_k w_k(t). \tag{3.15}$$

Since

$$w(x,t) = \sum_{k=1}^{\infty} w_k(t)\sqrt{2}\sin(\omega_k x),$$

we have that

$$y(t) = \frac{1}{2} \sum_{k=1}^{\infty} w_k(t)\sqrt{2}\, \omega_k \cos(\omega_k x)\Big|_{x=0} = \frac{1}{2} \frac{\partial w}{\partial x}(0,t). \tag{3.16}$$

It may be independently verified, using the method of characteristics, that y(t) as given by (3.16) does indeed satisfy the functional equation (3.12) when w(x,t), u(t) together satisfy (3.1), (3.2).

It is natural to ask why one might not use, instead of (3.12), the equation

$$y(t+1) + y(t-1) = u(t+1), \tag{3.17}$$

since a feedback relation such as

$$u(t+1) = \int_{-1}^{1} k(s)y(t+s)ds$$

is causal for (3.17) whereas

$$u(t) = \int_{-1}^{1} k(s+1)y(t+s)ds,$$

which effects the same closed loop dynamics in (3.12), is not causal. As a matter of fact, (3.17) can be achieved but, as is shown in [12], with the observation

$$y(t) = \frac{1}{2} \frac{\partial w}{\partial t}(1,t) + \frac{1}{2} u(t) = \frac{1}{2} \left(\frac{\partial w}{\partial t}(1,t) + \frac{\partial w}{\partial t}(1,t)\right) \tag{3.18}$$

which, it turns out, cannot be expressed in the form (3.15) but, rather, takes the form

$$y(t) = \sum_{k=1}^{\infty} \hat{g}_k w_k(t) + \frac{1}{2} u(t). \tag{3.19}$$

One can see, in fact, that (3.17) cannot be the control canonical form for (3.1), (3.2) if that system is to be transformable into the control canonical form by a transformation which preserves the transfer function of the process, for the transfer function of (3.17) is easily computed to be

$$\hat{T}(\lambda) = \frac{e^{\lambda}}{e^{\lambda}+e^{-\lambda}},$$

which does not have a partial fraction decomposition of the form (2.10). But we can write

$$\hat{T}(\lambda) = \frac{\frac{1}{2}e^{\lambda} - \frac{1}{2}e^{-\lambda}}{e^{\lambda}+e^{-\lambda}} + \frac{1}{2}$$

$$= \frac{1}{2}\tanh\lambda + \frac{1}{2}.$$

The constant $\frac{1}{2}$ corresponds to the term $\frac{1}{2}u(t)$ in (3.19), and $\frac{1}{2}\tanh\lambda$ does have a partial fraction decomposition leading to the coefficients $\hat{g}_k$ in (3.19) which correspond to the observation (3.18). Further details are given in [12].

Another, related, example concerns the heat equation

$$\frac{\partial w}{\partial t} - \frac{\partial^2 w}{\partial x^2} = 0, \tag{3.20}$$

with the same boundary conditions as (3.2):

$$w(0,t) = 0, \quad \frac{\partial w}{\partial x}(1,t) = u(t). \tag{3.21}$$

With the μ_k, ϕ_k as in (3.4), (3.5), the system (3.20), (3.21) is equivalent to

$$w_k'(t) = -\mu_k w_k(t) + \gamma_k u(t), \quad k = 1,2,\ldots \tag{3.22}$$

with (cf. (3.6))

$$\gamma_k = (-1)^{k-1}\sqrt{2}. \tag{3.23}$$

For

$$y(t) = \sum_{k=1}^{\infty} h_k w_k(t)$$

the transfer function is

$$T(\lambda) = \sum_{k=1}^{\infty} \frac{h_k \gamma_k}{\gamma - \mu_k} .$$

An entire function having zeros at the points $-\mu_k$, $k = 1,2,3,\ldots$, is given by the convergent infinite product

$$q(\lambda) = \prod_{k=1}^{\infty} \left(1 + \frac{\lambda}{\mu_k}\right) = \cosh \lambda^{1/2}. \tag{3.24}$$

Using (2.12) again, the coefficients of the canonical observation are given by

$$h_k = \frac{1}{q'(-\mu_k)\gamma_k} = \frac{1}{\frac{1}{2}(-\mu_k)^{-1/2} \sinh(-\mu_k)^{1/2}(-1)^{k-1}\sqrt{2}}$$

$$= \frac{1}{\frac{1}{2}\left(\frac{1}{\pm i \frac{2k-1}{2}\pi}\right)\left(\frac{e^{\pm i \frac{2k-1}{2}\pi} - e^{\mp i \frac{2k-1}{2}\pi}}{2}\right)(-1)^{k-1}\sqrt{2}}$$

$$= \frac{1}{\frac{1}{(2k-1)\pi}\left(\sin \frac{2k-1}{2}\pi\right)(-1)^k\sqrt{2}} = \frac{(2k-1)\pi}{\sqrt{2}} .$$

Thus

$$y(t) = \sum_{k=1}^{\infty} \frac{(2k-1)\pi}{\sqrt{2}} w_k(t)$$

$$= \sum_{k=1}^{\infty} \frac{(2k-1)\pi}{2} \sqrt{2}\, w_k(t)$$

$$= \sum_{k=1}^{\infty} \frac{\partial}{\partial x} \left(w_k(t)\sqrt{2} \sin\left(\frac{(2k-1)\pi}{2} x\right)\right)\Big|_{x=0}$$

$$= \frac{\partial w}{\partial x}(0,t), \tag{3.25}$$

the same canonical observation as in the first example, except for a positive scalar factor.

The entire function $q(\lambda)$ shown in (3.24) is, formally, the characteristic function of the infinite order differential operator ($D = d/dt$)

$$q(D) = \prod_{k=1}^{\infty} \left(1 + \frac{D}{\mu_k}\right) \tag{3.26}$$

and the control canonical form for (3.20), (3.21) is the equation

$$q(D)y = u. \tag{3.27}$$

There arises, of course, the question of just what (3.27) means and in what sense (3.20), (3.21) is equivalent to (3.27). This, and other matters, will be treated in the next two sections.

We will conclude the present section with some brief references to other systems, the functional equations which constitute their control canonical forms, and reference material. In some cases the work on these systems is not yet complete.

The rigorous canonical reduction procedure for hyperbolic systems, of which the first example in this section may be considered a prototype, has been developed in a series of papers. The first, [11], considered two dimensional, first order linear symmetric hyperbolic systems

$$\frac{\partial}{\partial t}\begin{pmatrix} w \\ v \end{pmatrix} = A(x)\frac{\partial}{\partial x}\begin{pmatrix} w \\ v \end{pmatrix} + B(x)\begin{pmatrix} w \\ v \end{pmatrix} + g(x)u(t) \qquad 0 \le x \le 1, \quad t \ge 0 \tag{3.28}$$

with boundary conditions

$$\alpha_0 w(0,t) + \beta_0 v(0,t) = 0, \quad \alpha_1 w(1,t) + \beta_1 v(1,t) = 0 \tag{3.29}$$

and showed that, with certain restrictions on $A(x)$, α_0, β_0, α_1, β_1, such a system could be mapped by a one-to-one linear transformation (more about this in Section 4) to a system described by the zero order neutral equation

$$y(t+T) + c_0 y(t) + \int_0^T c(s)y(t+s)ds = u(t) \tag{3.30}$$

wherein the delay T depends on the wave propagation time associated with (3.28), c_0 depends on α_0, β_0, α_1 and β_1, and the delay kernel c(s) depends on dA(x)/dx and B(x). This work was extended in [16] to cover what we call "augmented" hyperbolic systems which consist of a partial differential equation (3.28), finite dimensional systems ($z_0 \in E^{n_0}$, $z_1 \in E^{n_1}$)

$$\frac{dz_0}{dt} = A_0 z_0 + a_0 w(0,t) + b_0 v(0,t) + g_0 u(t), \tag{3.31}$$

$$\frac{dz_1}{dt} = A_1 z_1 + a_1 w(1,t) + b_1 v(1,t) + g_1 u(t) \tag{3.32}$$

coupled to (3.28), and coupling boundary conditions

$$\alpha_0 w(0,t) + \beta_0 v(0,t) + \gamma_0^* z_0(t) = 0, \tag{3.33}$$

$$\alpha_1 w(1,t) + \beta_1 v(1,t) + \gamma_1^* z_1(t) = 0. \tag{3.34}$$

Again there are certain fairly minimal restrictions but, when these are fulfilled, and with $n = n_0 + n_1$, the system (3.28), (3.31) - (3.34) can be transformed into the canonical system

$$\frac{d^n y}{dt^n}(t+T) + \sum_{k=0}^{n} c_k \frac{d^k y}{dt^k}(t) + \int_0^T c(s) \frac{d^n y}{dt^n}(t+s)ds = u(t). \tag{3.35}$$

The paper also considers "deficient" hyperbolic systems whose control canonical forms are neutral functional equations of negative order. An example of such an equation is

$$\int_0^T c(s)y(t+s)ds = u(t) \tag{3.36}$$

where, if the system has order $-n$, $c \in H^n[0,T]$ with derivatives of order $\leq n - 2$ vanishing at 0 and T, and

$$\frac{d^{n-1}c}{ds^{n-1}}(0) \neq 0, \quad \frac{d^{n-1}c}{ds^{n-1}}(T) \neq 0.$$

Unique solutions exist in H_{loc}^{-n}. The wave equation (3.1) with boundary conditions

$$w(0,t) = 0, \quad w(1,t) = u(t)$$

leads to the (-1) order equation

$$\int_0^T y(t+s)ds = u(t).$$

Examples of all negative orders occur in connection with control of the wave equation in a disc in $\mathbb{R}^2$.

It is possible also to permit the control u(t) to appear in the boundary conditions (3.29) or (3.33), (3.34) rather than, or in addition to, appearing in (3.28), (3.31), (3.32). The control canonical forms are unchanged; in fact, one of the important facts about the control canonical form is that it is independent of the precise form of control input - but the canonical observation does change with the input, as does the transformation which reduces the original system to its canonical form.

One of the author's former students, R. G. Teglas, has studied systems of hyperbolic equations with $w \in E^{2n}$:

$$\frac{\partial w}{\partial t} = A(x) \frac{\partial w}{\partial x} + B(x)w = 0 \tag{3.37}$$

and boundary conditions

$$C_0 w(0,t) = 0, \quad C_1 w(1,t) = b_1 u(t) \tag{3.38}$$

wherein C_0, C_1 are $n \times 2n$ matrices and $b \in E^n$. With appropriate restrictions and assumptions of observability and controllability type, Teglas has shown in his thesis [17] that a sub-class of these systems can be reduced to neutral functional equations involving multiple discrete delays and a distributed delay:

$$\sum_{k=1}^{N} c_k y(t+kh) + \int_0^{T(=Nh)} c(s)y(t+s)ds = u(t), \quad c_0 \neq 0, \quad c_N \neq 0. \tag{3.39}$$

Augmentation of such systems would lead to higher order equations with multiple delays.

Another of the author's students, L. F. Ho, has completed a study [5] of systems of the form (3.28) wherein the boundary conditions replacing (3.29) are degenerate - no "energy" is reflected from one or more of the boundaries $x = 0$, or $x = 1$. An example derived from a simulation of a conveyor system in a gas kiln is studied in some detail. The resulting canonical system is a retarded equation, typically having the form

$$y(t+T) + \int_0^T c(s)y(t+s)ds = u(t). \tag{3.40}$$

There appears to be a very extensive collection of systems which lead to canonical forms of Volterra equation type:

$$y(t) + \int_0^\infty c(s)y(t-s)ds = u(t)$$

with the kernel c having various properties depending on the particular system under consideration. The equation for linear surface waves on a liquid, studied by R. M. Reid and the author ([15], [10]), leads to a kernel $c(s)$ which behaves, asymptotically, like $\sin(s^2)$, not decaying to zero but oscillating ever more rapidly as $s \to \infty$. This equation is not yet well understood.

4. STATE SPACES FOR CANONICAL FORMS; TRANSFORMATION TO CONTROL CANONICAL FORM

Up to this point in the paper we have offered what is basically a plausibility argument for considering the system (2.1) to be, in some rather vague sense, equivalent to the system represented by the functional equation (2.9). But, if a rigorous theory is to be developed, we need to be able to describe a state space, E, for (2.9) and describe a transformation $Q: H \to E$ which carries the system (2.1) into (2.9).

In the case of the system (3.8), equivalent to the wave equation (3.1), (3.2), there is no real difficulty; the details have already appeared in [11] and [12]. An appropriate state space to use for the neutral equation (3.12) is $L^2[-1,1]$. The relevant semigroup (group in this case) is that generated by the operator

$$\mathcal{D} = \frac{\partial}{\partial s} \tag{4.1}$$

on the domain consisting of all y in $H^1[-1,1]$ which satisfy the boundary condition $y(1) + y(-1) = 0$. Solutions of (3.12) are re-expressed as functions of two variables

$$y(t+s) = y(t,s), \quad -cs < t < cs, \quad -1 \le s \le 1,$$

and $y(t,s)$ is interpreted as a generalized solution of

$$\frac{\partial y}{\partial t} = \frac{\partial y}{\partial s} = \mathcal{D}y \tag{4.2}$$

with boundary conditions

$$y(t,1) + y(t,-1) = u(t). \tag{4.3}$$

The exponential functions $e^{\lambda s}$ which satisfy the homogeneous condition conformable with (4.3), i.e.,

$$e^{\lambda} + e^{-\lambda} = 0 \quad (\text{or } e^{2\lambda} = -1),$$

are precisely $e^{\pm i(\frac{2k-1}{2})\pi s}$, $k = 1,2,3,\ldots$. These form a uniform, or Riesz, basis for $L^2[-1,1]$ as is well known; in fact, dividing by $\sqrt{2}$ we obtain an orthonormal basis. If one writes

$$y(t,s) = \sum_{k=1}^{\infty} [y_k(t)e^{i(\frac{2k-1}{2})\pi s} + y_{-k}(t)e^{-i(\frac{2k-1}{2})\pi s}],$$

then it is not hard to see that

$$y_k'(t) = i(\frac{2k-1}{2})\pi\, y_k(t) + \frac{1}{p'(i(\frac{2k-1}{2})\pi)} u(t) \tag{4.4}$$

$$y_{-k}'(t) = -i(\frac{2k-1}{2})\pi\, y_{-k}(t) + \frac{1}{p'(i(\frac{2k-1}{2})\pi)} u(t). \tag{4.5}$$

If we start with the system (3.8), the state space is ℓ^2, indexed by $k = \pm 1, \pm 2, \pm 3, \ldots$, and the map

$$\eta_{\pm k}(t) = b_{\pm k} p'(\pm i(\frac{2k-1}{2})\pi) y_{\pm k}(t)$$

carries (3.8) into (4.4), (4.5). In terms of basis elements, if we let $e_{\pm k}$ be the obvious basis elements for ℓ^2, the transformation may be described by $Q: \ell^2 \to L^2[-1,1]$ with the property

$$Q: e_{\pm k} \to \frac{1}{b_{\pm k} p'(\pm i(\frac{2k-1}{2})\pi)} e^{\pm i(\frac{2k-1}{2})s} . \tag{4.6}$$

In the given example (3.1), (3.2), the $b_{\pm k}$, $p'(\pm i(\frac{2k-1}{2})\pi)$ are bounded and bounded away from zero so that Q is bounded and boundedly invertible. In examples involving distributed, rather than boundary, control in (3.1), (3.2), Q becomes an unbounded (albeit one-to-one) operator because the $b_{\pm k}$ tend to zero as $k \to \infty$.

In order to have a general theory, one requires a Hilbert space E, whose elements may be identified as scalar functions or distributions, with the property that E includes all exponentials and polynomials. It should be possible to define the differentiation operator, $\mathcal{D}$, on a dense subspace of E, and the set of all $f \in E$ such that $\mathcal{D}f \in E$ and $\langle f,p \rangle = 0$ should define a domain for $\mathcal{D}$ such that $\mathcal{D}$ has the complex numbers λ_k as eigenvalues, the corresponding eigenvectors being the exponential functions $e^{\lambda_k s}$, since $\mathcal{D}(e^{\lambda_k s}) = \lambda_k e^{\lambda_k s}$. The transformation from the original system (2.1), equivalently (2.7), to (2.9) is effected by the map (cf. 4.6))

$$Q: e_k \to \frac{1}{b_k p'(\lambda_k)} e^{\lambda_k s} . \tag{4.7}$$

The operator Q maps ℓ^2 (indexed by $k = 1,2,3,\ldots$ with e_k the "obvious" basis) to E. In order to be able to say that Q is one-to-one with dense range, the $e^{\lambda_k s}$ should be complete and strongly linearly independent (see [14]) for strong linear independence) in E at least; preferably, the $e^{\lambda_k s}$ should form a uniform basis for E. In the latter case it will be possible to conclude that Q is bounded and boundedly invertible if the b_k, $p'(\lambda_k)$ are bounded and bounded away from zero.

The requirements just elaborated for E in some cases make the identification and/or construction of E a rather challenging task. To illustrate what can be involved in such an undertaking, we will use the final section of this paper to describe the construction and transform representation of E in the case of the heat conduction system (3.20), (3.21), for which we have made a plausible case that a functional equation of the form (3.26), (3.27) should act as the control canonical form. In the process we offer, we believe for the first time, a state space (i.e., semigroup) formulation of differential equations of infinite order. An extensive classical literature exists for such spaces. See [1] and references listed there.

5. STATE SPACES FOR A CLASS OF DIFFERENTIAL EQUATIONS OF INFINITE ORDER

In this section, we propose to discuss differential equations

$$q(D)y = 0, \tag{5.1}$$

or

$$q(D)y = u, \tag{5.2}$$

where q(D), rather than being a polynomial in D, is an entire function in a certain class to be specified below. A rather more general treatment than what we will give here is possible and will, we hope, appear elsewhere in the near future.

There is a substantial literature (see [1], e.g.) on differential equations of infinite order, but it seems safe to say that this literature has largely ignored the question of state space representation or, to say the same thing differently, semi-group representation. The restriction to finite dimensions is the familiar replacement of the differential equation

$$D^n y + a_1 D^{n-1} y + \dots + a_{n-1} Dy + a_n y = u$$

by the first order system, with $y_1(t) \equiv y(t)$,

$$Dy_k(t) = y_{k+1}(t), \quad k = 1,2,\dots,n-1$$

$$Dy_n(t) = - a_1 y_n(t) - \dots - a_n y_1(t) + u(t).$$

We are concerned with differential equations (5.1), (5.2) wherein $q(\lambda)$ is given by a convergent infinite product

$$q(\lambda) = \prod_{k=1}^{\infty} \left(1 + \frac{\lambda}{\mu_k}\right), \tag{5.3}$$

the μ_k having the form

$$\mu_k = \left(\frac{2k-1}{2}\right)^2 \pi^2 + k\varepsilon_k \tag{5.4}$$

with

$$\sum_{k=1}^{\infty} |\varepsilon_k|^2 < \infty. \tag{5.5}$$

The function $q(\lambda) = \cosh \lambda^{1/2}$ in (3.24) is a member of this family with the ε_k being identically zero.

There is a close relationship between functions (5.3) and the functions

$$p(z) = \prod_{k=1}^{\infty} \left(1 + \frac{z^2}{\mu_k}\right) = \prod_{k=1}^{\infty} \left(1 + \frac{z^2}{\omega_k^2}\right) \tag{5.6}$$

where

$$\omega_k = \left(\frac{2k-1}{2}\right)\pi + \hat{\varepsilon}_k \tag{5.7}$$

and

$$\prod_{k=1}^{\infty} |\hat{\varepsilon}_k|^2 < \infty. \tag{5.8}$$

The functions (5.6) have played an important role in the study of nonharmonic Fourier series ([1], [9], [16]). From this work (see particularly [8], [9]) we are able to extract a number of important properties of the functions $q(\lambda)$ and $p(z)$. We will just give a summary of those properties here. If we denote by Γ_ρ the vertical line in the complex plane

$$\Gamma_\rho = \{z \mid \mathrm{Re}(z) = \rho\}, \tag{5.9}$$

then it is known that there are positive constants m and M such that

$$me^{|Re(z)|} = me^{\rho} \leq |p(z)| \leq Me^{\rho} = Me^{|Re(z)|} \tag{5.10}$$

for $z \in \Gamma_{\rho}$.

It is also well known that the derivatives of p(z) at the points $\pm i\omega_k = \pm i(\frac{2k-1}{2})\pi$, k = 1,2,3,..., are bounded and bounded away from zero.

These properties of p(z) can be used to establish that the functions $e^{\pm i\omega_k t}$ form a uniform (Riesz) basis for the space $L^2[-1,1]$. The completeness follows from the observation that if $f \in L^2[-1,1]$ and has the Fourier transform (except for a trivial change of variable)

$$(Ff)(z) = \hat{f}(z) = \int_{-1}^{1} e^{zt} f(t)dt, \tag{5.11}$$

and if C_k denotes the square of side length $2k\pi$ centered at the origin and oriented in the positive direction, then the properties of Fourier transforms of functions $f \in L^2[-1,1]$ together with (5.10) and easily obtained estimates for 1/p(z) on the horizontal lines Im(z) = $k\pi$, where k is an integer, enable one to see that

$$\lim_{k\to\infty} \frac{1}{2\pi i} \int_{C_k} \frac{1}{\zeta - z} \frac{\hat{f}(\zeta)}{p(\zeta)} d\zeta = 0,$$

and hence

$$\frac{\hat{f}(z)}{p(z)} = -\sum_{k=1}^{\infty} [\hat{f}(i\omega_k) \text{Res}(\frac{1}{p(\zeta)})_{\zeta = i\omega_k} + \hat{f}(-i\omega_k)\text{Res}(\frac{1}{p(\zeta)})_{\zeta=-i\omega_k}]$$

for z not equal to any of the $\pm i\omega_k$. Thus if

$$\int_{-1}^{1} e^{\pm i\omega_k t} f(t)dt = \hat{f}(\pm i\omega_k) = 0$$

for all k, then $\hat{f}(z) \equiv 0$, which implies f = 0 in $L^2[-1,1]$. The strong linear independence of the functions $e^{\pm i\omega_k t}$ follows from the fact that the functions

$$p_k(z) = \frac{p(z)}{p'(i\omega_k)(z-i\omega_k)}$$

$$p_{-k}(z) = \frac{p(z)}{p'(-i\omega_k)(z+i\omega_k)}$$

may be seen to be Fourier transforms of functions g_k, g_{-k}, respectively, in $L^2[-1,1]$ with the property that for $k,\ell = 1,2,3,\ldots$

$$\int_{-1}^{1} e^{i\omega_k t} g_\ell(t)dt = \delta_{k,\ell}, \quad \int_{-1}^{1} e^{-i\omega_k t} g_\ell(t)dt = 0,$$

$$\int_{-1}^{1} e^{-i\omega_k t} g_{-\ell}(t)dt = \delta_{k,\ell}, \quad \int_{-1}^{1} e^{i\omega_k t} g_{-\ell}(t)dt = 0.$$

That the $e^{\pm i\omega_k t}$ actually form a Riesz basis can be derived from the boundedness above and below of the $p'(\pm i\omega_k)$ (see [14]).

Since there is such a strong relationship between the function $p(z)$, the functions $e^{\pm i\omega_k t}$, and the space $L^2[-1,1]$, it is perhaps not surprising that a comparable relationship should exist between the function $q(\lambda)$, the exponential functions $e^{-\mu_k t}$ associated with the zeros μ_k of $q(\lambda)$ and another Hilbert space which we will define subsequently.

We observe that the mapping

$$\lambda = z^2 \tag{5.12}$$

transforms both Γ_ρ and $\Gamma_{-\rho}$, as given by (5.9), onto the curve in the λ ($= \mu+i\nu$) plane

$$C_\rho = \{\mu + i\nu \,|\, \nu^2 + 4\rho^2\mu - 4\rho^4 = 0\}, \tag{5.13}$$

a parabola opening to the left, crossing the real (μ) axis at $\mu = \rho^2$ and crossing the imaginary (ν) axis at $\nu = \pm 2\rho^2$, collapsing to the negative real axis as ρ tends to zero. Any integral over Γ_ρ, say

$$\int_{\Gamma_\rho} \hat{f}(z)dz,$$

transforms under (5.12) to

$$\frac{1}{2}\int_{C_\rho} \hat{f}(\lambda^{1/2}) \frac{d\lambda}{\lambda^{1/2}} .$$

If f(z) is an even entire function of z, then $f(\lambda^{1/2})$ is an entire function of λ and, very conveniently,

$$\int_{\Gamma_\rho} \hat{f}(z)dz + \int_{\Gamma_{-\rho}} \hat{f}(z)dz = \int_{C_\rho} \hat{f}(\lambda^{1/2}) \frac{d\lambda}{\lambda^{1/2}} . \tag{5.14}$$

Now it is well known that the set of all Fourier transforms (5.12) corresponding to $f \in L^2[-1,1]$ is characterized by f(z) being entire, square integrable on every vertical line in the complex plane and satisfying an inequality

$$|\hat{f}(z)| \le Me^{|\mathrm{Re}(z)|} . \tag{5.15}$$

Moreover, if for $\rho > 0$ we define

$$\|\hat{f}\|_\rho = \int_{\Gamma_\rho} |\hat{f}(z)|^2 |dz| + \int_{\Gamma_{-\rho}} |\hat{f}(z)|^2 |dz| , \tag{5.16}$$

then the map (cf. (5.11))

$$F: f \in L^2[-1,1] \to \hat{f}$$

is an algebraic and topological isomorphism (the norms $\|\cdot\|_\rho$ corresponding to different ρ are equivalent in that they generate the same topology). Put another way, F (the Fourier operator) is bounded and boundedly invertible relative to $\|f\|_{L^2[-1,1]}$ and $\|\hat{f}\|_\rho$.

The space $L^2[-1,1]$ is the direct orthogonal sum of subspaces consisting of even and odd functions, respectively, and the corresponding transforms are even and odd, respectively. Let $H(\rho,e)$ denote the subspace of even $\hat{f}$ as described above equipped with the norm $\|\cdot\|_\rho$. Then $H(\rho,e)$ is a Hilbert space. The functions in $H(\rho,e)$ do not depend on the choice of ρ. When the particular norm $\|\cdot\|_\rho$ is not important we will refer to the class of functions as H_e. If $\hat{f} \varepsilon H_e$ and

$$\phi(\lambda) = (G\hat{f})(\lambda) \equiv \hat{f}(\lambda^{1/2}), \tag{5.17}$$

then $\phi(\lambda)$ is an entire function of λ, and, for each $\rho > 0$,

$$\int_{C_\rho} |\phi(\lambda)|^2 \left|\frac{d\lambda}{\lambda^{1/2}}\right| = \|\hat{f}\|_\rho < \infty. \tag{5.18}$$

We define Φ as

$$\Phi = \{\phi \mid f(z) = \phi(z^2) \varepsilon H_e\} \tag{5.19}$$

and

$$\|\phi\|_\rho = \|f\|_\rho.$$

Equipped with this norm we refer to Φ as Φ_ρ. It may be seen then that Φ_ρ is a Hilbert space and consists of entire functions $\phi(\lambda)$ for which

$$\int_{C_\rho} |\phi(\lambda)|^2 \left|\frac{d\lambda}{\lambda^{1/2}}\right| < \infty \tag{5.20}$$

for every $\rho > 0$ and

$$|\phi(\lambda)| = |f(\lambda^{1/2})| \leq M_\phi e^\rho, \quad \lambda \varepsilon C_\rho. \tag{5.21}$$

Since the C_ρ expand away from the origin like ρ^2, this says that $\phi(\lambda)$ has growth of order 1/2, type 1, as $\lambda \to \infty$.

Now consider a class of meromorphic functions $\psi(\lambda)$ described as follows: the poles of $\psi(\lambda)$ all lie inside, i.e., to the left of, some C_{ρ_0}, $\rho_0 > 0$, and for any $\rho > \rho_0$ we have

$$\int_{C_\rho} |\psi(\lambda)|^2 |\lambda^{1/2} d\lambda| < K_\psi \tag{5.22}$$

where K_ψ is a positive constant independent of ρ. If we let

$$\theta(\lambda) = \psi(\lambda)\lambda^{1/2},$$

then

$$\int_{C_\rho} |\theta(\lambda)|^2 \left|\frac{d\lambda}{\lambda^{1/2}}\right| = \int_{C_\rho} |\psi(\lambda)|^2 |\lambda^{1/2} d\lambda| < K_\psi . \tag{5.23}$$

In addition to (5.22) (equivalently, (5.23)), we require that

$$h(z) = \theta(z^2) = \psi(z^2)z \tag{5.24}$$

should have the property that

$$|h(z)| \le \frac{B}{|\rho|^{1/2}}, \quad z \in \Gamma_\rho, \quad |\rho| > \rho_0, \tag{5.25}$$

for some $B > 0$, which is the same as saying that

$$|\theta(\lambda)| \le \frac{B}{|\rho|^{1/2}}, \quad \lambda \in C_\rho, \quad |\rho| > \rho_0. \tag{5.26}$$

The collection of all meromorphic functions ψ satisfying these requirements will be designated by the symbol Ψ.

We next group the elements of Ψ into equivalence classes by saying that ψ_1 and ψ_2 in Ψ are equivalent if

$$\int_{C_\rho} \phi(\lambda)\psi_1(\lambda)d\lambda = \int_{C_\rho} \phi(\lambda)\psi_2(\lambda)d\lambda \tag{5.27}$$

for all $\phi \in \Phi$. The equivalence class containing a given $\psi \in \Psi$ will be denoted by $\{\psi\}$. The set of all such equivalence classes may easily be seen to form a vector space, with the obvious definitions of addition and scalar multiplication, and will be denoted by E.

We define, for $\psi \in \Psi$,

$$||\psi||_\rho = \sup_{||\phi||_\rho = 1} \left|\int_{C_\rho} \phi(\lambda)\psi(\lambda)d\lambda\right|. \tag{5.28}$$

It is clear that if ψ_1 and ψ_2 are equivalent then $||\psi_1||_\rho = ||\psi_2||_\rho$ and $||\psi_1 - \psi_2||_\rho = 0$. Hence, we may define a norm in E by setting

$$||\{\psi\}||_\rho = ||\psi||_\rho . \tag{5.29}$$

Equipped with this norm E becomes a normed linear space E_ρ. That E_ρ is, in fact, a Hilbert space is implied by the following theorem.

THEOREM. E_ρ is the dual space to Φ_ρ with

$$\langle\phi,\{\psi\}\rangle_\rho = \frac{1}{2\pi i}\int_{C_\rho} \phi(\lambda)\psi(\lambda)d\lambda, \quad \psi \in \{\psi\}. \tag{5.30}$$

Proof. That $E_\rho \subseteq \Phi_\rho^*$ follows from (5.20), (5.23), (5.30) since

$$\left|\int_{C_\rho} \phi(\lambda)\psi(\lambda)d\lambda\right| < \int_{C_\rho} |\phi(\lambda)|\ |\psi(\lambda)|\ |d\lambda|$$

$$= \int_{C_\rho} |\phi(\lambda)|\ |\theta(\lambda)|\ \left|\frac{d\lambda}{\lambda^{1/2}}\right|$$

$$\leq \left(\int_{C_\rho} |\phi(\lambda)|^2 \left|\frac{d\lambda}{\lambda^{1/2}}\right|\right)^{1/2} \left(\int_{C_\rho} |\theta(\lambda)|^2 \left|\frac{d\lambda}{\lambda^{1/2}}\right|\right)^{1/2}.$$

To complete the proof it is only necessary to show that every linear functional on Φ_ρ can be expressed in the form (5.30) for an appropriate $\psi \in \Psi$.

We have seen that Φ_ρ is isomorphic to the even subspace of $L^2[-1,1]$, the isomorphism being given by the map

$$GF\colon L^2[-1,1] \to \Phi$$

with G,F described by (5.17), (5.11) respectively. The space $L_e^2[-1,1]$ (the even subspace of $L^2[-1,1]$) is its own dual, so an arbitrary linear functional, ℓ, on that space may be represented as

$$\ell(f) = \int_{-1}^{1} f(t)g(t)dt$$

for some $g \in L_e^2[-1,1]$ uniquely associated with ℓ. Let g be extended to $(-\infty,\infty)$ by solving the functional equation (cf. (3.17))

$$g(t+1) + g(t-1) = 0$$

with the values of g on [-1,1] as initial state. The Laplace transform of g may be defined by

$$\mathcal{L}(g)(z) = \int_0^\infty e^{-zt} g(t)dt, \quad \mathrm{Re}(z) > 0,$$

and by

$$\mathcal{L}(g)(z) = -\int_{-\infty}^0 e^{-zt} g(t)dt, \quad \mathrm{Re}(z) < 0.$$

It is shown in [13] that $\mathcal{L}(g)(z)$, so defined, extends to a meromorphic function $\tilde{g}(z)$ with poles at the points $\pm i(\frac{2k-1}{2})\pi$, and that, with $\hat{f}(z) = (\mathcal{F}(f))(z)$ (cf. (5.11))

$$\int_{-1}^1 f(t)g(t)dt = \frac{1}{2\pi i}\left(\int_{\Gamma_\rho} + \int_{\Gamma_{-\rho}}\right)\hat{f}(z)\tilde{g}(z)dz \tag{5.31}$$

for any $\rho > 0$. The fact that both f and g are even in $L^2[-1,1]$ shows very readily that $\hat{f}(z)$ is an even entire function and $\tilde{g}(z)$ is an odd meromorphic function of z. With

$$\phi(\lambda) = \hat{f}(\lambda^{1/2}) \tag{5.32}$$

$$\psi(\lambda) = \tilde{g}(\lambda^{1/2})/\lambda^{1/2}, \tag{5.33}$$

$\phi(\lambda)$ and $\psi(\lambda)$ are entire and meromorphic, respectively, and (5.14), (5.31) give

$$\int_{-1}^1 f(t)g(t)dt = \frac{1}{2\pi i}\int_{C_\rho} \phi(\lambda)\psi(\lambda)d\lambda.$$

We can write this as

$$\langle f,g\rangle_{L_e^2[-1,1]} = \langle(\mathcal{GF})f, \{\psi\}\rangle_\rho$$

from which it follows that $g = (\mathcal{GF})^*\{\psi\}$, or $\{\psi\} = ((\mathcal{GF})^*)^{-1}g$.

Since GF is an isomorphism, so is $((GF)^*)^{-1}$. In particular, it is onto: its range is Φ_ρ^*, and this completes the proof.

COROLLARY. The bilinear form $\langle\phi,\{\psi\}\rangle_\rho$ defined by (5.30) is independent of ρ, provided ρ is chosen large enough so that the parabolic contour C_ρ encloses all of the poles of the particular $\psi(\lambda)$ under consideration.

This follows from the fact, established in [14], that if $\hat{f}$, $\tilde{g}$ are related to ϕ,ψ by (5.32), (5.33), then the integral

$$\frac{1}{2\pi i}\left(\int_{\Gamma_\rho} + \int_{\Gamma_{-\rho}}\right)(\hat{f}(z)\tilde{g}(z))dz$$

is independent of ρ.

COROLLARY. Let $\{\mu_k\}$ be a sequence of the form (5.4) and let $q(\lambda)$ be given by the infinite product (5.3). Then the functions

$$\phi_k(\lambda) = \frac{q(\lambda)}{q'(-\mu_k)(\lambda+\mu_k)} \tag{5.34}$$

form a uniform (Riesz) basis for Φ and the equivalence classes $\{\psi_k\}$, where

$$\psi_k(\lambda) = \frac{1}{\lambda+\mu_k} \tag{5.35}$$

form the dual, biorthogonal uniform basis for E_ρ, i.e.,

$$\langle\phi_k,\{\psi_\ell\}\rangle = \delta_{k\ell}. \tag{5.36}$$

Proof. From [8], [9], [16] it may be inferred that the functions (cf. (5.7))

$$g_k(t) = \frac{e^{i\omega_k t} + e^{-i\omega_k t}}{2}$$

form a uniform basis for $L_e^2[-1,1]$. The corresponding Laplace transforms are

$$\tilde{g}_k(z) = \frac{1}{2}\left(\frac{1}{z-i\omega_k} + \frac{1}{z+i\omega_k}\right) = \frac{z}{z^2+\mu_k}$$

and the corresponding $\psi_k(\lambda)$ under (5.33), equivalently (5.24), are just the functions (5.35). Since we have already seen that $\{\psi_k\} = ((GF)^*)^{-1}g_k$ ((5.33) ff.) and that $(GF^*)^{-1}$ is an isomorphism, it follows that the equivalence classes $\{\psi_k\}$ form a uniform basis for E_ρ.

Since the residue of $\psi_k(\lambda)$ at $\lambda = -\mu_k$ is 1 and the $\phi_k(\lambda)$ defined by (5.34) are such that

$$\psi_k(-\mu_\ell) = \delta_{k\ell},$$

it is not hard to see that the $\phi_k \in \Phi$ and

$$\langle\phi_k, \{\psi_\ell\}\rangle = \frac{1}{2\pi i}\int_{C_\rho} \phi_k(\lambda)\psi_\ell(\lambda)\,d\lambda = \delta_{k\ell}, \tag{5.37}$$

which establishes the ϕ_k as the dual basis to $\{\psi_k\}$ in Φ_ρ.

PROPOSITION. If $\{\psi\} \in E$, there is a function $f \in L^2_{loc}[0,\infty)$ such that ψ is the Laplace transform of $e^{-rt}f(t)$ (which, in turn, is an element of $L^2[0,\infty)$) for some real r.

Proof. Since (5.26) obviously implies that $\psi(\lambda)$ is bounded in some right half plane, a familiar theorem (see [4], e.g.) allows us to reach the above conclusions if we can show that the L^2 norm of ψ is uniformly bounded on all vertical lines

$$\text{Re}(\lambda) = \mu \tag{5.38}$$

for sufficiently large μ. We note that for any real μ, C_ρ intersects the line (5.38) at the points defined by the defining equation in (5.13). Solving that equation for ρ gives

$$2\rho^2 = 4\mu + 4\sqrt{\mu^2+\nu^2} = 4\mu + 4|\lambda|,$$

and (5.26) then implies that

$$|\psi(\mu+i\nu)| = |\theta(\lambda)/\lambda^{1/2}| < \frac{B}{|\rho|^{1/2}|\lambda|^{1/2}} = \frac{B\sqrt{2}}{[4\mu+4|\lambda|]^{1/4}|\lambda|^{1/2}} . \tag{5.39}$$

Since the function on the right hand side of (5.39) has uniformly bounded L^2 - norm, so does ψ. Of course, we have to take $\mu \geq r$ for some real r to avoid the singularities of ψ. Then, if we define

$$f(t) = \frac{1}{2\pi i} \int_{\mu-i\infty}^{\mu+i\infty} e^{\lambda t}\psi(\lambda)d\lambda$$

(in the usual l.i.m. sense), we have, as is well known, $e^{-rt}f(t) \in L^2[0,\infty)$, and the proposition is proved.

We now define $\hat{E}$ to be the set of all functions $f(t)$, as defined above, corresponding to ψ for which $\{\psi\} \in E$. Two functions, f_1 and f_2, in $\hat{E}$ are equivalent if their transforms ψ_1 and ψ_2 are equivalent, i.e., $\{\psi_1\} = \{\psi_2\}$, and E is the linear space of such equivalence classes. We define ($Lf = \psi$)

$$||f||_\rho = ||Lf||_\rho \tag{5.40}$$

and thereby obtain a Hilbert space E of equivalence classes of functions f, equipped with a particular norm.

Since the map from f to $\{Lf\}$ is bounded, by definition essentially, and since the equivalence classes corresponding to the functions (5.35) form a uniform basis for E, we conclude that the corresponding functions

$$p_k(t) = e^{-\mu_k t}, \quad k = 1,2,3,\ldots, \tag{5.41}$$

form a uniform basis for the space E.

What we have done, through the space Φ, is to define a Hilbert space of distributions whose Fourier transforms are the elements of Φ. The fact that the entire functions in Φ have exponential growth of order less than one implies that the support of these distributions is just the set $\{0\}$. The distributions in E^*, the dual space to E, are certain infinite linear combinations of $\delta_{\{0\}}$, $\delta'_{\{0\}}$, $\delta''_{\{0\}}$, etc. The coefficients are computable using the Taylor expansion of the corresponding Fourier transform $\phi(\lambda)$ at $\lambda = 0$. Individual distributions $\delta_{\{0\}}$, $\delta'_{\{0\}}$, $\delta''_{\{0\}}, \ldots$, etc., do not belong to E^* because the corresponding transforms $1, \lambda, \lambda^2, \ldots$, etc. do not lie in Φ - a consequence of the fact that these functions do not satisfy (5.20).

Each function $q(\lambda)$ of the form (5.3), the μ_k having the form (5.4), leads to a set of exponential functions (5.41) for which the corresponding equivalence classes form a uniform basis for the space E. Whatever the sequence $\{\mu_k\}$, within the limits of (5.4), every equivalence class which is an element of $\mathcal{E}$, call it $\{\psi\}$, will contain exactly one function of the form

$$\tilde{\psi}(\lambda) = \sum_{k=1}^{\infty} \frac{\alpha_k}{\lambda + \mu_k}$$

with

$$\sum_{k=1}^{\infty} |\alpha_k|^2 < \infty.$$

The α_k are the expansion coefficients of $\{\psi\}$ with respect to the elements $\{\frac{1}{\lambda+\mu_k}\}$ in $\mathcal{E}$ or, equivalently, the expansion coefficients of $\{f\}$ (f = inverse Laplace transform of ψ) with respect to $\{e^{-\mu_k t}\}$. The different representations of a given equivalence class in $\mathcal{E}$ thus correspond to different coordinate systems in terms of which a given element of $\mathcal{E}$ may be expressed. As we will see, these different representations of the same element in $\mathcal{E}$ (equivalently, $\mathcal{E}$) are useful in defining certain linear operators on these spaces.

All we have done here is pointless unless we can show that the space $\mathcal{E}$ enables us to define a semigroup on $\mathcal{E}$ for which the corresponding state trajectories yield, in a very natural way, solutions of the differential equation $q(D)y = 0$ with q described by (5.3), (5.4). We will conclude our work by making what we hope is a convincing case for this.

We begin by noting, informally, what a scalar functional equation corresponds to in terms of semigroup theory. It corresponds to a semigroup whose generator is the infinitesimal shift operator

$$D = \frac{\partial}{\partial s} \tag{5.42}$$

densely defined on a Hilbert (or Banach) space E of scalar functions of the type discussed in this section or, more generally, in the immediately preceding section. Different scalar functional equations are obtained by specifying different domains for this operator, provided the domain of D so selected is such that D does, indeed, generate a semigroup on E. What is usually referred to as the functional equation is an unbounded linear functional, ℓ, on E, with domain $\mathcal{D}(\ell)$ dense in E, such that $\{e \in \mathcal{D}(\ell) \subset E | \ell(e) = 0\}$ is still dense in E and constitutes a domain for (5.42) such that, with domain specification, D generates a semigroup on E. Thus the linear functional $\ell = \delta_{(1)} + \delta_{(-1)}$ is defined on the dense domain $H^1[-1,1] \subseteq L^2[-1,1]$ and the dense domain in $L^2[-1,1]$ consisting of those $w \in H^1[-1,1]$ for which $\ell(w) = w(1) + w(-1) = 0$ is a domain for D in $L^2[-1,1]$(= E in this case) for which D generates a strongly continuous group, S(t), in E. What we usually call a solution of the functional equation

$$w(t+1) + w(t-1) = 0$$

is obtained by selecting an initial $w_0 \in L^2[-1,1]$, letting

$$w(t,\cdot) = S'(t)w_0, \quad t \geq 0,$$

and then letting

$$w(t) = w(t,0), \quad t \geq 0.$$

In the case of the space E being discussed in this section, it is easiest to work in the context of the transformation spaces $\mathcal{E}$ and Φ for E and E^*, respectively. The differentiation operator, D, on E is represented transformationally as multiplication by λ. The fact that it is an unbounded operator corresponds to the complementary fact that $\lambda\psi(\lambda)$ does not, in general, belong to Ψ when ψ does. The operator D is defined at the point $\{f\} \in E$ just in case each ψ in the equivalence class of $\{\mathcal{L}f\}$ in $\mathcal{E}$ can be written in the form

$$\psi(\lambda) = \frac{c}{\lambda} + \tilde{\psi}(\lambda) \tag{5.43}$$

where c is a complex constant and

$$\lambda\tilde{\psi}(\lambda) \in \Psi. \tag{5.44}$$

It may then be seen that $\{\mathcal{L}Df\} = \{\lambda\tilde{\psi}(\lambda)\}$. From the property (5.10) of the corresponding p(z), $q(\lambda)$ is bounded and bounded away from zero on each C_ρ with $\rho > 0$. It follows that the integral

$$\int_{C_\rho} q(\lambda)\tilde{\psi}(\lambda)d\lambda$$

converges absolutely. The domain of D, appropriate to the functional equation (5.1) may be described as follows: it is the set $\mathcal{D}(D) \subseteq E$ whose elements have Laplace transform classes $\{\psi\}$ such that (5.43), (5.44) are true and, letting K be a small circle about $\lambda = 0$ lying inside C_ρ, chosen to exclude the zeros of $q(\lambda)$,

$$\int_K q(\lambda)\frac{c}{\lambda}\, d\lambda + \int_{C_\rho} q(\lambda)\tilde{\psi}(\lambda)d\lambda$$

$$\left(= 2\pi i\, q(0)c + \int_{C_\rho} q(\lambda)\tilde{\psi}(\lambda)d\lambda\right) = 0. \tag{5.45}$$

Now let $\psi(\lambda)$ be that element in $\{Lf\}$ expressible as

$$\psi(\lambda) = \sum_{k=1}^{\infty} \alpha_k \left(\frac{1}{\lambda+\mu_k} \right), \quad \sum_{k=1}^{\infty} |\alpha_k|^2 < \infty. \tag{5.46}$$

Since the $\{\frac{1}{\lambda+\mu_k}\}$ form a uniform basis for E and $\{\frac{1}{\lambda}\}$ belongs to E, there are square summable coefficients β_k such that

$$\{\frac{c}{\lambda}\} = \{ \sum_{k=1}^{\infty} \beta_k \frac{1}{\lambda+\mu_k} \}.$$

Then $\tilde{\psi}(\lambda) = \psi(\lambda) - \frac{c}{\lambda}$ is equivalent to $\sum_{k=1}^{\infty} \left(\frac{\alpha_k - \beta_k}{\lambda+\mu_k} \right)$ and, since $q(-\mu_k) = 0$ for $k = 1,2,3,\ldots$, it follows (since $q(\lambda)/\lambda \in \Phi$) that

$$\int_{C_\rho} q(\lambda)\tilde{\psi}(\lambda)d\lambda = \int_{C_\rho} q(\lambda) \sum_{k=1}^{\infty} \left(\frac{\alpha_k - \beta_k}{\lambda+\mu_k} \right) d\lambda = 0$$

if $\{f\} \in \mathcal{D}(D)$. But then (5.45) shows that $c = 0$ because $q(0) = 1$. Hence $\psi(\lambda) = \tilde{\psi}(\lambda)$ and it must be true that $\lambda\psi(\lambda) \in \Psi$ and $\{\lambda\psi(\lambda)\} = \{Lf\}$. Note, however, that this result is only true for the particular element (5.46) in $\{Lf\}$, not for every element of Ψ in that class. Thus $L(\mathcal{D}(D))$ consists of all equivalence classes in E containing an element (5.46) with $\lambda\psi(\lambda) \in \Psi$, and for each $f \in \mathcal{D}(D)$, $\{Lf\}$ contains the element $\lambda\psi(\lambda)$ which, it may be shown, is expressible as

$$\lambda\psi(\lambda) = \sum_{k=1}^{\infty} [-\mu_k\alpha_k(\frac{1}{\lambda+\mu_k})].$$

Returning to E, we see that the differentiation operator, so defined, has the equivalence classes $\{e^{-\mu_k s}\}$ as eigenvectors and

$$D\{e^{-\mu_k s}\} = \{-\mu_k e^{-\mu_k s}\}$$

- to no one's amazement. The semigroup S(t) generated by D, so defined, is described by

$$S(t)\{e^{-\mu_k s}\} = \{e^{-\mu_k t} e^{-\mu_k s}\} = \{e^{-\mu_k (t+s)}\}$$

which extends continuously to all of E, relative to the topology of E because the $e^{-\mu_k t}$ are bounded and the $\{e^{-\mu_k s}\}$ form a uniform basis for E.

Transformationally the action of S(t) may be described as follows. If $\{f\} \in E$ has the transform class $\{Lf\} = \{\psi\}$ with ψ of the form (5.46), then

$$\{LS(t)f\} = \{\psi(t,\cdot)\},$$

where, for sufficiently large ρ,

$$\psi(t,\nu) = \frac{1}{2\pi i} \int_{C_\rho} \frac{e^{\lambda t}\psi(\lambda)}{\nu-\lambda}\, d\lambda \tag{5.47}$$

for ν external to C_ρ. It is not hard to show that

$$\psi(t,\lambda) = \sum_{k=1}^{\infty} \frac{\alpha_k e^{-\mu_k t}}{\lambda+\mu_k}.$$

The initial conditions necessary and sufficient to specify a solution of $q(D)y = 0$ are not the values and derivatives $y(0)$, $y'(0)$, $y''(0)$,... etc.

In general, these are not defined. Rather, it is necessary and sufficient to specify all of the integrals

$$\langle\phi,\psi\rangle = \frac{1}{2\pi i}\int_{C_\rho} \phi(\lambda)\psi(\lambda)d\lambda, \quad \phi \in \Phi \tag{5.49}$$

in order to specify $\{\psi\}$. Expanding ϕ in Maclaurin series

$$\phi(\lambda) = \sum_{k=0}^{\infty} \phi_k\lambda^k,$$

it may be useful to think of (5.49) as corresponding to a certain linear combination of derivatives, i.e.,

$$\sum_{k=0}^{\infty} \phi_k y^{(k)}(0).$$

One can show that if $f \in \hat{E}$ and $\psi = Lf$ is its Laplace transform, the usual Laplace inversion formula

$$f(t) = \frac{1}{2\pi i} \lim_{A\to\infty} \int_{\mu-iA}^{\mu+iA} e^{(\lambda)t}\psi(\lambda)d\lambda,$$

valid for μ sufficiently large, can be replaced by

$$f(t) = \frac{1}{2\pi i}\int_{C_\rho} e^{\lambda t}\psi(\lambda)d\lambda, \quad t > 0. \tag{5.50}$$

Along similar lines it is possible to show that for $\{f\} \in E$, if ψ is the representative of $\{Lf\}$ of the form (5.46), the scalar function $y(t)$ which is the corresponding solution of $q(D)y = 0$ is given by

$$y(t) = \frac{1}{2\pi i}\int_{C_\rho} e^{\lambda t}\psi(\lambda)d\lambda, \quad t > 0.$$

(This may be undefined for $t = 0$.) The derivatives

$$(D^n y)(t) = \frac{1}{2\pi i} \int_{C_\rho} \lambda^n e^{\lambda t} \psi(\lambda) d\lambda$$

are all defined for $t > 0$ and, in fact, with

$$q(\lambda) = \prod_{k=1}^{\infty} \left(1 + \frac{\lambda}{\mu_k}\right),$$

the partial product differential polynomials

$$(q_k(D)y)(t) = \left(\prod_{k=1}^{K} \left(1 + \frac{1}{\mu_k} D\right)y\right)(t)$$

$$= \frac{1}{2\pi i} \int_{C_\rho} q_K(\lambda) e^{\lambda t} \psi(\lambda) d\lambda$$

are defined for $K = 1,2,3,\ldots$ and

$$\lim_{K\to\infty} (q_K(D)y)(t) = (q(D)y)(t) = 0$$

uniformly for t in any interval $[\tau,\infty)$ with $\tau > 0$ - thus y satisfies $q(D)y = 0$ in a very strong sense for positive t.

Finally, there is the matter of the inhomogeneous equation (5.2), i.e.,

$$q(D)y = u.$$

It is clearly enough to discuss the solution corresponding to the initial state $\{0\}$ in E.

We proceed formally, at first, in the sense that we assume that a solution $y(t)$ of (5.2), with initial state equal to zero in E, exists and has the usual properties. Let $T > 0$ and let $y_T(t)$ be the solution of (5.2), still with zero initial state, corresponding to the same $u(t)$ for $0 \le t \le T$ but satisfying

$$q(D)y_T(t) = 0, \quad t > T.$$

Then for $t > T$ we should have

$$y_T(t) = \sum_{k=1}^{\infty} q_k e^{-\mu_k(t-T)},$$

indicating that the state at time T is the one which, in E, corresponds to the equivalence class of the function $\sum_{k=1}^{\infty} q_k e^{-\mu_k(t-T)}$. The Laplace transform of $y_T(t)$ is

$$\hat{y}_T(\lambda) = \int_0^T e^{-\lambda t} y_T(t)dt + \sum_{k=1}^{\infty} q_k e^{\mu_k T} \frac{1}{\lambda+\mu_k}.$$

From this the desired transform

$$\eta(\nu,T) = \sum_{k=1}^{\infty} \frac{\mu_k}{\nu+\mu_k}$$

can be obtained, at least for ν outside C_ρ, ρ chosen large enough so that the zeros of q lie inside C_ρ, from the formula

$$\eta(\nu,T) = \frac{1}{2\pi i} \int_{C_\rho} \frac{e^{\lambda T}\hat{y}_T(\lambda)}{\nu-\lambda} d\lambda.$$

Since the solution of $q(D)y = u$ which corresponds to $u = \delta_{(t)}$ has the transform $e^{-\lambda t}/q(\lambda)$, we see that we must have

$$\hat{y}_T(\lambda) = \frac{1}{q(\lambda)} \int_0^T e^{-\lambda t} u(t)dt$$

so that

$$\eta(\nu,T) = \frac{1}{2\pi i} \int_{C_\rho} \frac{1}{(\nu-\lambda)q(\lambda)} \int_0^T e^{\lambda(T-t)} u(t)dtd\lambda. \tag{5.51}$$

From this, assuming the convergence of the integral (5.51) and the sum below, one can derive the alternate expression

$$\eta(\nu,T) = \sum_{k=1}^{\infty} \frac{1}{q'(-\mu_k)(\nu+\mu_k)} \int_0^T e^{-\mu_k(T-t)} u(t)dt, \tag{5.52}$$

consistent with the fact that $q(D)y = u$ is equivalent to the infinite dimensional first order system

$$\dot{w}_k = -\mu_k w_k + \frac{1}{q'(-\mu_k)} u, \quad k = 1,2,3,\ldots .$$

It is possible to show, using an argument which appears in [6], that (5.52) is not, in general, convergent for $u \varepsilon L^2[0,T]$. If u is measurable and bounded on [0,T] we have, for some $M > 0$,

$$\left| \frac{1}{q'(-\mu_k)} \int_0^T e^{-\mu_k(T-t)} u(t)dt \right| \le Mk^{-1}$$

and the convergence of (5.52) then follows from the uniform basis property of the elements $\{\frac{1}{y+\mu_k}\}$ in E.

ACKNOWLEDGEMENTS

This research was supported in part by the Air Force Office of Scientific Research under Grant AFOSR 79-0018. Part of this material appeared originally in the Proceedings of the Third IMA Conference on Control Theory, Sheffield, 1980.

REFERENCES

1. Boas, R., *Entire Functions*, Academic Press, New York, 1954.

2. Dunford, N. and J. T. Schwartz, *Linear Operators, Part I: General Theory*, Interscience Pub., New York, 1958.

3. Duren, P. L., *Theory of H^p-Spaces*, Academic Press, New York, 1970.

4. Hoffman, K., *Banach Spaces of Analytic Functions*, Prentice Hall, Englewood Cliffs, N. J., 1962.

5. Ho, L. F., Thesis, Department of Mathematics, University of Wisconsin, Madison, August, 1981.

6. Ho, L. F., and D. L. Russell, Admissible input elements for linear systems in Hilbert Space and a Carleson measure criterion, *SIAM J. Cont. Opt.*, to appear.

7. Lee, R. C. K., *Optimal Estimation, Identification and Control*, M. I. T. Press, Cambridge, 1964.

8. Levinson, N., *Gap and Density Theorems*, Amer. Math. Soc. Colloq. Pub., Vol. 26, Providence, 1940.

9. Paley, R. E. A. C., and N. Wiener, *Fourier Transforms in the Complex Domain*, Amer. Math. Soc. Colloq. Pub., Vol. 19, Providence, 1934.

10. Reid, R. M., Thesis, Department of Mathematics, University of Wisconsin, Madison, August, 1979.

11. Russell, D. L., Canonical forms and spectral determination for a class of hyperbolic distributed parameter control systems, *J. Math. Anal. Appl.* 62 (1978), 186-225.

12. Russell, D. L., Closed-loop eigenvalue specification for infinite dimensional systems: augmented and deficient hyperbolic cases, Technical Summary Report #2021, Mathematics Research Center, University of Wisconsin, Madison, August, 1979.

13. Russell, D. L., Control canonical structure for a class of distributed parameter systems, *Proc. Third IMA Conference on Control Theory*, Sheffield, September, 1980.

14. Russell, D. L., Uniform bases of exponentials, neutral groups, and a transform theory for $H^m[a,b]$, Technical Summary Report #2149, Mathematics Research Center, University of Wisconsin, Madison, July, 1980.

15. Russell, D. L., and R. M. Reid, Water waves and problems of infinite time control, *Proc. IRIA International Symposium on Syst. Anal. & Optim.*, Rocquencourt, December, 1978.

16. Schwartz, L., *Étude des sommes d'exponentielles*, Hermann, Paris, 1950.

17. Teglas, R. G., Thesis, Department of Mathematics, University of Wisconsin, Madison, June, 1981.

EXPONENTIAL DICHOTOMIES IN EVOLUTIONARY EQUATIONS: A PRELIMINARY REPORT

Robert J. Sacker

Department of Mathematics
University of Southern California
Los Angeles, California

George R. Sell

School of Mathematics
University of Minnesota
Minneapolis, Minnesota

1. INTRODUCTION

In this lecture we want to present some preliminary results on the general theory of linear evolutionary equations. While the theory we discuss here concerns linear equations only, our investigations should be viewed as the first step in a somewhat broader program that includes the local behavior of nonlinear evolutionary equations near an invariant finite dimensional manifold.

In order to motivate our theory it is helpful to review several problems which arise in functional differential equations (FDE) and in parabolic partial differential equations (PPDE). Let us begin with some linear FDEs. First consider

$$\dot{x}(t) = \sum_{i=1}^{k} A_i(t)x(t-r_i). \tag{1}$$

Here we have $x \in \mathbb{R}^n$, $A_i: \mathbb{R} \to \mathrm{Hom}(\mathbb{R}^n, \mathbb{R}^n)$ is a real $(n \times n)$-matrix valued function that is periodic in t (perhaps with period that depends on i) and $r_i \geq 0$. Equation (1) is a special case of the combined FDE - ordinary differential equation (ODE) given by

$$\dot{x}(t) = \sum_{i=1}^{k} A_i(\theta)x(t-r_i), \quad \dot{\theta} = F(\theta) \tag{2}$$

where $x \in \mathbb{R}^n$, $\theta \in M$ and M is some compact manifold, $A_i: M \to \mathrm{Hom}(\mathbb{R}^n, \mathbb{R}^n)$ is continuous and $r_i \geq 0$. In the case where $A_i(t)$ is periodic in t in equation (1), then equation (1) is included in equation (2) when M is a torus with a suitable twist flow. Another special case of equation (2), which is of particular interest here, is

$$\begin{aligned} \dot{x}(t) &= a(r,\theta)x(t-1) + b(r,\theta)x(t) \\ \dot{r} &= r(1-r), \quad \dot{\theta} = \pi/2 \end{aligned} \tag{3}$$

where $M = \{(r,\theta): 0 \leq r \leq 1\}$ is the unit disc in $\mathbb{R}^2$ represented in polar coordinates.

The basic problem in the study of these equations is to give a qualitative description of the solution in terms of exponential growth rates, cf. [12 - 14].

Let us now look at the nonlinear FDE

$$\dot{x}(t) = f(a, x_t) \tag{4}$$

where a is a real parameter and f satisfies certain technical conditions to insure the local existence and uniqueness of the solutions of the initial value problem associated with (4), as well as the complete continuity of the solution operator for $t \geq r \geq 0$. Assume that M_0 is a compact invariant manifold for equation (4) when $a = a_0$. We are interested in sufficient conditions on the semi-flow near M_0 in order that there exists a family of compact invariant manifolds M_a that varies continuously in a with $M_{a_0} = M_0$. Such a result, if valid, would be a generalization of theorems of Sacker [10], Fenichel [1] and Pugh and Shub [9].

In the area of PPDE, the Navier - Stokes equations for fluid flow represent the type of equations that concern us. One problem of especial interest arises in the study of the Couette flow, which is the flow of a fluid between two coaxial cylinders where the outer cylinder is fixed and the inner cylinder rotates with an angular velocity ω. It is known that the behavior of the flow changes radically as ω crosses various critical values

$\omega_1,\ldots,\omega_4$ where $0 < \omega_1 < \omega_2 < \omega_3 < \omega_4$, cf. [2,7,17]. In the region $0 < \omega < \omega_1$ one observes the classical Couette flow. The Taylor's cells appear in the region $\omega_1 < \omega < \omega_2$. This state is represented as a stationary solution of the Navier - Stokes equations. In the region $\omega_2 < \omega < \omega_3$ one notices a periodic behavior in the flow. Most experts believe that the Couette flow goes through a Hopf bifurcation at $\omega = \omega_2$. In the region $\omega_3 < \omega < \omega_4$ it appears that there may be an invariant 2 - dimensional torus $M = T^2$ in the Couette flow. One mathematical problem which naturally arises is to study the behavior of M as ω crosses ω_4. Some questions one would like to have the answer to are: (1) Is there a strange attractor for $\omega > \omega_4$? (2) Does there exist a Hopf - Landau bifurcation $T^2 \to T^3$ as ω crosses ω_4? (See [16, 17] for more details.)

The dynamics represented by all the equations above can be summarized in terms of a skew - product semi - flow acting on a Banach space where the solution operator is completely continuous for $t \geq r$, $r \geq 0$. Even though the dynamical systems here are semi - flows on an infinite dimensional Banach space, we can obtain essential insight into the methods needed for analyzing these systems by examining the corresponding theories for ODE on finite dimensional spaces. In every such case the key to understanding these equations is the existence of exponential dichotomies, cf. [1,5,9 - 16]. For the nonlinear equations there is an added problem connected with the theory of linearization in the vicinity of an invariant manifold, [4,6,8,15]. We shall postpone a discussion of this linearization theory for a later report and concentrate in this lecture on the question of finding necessary and sufficent conditions for the existence of an exponential dichotomy for a linear evolutionary equation. The point of view which we adopt and which we next describe is, in our opinion, broad enough to apply to all the problems described above. Proofs will appear elsewhere.

2. GENERAL DISCUSSION

Let X denote a fixed Banach space and let M be a compact, connected Hausdorff space. Assume that we are given a (two - sided) flow $\sigma(\theta,t) = \theta\cdot t$ on M. This means that (i) $\sigma: M \times \mathbb{R} \to M$ is continuous, (ii) $\sigma(\theta,0) = \theta$ and (iii) $\sigma(\sigma(\theta,t),s) = \sigma(\theta,t+s)$. Next we assume that

$$\pi(x,\theta,t) = (\Phi(\theta,t)x,\theta\cdot t)$$

is a linear skew - product semi - flow on $E = X \times M$ where Φ is a linear operator that is *uniformly completely continuous*. This means that for each $\theta \in M$ there is a neighborhood U of θ and a time $T_\theta \geq 0$ such that for any $\rho > 0$ the set

$$\pi(D_\rho \times U,t) = \{(\Phi(\hat{\theta},t)x,\theta\cdot t): \hat{\theta} \in U, |x| \leq \rho\}$$

is relatively compact in E for each $t \geq T_\theta$. By using the compactness of M and the continuity of π it is not difficult to see that

$$T \equiv \sup \{T_\theta: \theta \in M\} < +\infty.$$

We shall say that π has an exponential dichotomy over M if there is a continuous mapping $\hat{P}: E \to E$ and constants $K > 0$, $\alpha > 0$ such that

(i) $\hat{P}(x,\theta) = (P(\theta)x,\theta)$,

where $P(\theta)$ is a linear projection on X,

(ii) $|\Phi(\theta\cdot s,t-s)P(\theta\cdot s)| \leq Ke^{-\alpha(t-s)}$

for all $s \in \mathbb{R}$ and all $(t-s) \geq 0$, and

(iii) $|\Phi(\theta\cdot s,t-s)[I - P(\theta\cdot s)]| \leq Ke^{-\alpha(s-t)}$,

for all $s \in \mathbb{R}$ and all $(t-s) \leq 0$.

Since π is assumed to be a semi - flow, i.e., $\Phi(\theta,t)$ is defined for $t \geq 0$, the inequality (iii) above needs explanation. First of all, define

$$R = \{(x,\theta) \in E: P(\theta)x = x\}$$

$$N = \{(x,\theta) \in E: P(\theta)x = 0\}.$$

Then the inequality (iii) requires that one has $N \subseteq \mathrm{N}$, where we define N to be the set of all $(x,\theta) \in E$ such that there is some negative continuation ϕ through (x,θ) defined for all $t \leq 0$ (i.e., there is a continuous function

ϕ: $(-\infty,0] \to X$ such that $\phi(0) = x$ and $\Phi(\theta\cdot s,t)\phi(s) = \phi(s+t)$ for all $s \leq 0$ and all t, $0 \leq t \leq -s$.) Let $N(\theta) = \{x \in X: (x,\theta) \in N\}$.

We now define the following five subsets of E:

U is the set of $(x,\theta) \in N$ such that there is some negative continuation ϕ defined for all $t \leq 0$ and satisfying $|\phi(t)| \to 0$ as $t \to -\infty$.

B^- is the set of $(x,\theta) \in N$ such that there is some negative continuation ϕ defined for all $t \leq 0$ and satisfying $\sup_{t\leq 0} |\phi(t)| < +\infty$.

$$B^+ = \{(x,\theta) \in E: \sup_{t\geq 0} |\Phi(\theta,t)x| < +\infty\}$$

$$S = \{(x,\theta) \in E: |\Phi(\theta,t)x| \to 0 \text{ as } t \to +\infty\}$$

$$B = B^+ \cap B^-.$$

In the theory of ODE on finite dimensional spaces we recall that a necessary condition for π to have an exponential dichotomy on M is that $B = \{0\} \times M$; see [11]. Also one should note that in some cases (e.g., when the flow on M is minimal) $B = \{0\} \times M$ is a sufficient condition for an exponential dichotomy on M. For the remainder of this lecture we shall assume that

$$B = \{0\} \times M$$

in addition to the assumptions stated earlier. An immediate consequence of this is that $S \cap U = \{0\} \times M$ since $S \cap U \subseteq B$.

THEOREM 1. The following statements are valid:

(A) $B^+ \subseteq S$, $B^- \subseteq U$.

(B) S is closed and the decay rate in S is uniformly exponential.

(C) U is closed, the decay rate in U is uniformly exponential and $\dim U(\theta) \leq \operatorname{codim} S(\theta) < +\infty$ for every $\theta \in M$.

(D) For each $\theta \in M$ the function $\operatorname{codim} S(\theta\cdot t)$ is nondecreasing in t.

The first surprise in this infinite dimensional theory is that the function $\operatorname{codim} S(\theta\cdot t)$ need not be constant in t. The example illustrating this phenomenon is a variation of an example of Winston and Yorke [3, 18]. Specifically we let

$$\dot{x}(t) = a(r,\theta)x(t - 1) + b(r,\theta)x(t)$$

where $\dot{r} = r(r - 1)$, $\dot{\theta} = \pi/2$ in the unit disc

$$M = \{(r,\theta)\colon 0 \le r \le 1, \quad 0 \le \theta \le 2\pi\},$$

$b(r,\theta) \equiv 10$ for $r < 0.3$, $b(r,\theta) = 0$ for $r > 0.7$, and $a(r,\theta)$ is a C^1-function defined on M satisfying

(i) $a(r,\theta) = 0$, for $|r| \le 0.3$.
(ii) $a(r,\theta) = 0$, for $|r| \ge 0.7$ and $(\frac{\pi}{2} \le \theta \le \pi$ or $\frac{3\pi}{2} \le \theta \le 2\pi)$.
(iii) $a(r,\theta) = -2\sin^2 2\theta$, for $|r| > 0.7$ and $(0 \le \theta \le \frac{\pi}{2}$ or $\pi \le \theta \le \frac{3\pi}{2})$.

It is not hard to verify that for r small one has codim $S(r,\theta) = 1$ while for r large one has codim $S(r,\theta) = 0$; cf. [3,18] for details.

It would be interesting to obtain a better understanding of the discontinuity set of codim $S(\theta)$, both in this example and in general.

THEOREM 2. Fix $\theta \in M$ and define $\ell = \ell(\theta)$ by

$$\ell = \lim_{t\to+\infty} \text{codim } S(\theta\cdot t).$$

Then $\ell < +\infty$ and for all $\hat{\theta} \in \Omega(\theta)$, the ω-limit set of θ, one has codim $S(\hat{\theta}) = \dim U(\hat{\theta}) = \ell$ and $X = S(\hat{\theta}) \oplus U(\hat{\theta})$. In particular, if $M = \Omega(\theta)$ (e.g., if M is a minimal set) then π has an exponential dichotomy over M.

The main conclusion of the next theorem is that there exists a full complement to $S(\theta)$ with the property that for every initial vector x in this complement there exists a negative continuation defined for all $t \le 0$. Since codim $S(\theta)$ is finite, this complement is finite dimensional.

THEOREM 3. Fix $\theta \in M$ with the property that codim $S(\theta\cdot t)$ = constant for all $t \le 0$. Then there is a linear subspace $K(\theta)$ with $K(\theta) \subset N(\theta)$, $(U(\theta) \oplus S(\theta)) \cap K(\theta) = \{0\}$ and $X = S(\theta) \oplus K(\theta) \oplus U(\theta)$. (Note that codim $S(\theta) = \dim K(\theta) + \dim U(\theta)$.)

The next theorem describes the behavior of π over α-limit sets. The major difficulty in proving this result, as with many of our theorems, is the fact that $\Phi(\theta,t)x$ need not be defined for $t \le 0$.

THEOREM 4. Fix $\theta \in M$ and define $k = k(\theta)$ and $m = m(\theta)$ by

$$k = \lim_{t\to-\infty} \operatorname{codim} \mathcal{S}(\theta\cdot t), \quad m = \dim \mathcal{U}(\theta).$$

Then $m \leq k$ and for all $t \in \mathbb{R}$ one has $\dim \mathcal{U}(\theta\cdot t) = m$, and for all $\hat{\theta} \in A(\theta)$, the α-limit set of θ, one has $\operatorname{codim} \mathcal{S}(\hat{\theta}) = \dim \mathcal{U}(\hat{\theta}) = m$ and $X = \mathcal{S}(\hat{\theta}) \oplus \mathcal{U}(\hat{\theta})$.

The next result, which is our main theorem, summarizes everything stated above. The reader should compare this with the finite dimensional theory, [11].

THEOREM 5. Assume that $\mathcal{B} = \{0\} \times M$. Then either π admits exponential dichotomy over M or the flow on M has a gradient-like structure.

The alternative statement on the gradient-like flow on M means that (i) at least two of the sets

$$M_i \equiv \{\theta \in M: \operatorname{codim} \mathcal{S}(\theta\cdot t) = \dim \mathcal{U}(\theta\cdot t) = i, \text{ for all } t \in \mathbb{R}\}, i = 0,1,\ldots .$$

are nonempty, (ii) all the sets M_i are closed, compact and invariant, (iii) for every $\theta \in M$ one has $k \leq \ell$, $A(\theta) \subseteq M_k$, $\Omega(\theta) \subseteq M_\ell$ where

$$k \equiv \lim_{t\to-\infty} \operatorname{codim} \mathcal{S}(\theta\cdot t)$$

$$\ell \equiv \lim_{t\to+\infty} \operatorname{codim} \mathcal{S}(\theta\cdot t),$$

and $k = \ell$ if and only if $\theta \in M_k$.

ACKNOWLEDGEMENT

This research was supported in part by National Science Foundation grant MCS 79-01998, while the second author was visiting the University of Southern California.

REFERENCES

1. N. Fenichel, Persistence and smoothness of invariant manifolds for flows, *Indiana University Math. J.* 21 (1971/72), 193-226.

2. J. P. Gollub and H. L. Swinney, Onset of turbulence in a rotating fluid, *Phys. Rev. Letters* 35 (1975), 921 ff.

3. J. K. Hale, *Theory of Functional Differential Equations*. Springer-Verlag, New York, 1977.

4. D. Henry, *Geometric Theory of Semilinear Parabolic Equations*. Lecture Notes in Mathematics No. 840, Springer-Verlag, New York, 1980.

5. M. Hirsch, et al., *Invariant Manifolds*. Lecture Notes in Math. No. 583, Springer-Verlag, New York, 1977.

6. R. A. Johnson, Concerning a theorem of Sell. *J. Differential Equations* 30 (1978), 324-339.

7. D. Joseph, *Stability of Fluid Motions* I, II. Springer-Verlag, New York, 1976.

8. J. E. Marsden and M. McCracken, *The Hopf Bifurcation and Its Applications*. Springer-Verlag, New York, 1976.

9. C. C. Pugh and M. Shub, Linearization of normally hyperbolic diffeomorphisms and flows, *Invent. Math.* 10 (1970), 187-198.

10. R. J. Sacker, A perturbation theorem for invariant manifolds and Hölder continuity, *J. Math. Mech.* 18 (1969), 705-762.

11. R. J. Sacker and G. R. Sell, Existence of exponential dichotomies and invariant splittings I, II, *J. Differential Equations* 15 (1974), 429-458 and 22 (1976), 478-496.

12. R. J. Sacker and G. R. Sell, A spectral theory for linear differential systems, *J. Differential Equations* 27 (1978), 320-358.

13. R. J. Sacker and G. R. Sell, The spectrum of an invariant submanifold, *J. Differential Equations* 38 (1980), 135-160.

14. J. F. Selgrade, Isolated invariant sets for flows on vector bundles, *Trans. Amer. Math. Soc.* 203 (1975), 359-390.

15. G. R. Sell, The structure of a flow in the vicinity of an almost periodic motion, *J. Differential Equations* 27 (1978), 359-393.

16. G. R. Sell, Bifurcation of higher dimensional tori, *Arch. Rational Mech. Anal.* 67 (1979), 199-230.

17. G. R. Sell, Hopf - Landau bifurcation near strange attractors, *Proc. International Symposium on Synergetics*, W. Germany, 1981, to appear.

18. E. Winston and J. A. Yorke, Linear delay differential equations whose solutions become identically zero, *Rev. Roum. Math. Pures Appl.* 14 (1969), 885-887.

ON WEIGHTED MEASURES AND A NEUTRAL FUNCTIONAL DIFFERENTIAL EQUATION

Olof J. Staffans

Helsinki University of Technology
Institute of Mathematics
Otaniemi, Finland

1. INTRODUCTION

We study the linear, autonomous, neutral system of functional differential equations

$$\frac{d}{dt}(\mu * x(t) + f(t)) = \nu * x(t) + g(t), \quad t \in \mathbb{R}^+ \tag{1.1}$$

$$x(t) = \phi(t), \quad t \in \mathbb{R}^- . \tag{1.2}$$

Here $\mathbb{R}^+ = [0,\infty)$, $\mathbb{R}^- = (-\infty,0]$, μ and ν are matrix-valued measures on $\mathbb{R}^+$, finite with respect to a weight function, and f, g and ϕ are continuous and satisfy certain growth conditions as $t \to \pm\infty$. We give conditions which imply that solutions of (1.1), (1.2) can be decomposed into components with different exponential growth rate. Similar results have earlier been obtained by D. Henry [4] and by T. Naito [8], [9]. Henry studies the case when the delay is finite, i.e., μ and ν are supported on a finite interval $[0,r]$, and does not allow μ to have a singular part. Naito does permit the delay to be infinite, but he treats only the retarded equation, which one gets by replacing $\mu * x$ by x in (1.1). Our decomposition result applies to neutral (as well as retarded) equations with (finite or) infinite delay, and we do not assume that the singular part of μ vanishes. On the other hand, our initial function ϕ is less general than, e.g., the initial function in [9].

This work relies heavily on I. Gelfand's theory of the existence of inverses of some weighted measures. This theory, in a locally analytic setting, has recently been applied to equations similar to (1.1) in [5]. To give the reader an idea of the type of arguments that are used in the proof of our decomposition theorem, we first discuss the theory of locally analytic transforms of weighted L^1-functions, as presented in [5, Chapter 2].

2. WEIGHTED MEASURES AND L^1-FUNCTIONS

A continuous, strictly positive function ρ on R is called a *weight function on* $\mathbb{R}$, if

$$\rho(s + t) \leq \rho(s)\rho(t) \quad (s,t \in \mathbb{R}); \quad \rho(0) = 1. \tag{2.1}$$

The condition $\rho(0) = 1$ is not used in [5], but we have included it here because it is needed if one wants to apply [1, Satz 7], as we do in our proof of the decomposition theorem. Clearly, by (2.1), $\log \rho(t)$ is subadditive, and this fact implies that ρ_* and ρ^*, defined by

$$\begin{aligned} \rho_* &= -\inf_{t>0} \frac{\log \rho(t)}{t} = -\lim_{t\to\infty} \frac{\log \rho(t)}{t} \\ \rho^* &= -\sup_{t<0} \frac{\log \rho(t)}{t} = -\lim_{t\to-\infty} \frac{\log \rho(t)}{t} \end{aligned} \tag{2.2}$$

satisfy $-\infty < \rho_* \leq \rho^* < \infty$.

Let $L^1(\mathbb{R};\rho)$ be the set of locally integrable functions a on $\mathbb{R}$ such that

$$||a|| = \int_{\mathbb{R}} \rho(t)|a(t)|dt < \infty. \tag{2.3}$$

Let $M(\mathbb{R};\rho)$ be the set of locally finite measures μ on $\mathbb{R}$ such that

$$||\mu|| = \int_{\mathbb{R}} \rho(t)d|\mu|(t) < \infty. \tag{2.4}$$

Finally, let $V(\mathbb{R};\rho) = L^1(\mathbb{R};\rho) \oplus \delta$, where δ is the unit point mass at zero, i.e., $V(\mathbb{R};\rho)$ consists of measures which are sums of a point mass at the origin and an $L^1(\mathbb{R};\rho)$-function.

The three spaces $M(\mathbb{R};\rho)$, $V(\mathbb{R};\rho)$ and $L^1(\mathbb{R};\rho)$ are Banach algebras with convolution as the multiplication. The convolution of two functions $a, b \in L^1(\mathbb{R};\rho)$ is defined a.e. by

$$a * b(t) = \int_{\mathbb{R}} a(t-s)b(s)ds, \tag{2.5}$$

and the convolution of a measure $\mu \in M(\mathbb{R};\rho)$ and a function $a \in L^1(\mathbb{R};\rho)$ is defined a.e. by

$$\mu * a(t) = \int_{\mathbb{R}} a(t-s)d\mu(s). \tag{2.6}$$

One can define the convolution of two measures $\mu, \nu \in M(\mathbb{R};\rho)$ in different ways, e.g., one can use the Riesz representation theorem to define $\mu * \nu$ by

$$(\mu * \nu) * \phi(0) = \mu * (\nu * \phi)(0) \tag{2.7}$$

for every ϕ belonging to the predual of $M(\mathbb{R};\rho)$. For more details, see [10].

Every Banach algebra has a maximal ideal space [2]. The maximal ideal space of $V(\mathbb{R};\rho)$ can be identified with $\overline{\Pi} = \Pi \cup \{\infty\}$, where

$$\Pi = \{z \in \mathbb{C} \mid \rho_* \le \operatorname{Re} z \le \rho^*\}, \tag{2.8}$$

with the topology of the compactified complex plane. The Gelfand transform on Π can be identified with the bilateral Laplace transform

$$\hat{a}(z) = \int_{\mathbb{R}} e^{-zt}a(t)dt \quad (a \in L^1(\mathbb{R};\rho);\ z \in \Pi). \tag{2.9}$$

The maximal ideal space of $M(\mathbb{R};\rho)$ is larger than Π, but we can still define the bilateral Laplace transform $\hat{\mu}$ of a measure $\mu \in M(\mathbb{R};\rho)$ by

$$\hat{\mu}(z) = \int_{\mathbb{R}} e^{-zt} d\mu(t) \quad (z \in \Pi). \tag{2.10}$$

The integrals in (2.9) and (2.10) converge absolutely because of (2.2).

One defines the concept of a weight function on $\mathbb{R}^+$ in a similar way. A continuous, strictly positive function ρ on $\mathbb{R}^+$ is called a *weight function on* $\mathbb{R}^+$, if

$$\rho(s + t) \le \rho(s)\rho(t) \quad (s, t \in \mathbb{R}^+); \quad \rho(0) = 1. \tag{2.11}$$

Define

$$\rho_* = -\inf_{t>0} \frac{\log \rho(t)}{t} = -\lim_{t\to\infty} \frac{\log \rho(t)}{t}. \tag{2.12}$$

Then $-\infty < \rho_* \le \infty$. To avoid trivialities we assume throughout that $\rho_* < \infty$. Define $L^1(\mathbb{R}^+;\rho)$, $V(\mathbb{R}^+;\rho)$ and $M(\mathbb{R}^+;\rho)$ as above. They are still Banach algebras, and the only major change in the theory presented above is that one has to redefine Π by

$$\Pi = \{z \in \mathbb{C} \mid \text{Re } z \ge \rho_*\}, \tag{2.13}$$

and integrate over the interval where the functions and measures are defined (i.e., either over $\mathbb{R}^+$ or over $[0,t]$).

In the sequel, if we do not specify that we work on either $\mathbb{R}$ or $\mathbb{R}^+$, then both cases are included. In our notations we simply omit the domain of definition of our functions and measures, and write $L(\rho)$, $V(\rho)$ and $M(\rho)$.

Gelfand, Raikov and Shilov [2] call a continuous function ϕ locally analytic with respect to $L^1(\rho)$, if for every $z_0 \in \overline{\Pi}$, there exist a neighborhood U of z_0, functions $a_1, \ldots, a_k \in L^1(\rho)$, and a function ψ, analytic at $(\hat{a}_1(z_0),\ldots,\hat{a}_k(z_0))$, such that

$$\phi(z) = \psi(\hat{a}_1(z),\ldots,\hat{a}_k(z)) \quad (z \in U). \tag{2.14}$$

We shall refer to the preceding concept as local analyticity [2], to distinguish it from a related concept defined below, and referred to as local analyticity.

Applying Theorem 1 in [2,p. 82] we get the following fundamental theorem.

THEOREM A. Let ϕ be locally analytic [2]. Then $\phi(z) - \phi(\infty) = \hat{a}(z)$ for some $a \in L^1(\rho)$.

Theorem A has certain shortcomings. We would like to apply it to the following situation. The equation for the differential resolvent r of a measure μ is given by

$$\begin{aligned} r'(t) + \mu * r(t) &= 0, \quad t \in \mathbb{R}^+ \\ r(0) &= 1. \end{aligned} \tag{2.15}$$

Transforming (2.15) one gets

$$\hat{r}(z) = (z + \hat{\mu}(z))^{-1}, \tag{2.16}$$

and even if $z + \hat{\mu}(z) \neq 0$ $(z \in \Pi)$, Theorem A does not apply directly, because we have an explicit z and the transform of a measure on the right hand side of (2.16) rather than the transform of $L^1(\rho)$-functions.

Motivated by example (2.16), and other similar examples, we make the following definiton.

DEFINITION 2.1. We call a function ϕ locally analytic at a point $z_0 \in \Pi$, if there exists a neighborhood U of z_0, measures $\mu_1, \ldots, \mu_k \in M(\rho)$, and a function ψ, analytic at $(z_0, \hat{\mu}_1(z_0), \ldots, \hat{\mu}_k(z_0))$, such that

$$\phi(z) = \psi(z, \hat{\mu}_1(z), \ldots, \hat{\mu}_k(z)) \quad (z \in U). \tag{2.17}$$

We say that ϕ is locally analytic at infinity, if there exist a neighborhood U of infinity, functions $a_1, \ldots a_n \in L^1(\rho)$, measures $\mu_1, \ldots, \mu_k \in M(\rho)$, and a function ψ, analytic at $(0, 0, \ldots, 0)$, such that

$$\phi(z) = \psi(z^{-1}, \hat{a}_1(z), \ldots, \hat{a}_n(z), \hat{\mu}_1(z)/z, \ldots, \hat{\mu}_k(z)/z) \quad (z \in U). \tag{2.18}$$

Finally, we say that ϕ is locally analytic, if it is locally analytic everywhere.

Miller [7] has discovered a method which transforms (2.16) into a form acceptable by Theorem A, and it turns out that a similar method can be applied in the general case:

PROPOSITION 2.2. Local analyticity is equivalent to local analyticity [2].

In particular, with the new definition of local analyticity, Theorem A does apply to (2.16), and shows that $r \in L^1(\mathbb{R}^+;\rho)$, if $\mu \in M(\mathbb{R}^+;\rho)$ and $z + \hat{\mu}(z) \neq 0$ $(z \in \Pi)$.

3. THE NEUTRAL EQUATION

We now return to the neutral equation (1.1), (1.2).

Let ρ be a weight function on $\mathbb{R}^+$, with $\rho_* < \infty$, and such that

$$\rho(t)e^{\rho_* t} \text{ is nondecreasing.} \tag{3.1}$$

Let f, g and ϕ be continuous, $\mathbb{C}^n$-valued functions satisfying

$$\begin{aligned} f(t) &= o(e^{\rho_* t}) \quad (t \to \infty), \\ g(t) &= o(e^{\rho_* t}) \quad (t \to \infty), \\ \phi(t) &= o(\rho(-t)) \quad (t \to -\infty). \end{aligned} \tag{3.2}$$

Let μ and ν be matrix-valued measures, whose components belong to $M(\mathbb{R}^+;\rho)$, and suppose that

$$\mu \text{ is atomic at zero,} \tag{3.3}$$

i.e., μ has an invertible point mass at zero. Define

$$D(z) = z\hat{\mu}(z) - \hat{\nu}(z) \quad (\text{Re } z \geq \rho_*). \tag{3.4}$$

Following Wheeler [11] and Jordan and Wheeler [6], we define the determinant measure det μ of μ by computing the formal determinant of μ, but replacing all multiplications by convolutions. Split det μ into its discrete, singular and absolutely continuous parts,

$$\det \mu = (\det \mu)_d + (\det \mu)_s + (\det \mu)_a. \tag{3.5}$$

Define

$$\Omega = \{\lambda \geq \rho_* \mid \inf_{\omega \in \mathbb{R}} | (\det \mu)_d\hat{}(\lambda + i\omega) | > \| (\det \mu)_s \|_\lambda \}, \tag{3.6}$$

where

$$\| (\det \mu)_s \|_\lambda = \int_{\mathbb{R}^+} e^{-\lambda t} | (\det \mu)_s | (t). \tag{3.7}$$

We are now ready to state our decomposition theorem.

THEOREM 3.1. Let $[\alpha, \beta] \in \Omega$, and assume that $\det D(z) \neq 0$ on the lines $\mathrm{Re}\, z = \alpha$ and $\mathrm{Re}\, z = \beta$. Then the solution x of (1.1), (1.2) can be written as a unique sum $x = x_S + x_C + x_U$, where the "stable" component x_S is a solution of (1.1), the "central" and "unstable" components x_C and x_U are solutions of the homogeneous equation

$$\frac{d}{dt} (\mu * x(t)) = \nu * x(t) \quad (-\infty < t < \infty), \tag{3.8}$$

x_S and x_U satisfy

$$x_S(t) = o(e^{\alpha t}) \quad (t \to \infty), \tag{3.9}$$

$$x_U(t) = o(e^{(\beta+\varepsilon)t}) \quad (t \to -\infty), \quad x_U(t) = o(e^{dt}) \quad (t \to \infty), \tag{3.10}$$

for some $d \geq \rho_*$ and some $\varepsilon > 0$, and x_C is an exponential polynomial

$$x_C(t) = \sum_{j=1}^{m} p_j(t) e^{z_j t}, \tag{3.11}$$

where z_j $(1 \leq j \leq m)$ are the zeros of $\det D(z)$ in the strip $\alpha < \mathrm{Re}\, z < \beta$, and p_j are polynomials of degree at most one less than the order of the zero z_j. In particular, if $\det D(z) \neq 0$ for $\alpha \leq \mathrm{Re}\, z \leq \beta$, then $x_C = 0$. If β is sufficiently large, then $x_U = 0$.

The proof of Theorem 3.1 is given in [10]. First one rewrites (1.1) to get an equation on all of $\mathbb{R}$ and not just on $\mathbb{R}^+$, and to make all functions vanish for $t \le -1$. In particular, one has to redefine ϕ for $t \le -1$. This can be done, if one changes the functions f and g in a suitable way. On each vertical line Re $z = \lambda$ such that $\lambda \in \Omega$ and det $D(z) \neq 0$ on Re $z = \lambda$ one can invert $D(z)$ to get a "resolvent" r_λ such that

$$\hat{r}_\lambda(z) = D^{-1}(z) \quad (\text{Re } z = \lambda).$$

In general r_λ does not vanish for $t < 0$, but it can be used to define a solution x_λ to (1.1). This solution does not, in general, satisfy the initial condition (1.2). However, if λ is sufficiently large (i.e., $\lambda \ge d$, where d is the constant in (3.10)), then the resolvent r_λ does vanish for $t < 0$, and using this resolvent one constructs a solution x to (1.1) which does satisfy (1.2). Let $[\alpha,\beta] \subset \Omega$, as in Theorem 3.1, with det $D(z) \neq 0$ on the lines Re $z = \alpha$ and Re $z = \beta$. Then we get two solutions x_α and x_β of (1.1). Define

$$x_U = x - x_\beta, \quad x_C = x_\beta - x_\alpha, \quad x_S = x_\alpha.$$

Then $x = x_U + x_C + x_S$, the different components have the right order of growth, and they satisfy the right equations. The function $D^{-1}(z)$ can have at most finitely many poles in $\alpha < \text{Re } z < \beta$, and if one subtracts off these poles and then takes the inverse Laplace transform of the remainder, one gets a residual resolvent $r_{\alpha,\beta}$. The residual resolvent $r_{\alpha,\beta}$ gives the same contribution to x_β as it does to x_α, so its influence vanishes in the difference $x_C = x_\beta - x_\alpha$. This, in turn, implies that x_C is a exponential polynomial of the form given.

One can interpret Theorem 3.1 in a semigroup setting. The predual of $M(\mathbb{R}^+;\rho)$ can be identified with the set of continuous, complex functions ϕ on $\mathbb{R}^-$ satisfying

$$\lim_{t\to-\infty} \phi(t)/\rho(-t) = 0$$

(cf. (3.2)). This space turns out to be a space of fading memory type, and (1.1), (1.2) (with $f = g = 0$) generates a strongly continuous semigroup in this space. Theorem 3.1 says that this semigroup can be split into three parts, a "stable", a "central" and an "unstable" part, which have different exponential growth rates. The central and unstable parts of this semigroup can be extended to strongly continuous groups, and the central part is finite dimensional. For more details, see [10].

REFERENCES

1. I. Gelfand, Über absolut konvergente trigonometrische Reihen und Integrale, *Math. Sb.* 9 (1941), 51-66.

2. I. Gelfand, D. A. Raikov and G. E. Shilov, *Commutative Normed Rings*, Chelsea, New York, 1964.

3. J. K. Hale, *Theory of Functional Differential Equations*, Springer, New York, 1977.

4. D. Henry, Linear autonomous neutral functional differential equations, *J. Differential Equations* 15 (1974), 106-128.

5. G. S. Jordan, O. J. Staffans and R. L. Wheeler, Local analyticity in weighted L^1-spaces, and applications to stability problems for Volterra equations, *Trans. Amer. Math. Soc.*, to appear.

6. G. S. Jordan and R. L. Wheeler, Asymptotic behavior of unbounded solutions of linear Volterra integral equations, *J. Math. Anal. Appl.* 55 (1976), 596-615.

7. R. K. Miller, Asymptotic stability and perturbations for linear Volterra integrodifferential systems, in *Delay and Functional Differential Equations and Their Applications*, Academic Press, New York, 1972.

8. T. Naito, On autonomous linear functional differential equations with infinite retardations, *J. Differential Equations* 21 (1976), 297-315.

9. T. Naito, On linear autonomous retarded equations with an abstract phase space for infinite delay, *J. Differential Equations* 33 (1979), 74-91.

10. O. J. Staffans, On a neutral functional differential equation in a fading memory space, *J. Differential Equations*, to appear.

11. R. Wheeler, A note on systems of linear integrodifferential equations, *Proc. Amer. Math. Soc.* 35 (1972), 477-482.

PART TWO

A PARAMETER DEPENDENCE PROBLEM IN FUNCTIONAL DIFFERENTIAL EQUATIONS

Dennis W. Brewer

Department of Mathematics
University of Arkansas
Fayetteville, Arkansas

1. INTRODUCTION

We consider the differentiability with respect to a parameter p of solutions of the linear functional differential equation

$$x'(t) = L(p)x_t, \quad t \geq 0$$
$$x(0) = \eta, \quad x_0 = \phi, \tag{1.1}$$

where $x_t(s) = x(t + s)$, $t \geq 0$, $s \leq 0$, $\phi \in L^1(-\infty,0)$ with values in $\mathbb{C}^n$, $\eta \in \mathbb{C}^n$, and, for each parameter p in an open subset P of a normed linear parameter space $\mathcal{P}$, L(p) is a linear functional from a subset of $L^1(-\infty,0)$ into $\mathbb{C}^n$. By a solution of (1.1) we mean a function in $L^1(-\infty,t_1)$ for each $t_1 > 0$ which is absolutely continuous (a.c.) on compact subsets of $[0,\infty)$ and satisfies (1.1) a.e. on $[0,\infty)$.

The differentiability of the solution x(t;p) with respect to p will be demonstrated by transforming (1.1) to an abstract Cauchy problem in a state space and applying a differentiability result for solutions of abstract Cauchy problems. In Section 2 this abstract theorem is presented. Section 3 considers the application of this theorem to equation (1.1). A Volterra integro-differential equation with delay is discussed in Section 4 as an important special case of equation (1.1).

2. AN ABSTRACT RESULT

Consider the linear abstract Cauchy problem

$$\begin{aligned} u'(t) &= A(p)u(t), \quad t \geq 0 \\ u(0) &= u_0, \end{aligned} \tag{2.1}$$

where for every $p \in P$, $A(p)$ is a linear operator with domain $\mathcal{D}(A(p))$ in a Banach space X with norm $\|\cdot\|$. We assume

$A(p)$ generates a strongly continuous semigroup $S(t;p)$ on X, (2.2)

$\mathcal{D}(A(p)) = \mathcal{D}$ is independent of p, (2.3)

$\|S(t;p)x\| \leq Me^{\omega t}\|x\|$, $x \in X$, $t \geq 0$, $p \in P$ for some constants M and ω independent of p, x, and t. (2.4)

If $u_0 \in \mathcal{D}$, then the solution of (2.1) is given by $u(t) = S(t;p)u_0$. We consider the somewhat more general problem of differentiating $S(t;p)u_0$ with respect to p for any fixed $u_0 \in X$. All derivatives are in the sense of Fréchet.

Suppose $A(p) = A + B(p)$ where A is independent of p. Fixing $p_0 \in P$ and $T > 0$ we assume $B(p)$ satisfies the following hypothesis:

$$\left\{ \begin{array}{l} \text{for every } p \in P, \text{ there is a constant } K > 0 \text{ such that} \\ \displaystyle\int_0^T \|B(p)S(t;p_0)x\|\,dt \leq K\|x\|, \quad x \in \mathcal{D}. \end{array} \right. \tag{2.5}$$

This hypothesis implies that the linear mapping $x \to B(p)S(\cdot;p_0)x$ from $\mathcal{D}$ into $L^1(0,T;X)$ is bounded on $\mathcal{D}$. Let $F(p)$ denote the bounded linear extension of this mapping to $\overline{\mathcal{D}} = X$. Concerning $F(p)$ we assume

$$\left\{\begin{array}{l}\text{there is a closed subspace } Y \text{ of } X \text{ such that} \\ \text{i) } F(p)u_0 \in L^1(0,T;Y) \text{ for every } p \in P, \text{ and} \\ \text{ii) for every } \varepsilon > 0 \text{ there exists } \delta > 0 \text{ such that} \\ \qquad \|F(p_0+h)y - F(p_0)y\|_{L^1} \le \varepsilon\|y\| \\ \quad \text{whenever } y \in Y \text{ and } |h| \le \delta. \end{array}\right. \tag{2.6}$$

These conditions lead to the following result for the differentiability of $S(t;p)u_0$ with respect to p. This theorem is proved by the author in [1].

THEOREM 1. Suppose (2.2) - (2.6) hold. In addition, suppose

$$F(p)u_0 \text{ is differentiable with respect to p at } p_0. \tag{2.7}$$

Then for every $t \in [0,T]$, $S(t;p)u_0$ is differentiable with respect to p at p_0 and

$$D_pS(t;p_0)u_0 = \int_0^t S(t-s;p_0)[D_pF(p_0)u_0](s)ds, \quad 0 \le t \le T.$$

REMARK. A similar result is obtained in [3] under the hypothesis that B(p) is a bounded perturbation. In Section 4 we consider an example for which B(p) is unbounded but (2.5) holds. Hypothesis (2.5) is similar to one of the conditions employed by Hille and Phillips to obtain a perturbation series, namely,

$$\left\{\begin{array}{l} B(p)S(t;p_0) \text{ is bounded on } \mathcal{D} \text{ for } t > 0 \text{ and} \\ \\ \int_0^T \|B(p)S(t;p_0)\|_{\mathcal{D}}\, dt < \infty. \end{array}\right. \tag{2.8}$$

See [4] p. 394 and following. Note that (2.8) implies (2.5). The example given in Section 4 does not satisfy (2.8) but does satisfy (2.5).

A result similar to Theorem 1 for the nonhomogeneous analogue of (2.1) may also be found in [1].

3. THE FUNCTIONAL DIFFERENTIAL EQUATION

In [2] existence and uniqueness of solutions of (1.1) is provided by the following theorem:

THEOREM 2. (J. A. Burns and T. L. Herdman) Suppose

(i) the restriction of $L(p)$ to $BC(-\infty,0]$ is a bounded linear operator, (3.1)

(ii) if $t_1 > 0$, then for each $x \in L^1(-\infty,t_1)$, $t \to L(p)x_t$ defines a function a.e. on $(0,t_1)$ which depends only on the equivalence class of x, (3.2)

(iii) there is a continuous function Γ, independent of p, such that if $x \in L^1(-\infty,t_1)$, then $L(p)x_t$ belongs to $L^1(0,t_1)$ and (3.3)

$$\int_0^t |L(p)x_s|ds \le \Gamma(t)\int_{-\infty}^t |x(s)|ds, \quad 0 \le t \le t_1,$$

(iv) if $x \in L^1(-\infty,t_1) \cap BC(-\infty,t_1]$, then $L(p)x_t$ is continuous at zero from the right. (3.4)

Then (1.1) has a unique solution for every $(\eta,\phi) \in \mathbb{C}^n \times L^1(-\infty,0)$.

The functional differential equation (1.1) may be placed in the setting of (2.1) in a standard way. Let $X = \mathbb{C}^n \times L^1(-\infty,0)$ with

$$||(\eta,\phi)|| = |\eta| + ||\phi||_{L^1}.$$

For $(\eta,\phi) \in X$ define

$$S(t;p)(\eta,\phi) = (x(t),x_t)$$

where x is the solution of (1.1) with initial data (η,ϕ). Define an operator $A(p)$ in X by

$$\mathcal{D}(A(p)) = \mathcal{D} = \{(\eta,\phi) \in X\colon \phi \text{ is a.c. on compact subsets of } (-\infty,0], \\ \phi' \in L^1(-\infty,0),\ \eta = \phi(0)\},$$

$$A(p)(\eta,\phi) = (L(p)\phi,\phi'),$$

$$B(p)(\eta,\phi) = (L(p)\phi,0) \quad \text{for} \quad (\eta,\phi) \in \mathcal{D}.$$

LEMMA 1. Suppose (3.1) - (3.4) hold. Then for every $p \in P$, $S(t;p)$ is a strongly continuous semigroup on X generated by $A(p)$ and satisfying (2.2) - (2.4).

Proof. This result is essentially proved in [2]. The parameter independence required by (2.4) follows from the assumption that the function Γ of (3.3) is independent of p.

LEMMA 2. Suppose (3.1) - (3.4) hold. Then $S(t;p)$ and $B(p)$ as defined in this section satisfy (2.5) for any fixed $T > 0$.

Proof. Let $K = \Gamma(T)(1 + MTe^{\omega T})$. Then by (3.3)

$$\begin{aligned}\|F(p)(\eta,\phi)\|_{L^1} &= \int_0^T \|(L(p)x_t,0)\|dt = \int_0^T |L(p)x_t|dt \\ &\le \Gamma(T)\int_{-\infty}^T |x(t)|dt \le \Gamma(T)\|\phi\|_{L^1} + \Gamma(T)MTe^{\omega T}\|(\eta,\phi)\| \\ &\le K\|(\eta,\phi)\|.\end{aligned}$$

It is interesting to note that in this setting hypothesis (2.5) of the abstract theorem is a direct consequence of the estimate (3.3) used in [2] to obtain existence and uniqueness of solutions of (1.1).

LEMMA 3. Suppose (3.1) - (3.4) hold and in addition assume that $L(p)$ is continuous at p_0 in the sense that

$$\left\{\begin{array}{l}\text{for every } \varepsilon > 0, \text{ there exists } \delta > 0 \text{ such that} \\ \qquad \int_0^T |L(p_0+h)y_t - L(p_0)y_t|dt \le \varepsilon|\eta| \\ \text{for every } \eta \in \mathbb{C}^n \text{ and } |h| \le \delta, \text{ where } y \text{ is the solution} \\ \text{of (1.1) with } p = p_0 \text{ and initial data } (\eta,0).\end{array}\right. \tag{3.5}$$

Then (2.6) holds.

Proof. Let $Y = \mathbb{C}^n \times \{0\}$. Then $F(p)(\eta,\phi) = (L(p)x_t, 0) \in L^1(0,T;Y)$ and (3.5) is easily seen to be equivalent to (2.6) in this setting.

Fix $u_0 = (\xi,\psi) \in X$. Let z be the solution of (1.1) with $p = p_0$ and initial data (ξ,ψ). We are now in a position to apply Theorem 1 to equation (1.1).

THEOREM 3. Suppose (3.1) - (3.5) hold and, in addition, assume that

$$\begin{cases} \text{the mapping } p \to L(p)z_t \text{ from } P \text{ into } L^1(0,T) \\ \text{is differentiable with respect to } p \text{ at } p_0. \end{cases} \tag{3.6}$$

Then $S(t;p)u_0 = (x(t),x_t)$ is differentiable with respect to p at p_0 for every $t \in [0,T]$ and

$$D_pS(t;p_0)u_0 = \int_0^t S(t-s;p_0)(D_pL(p_0)z_s,0)ds, \quad 0 \le t \le T.$$

Proof. It is not difficult to show that if L(p) satisfies (3.6), then F(p) satisfies (2.7). In view of the preceeding lemmas, the theorem follows directly from Theorem 1.

Note that for $h \in P$, $[D_pS(t;p_0)u_0]h = (w(t),w_t)$, where w is a mild solution of the nonhomogeneous functional differential equation

$$w'(t) = L(p_0)w_t + [D_pL(p_0)z_t]h, \quad t \ge 0$$

$$w(0) = 0, \quad w_0 = 0.$$

4. A VOLTERRA INTEGRO-DIFFERENTIAL EQUATION

In this section we consider a Volterra integro-differential equation with delay

$$\begin{aligned} &x'(t) = ax(t) + bx(t-r) + \int_{-\infty}^t K(t-s;q)x(s)ds, \quad t \ge 0 \\ &x(0) = \eta, \quad x_0 = \phi, \end{aligned} \tag{4.1}$$

where a,b are constant $n \times n$ matrices, $r > 0$, $K(\cdot;q)$ is an $n \times n$ measurable matrix function on $(0,\infty)$ for each q in an open subset Q of $\mathbb{R}$, and $(\eta,\phi) \in \mathbb{C}^n \times L^1(-\infty,0)$. We consider the differentiability of solutions of (4.1) with respect to the parameter $p = (r,q) \in P$, where P is a bounded, open subset of $(0,\infty) \times Q$.

Note that (4.1) becomes a special case of (1.1) if one defines

$$L(p)\phi = a\phi(0) + b\phi(-r) + \int_{-\infty}^{0} K(-s;q)\phi(s)ds. \tag{4.2}$$

LEMMA 4. Suppose

$$\left\{ \begin{array}{l} \text{there is a constant C, independent of q, such that} \\ \int_0^\infty |K(s;q)|ds \le C, \quad q \in Q. \end{array} \right. \tag{4.3}$$

Then L(p) as defined by (4.2) satisfies (3.1) - (3.4) and therefore, in view of Section 3, also satisfies hypotheses (2.2) - (2.5) of Theorem 1.

Proof. This result is proved by Burns and Herdman in [2]. Hypothesis (4.2) and the boundedness of P provide the parameter independence required for the estimates.

REMARK. In this example

$$B(p)(\eta,\phi) = (a\phi(0) + b\phi(-r) + \int_{-\infty}^{0} K(-s;q)\phi(s)ds,0)$$

and

$$B(p)S(t;p_0)(\eta,\phi) = (ax(t) + bx(t-r) + \int_{-\infty}^{0} K(-s;q)x(t+s)ds,0),$$

where x satisfies (4.1) with $p = p_0 = (r_0,q_0)$. These mappings are both unbounded on $\mathcal{D}$ so that the bounded perturbation assumption of [3] does not hold for this example and the hypothesis of [4] for a perturbation series does not hold for this example. These mappings are bounded on a state space of continuous initial data with uniform norm, however in this case the definition of $\mathcal{D}(A(p))$ must include the parameter, so (2.3) is no longer true.

LEMMA 5. Fix $(r_0,q_0) \in P$ and suppose, in addition to (4.3), that

$$\left\{\begin{array}{l}\text{the mapping } q \to K(\cdot;q) \text{ from } Q \text{ into } L^1(0,\infty) \text{ is} \\ \text{differentiable with respect to } q \text{ at } q_0.\end{array}\right. \tag{4.4}$$

Then L(p) as defined by (4.2) satisfies (3.5).

Proof. In this setting (3.5) is true if for every $\varepsilon > 0$, there exists $\delta > 0$ such that

$$\int_0^T \left| by(t-r_0-h_1) - by(t-r_0) + \int_{-\infty}^0 [K(-s;q_0+h_2) - K(-s;q_0)]y(t+s)ds \right| dt \le \varepsilon|\eta|$$

for every $\eta \in \mathbb{C}^n$ and $|h_1| + |h_2| \le \delta$, where y is the solution of (4.1) with $(r,q) = (r_0,q_0)$ and initial data $(\eta,0)$. This estimate may be proved using standard arguments.

LEMMA 6. Suppose (4.3) and (4.4) hold and $u_0 = (\xi,\psi) \in \mathcal{D}$. Then L(p) as defined by (4.2) satisfies (3.6).

Proof. Hypothesis (3.6) holds in this setting if the mapping

$$G\colon (r,q) \to az(t) + bz(t-r) + \int_{-\infty}^0 K(-s;q)z(t+s)ds$$

from P into $L^1(0,T)$ is differentiable at $p_0 = (r_0,q_0)$, where z is the solution of (4.1) with $(r,q) = p_0$ and initial data (ξ,ψ). The differentiability of G with respect to r depends on the fact that $u_0 \in \mathcal{D}$, and the differentiability of G with respect to q may be proved using (4.4). In fact, one can show that for $(h_1,h_2) \in P$,

$$[D_pG(p_0)](h_1,h_2) = -bz'(t-r_0)h_1 + \int_{-\infty}^0 [D_qK(-s;q_0)]h_2z(t+s)ds.$$

Combining the preceding lemmas with Theorems 1 and 3 yields the following theorem concerning the differentiability of solutions of (4.1) with respect to (r,q).

THEOREM 4. Suppose (4.3) and (4.4) hold and $u_0 \in \mathcal{D}$. Let x be the solution of (4.1) with initial data u_0. Then $S(t;p)u_0 = (x(t),x_t)$ is differentiable with respect to p at p_0 for every $t \in [0,T]$ and

$$D_pS(t;p_0)u_0 = \int_0^t S(t-s;p_0)(-bz'(s-r_0) + \int_{-\infty}^0 D_qK(-\tau;q_0)z(s+\tau)d\tau,0)ds.$$

Furthermore, if $(h_1,h_2) \in P$, then $[D_pS(t;p_0)u_0](h_1,h_2) = (w(t),w_t)$, where w is a mild solution of the nonhomogeneous equation

$$w'(t) = aw(t) + bw(t-r_0) + \int_{-\infty}^t K(t-s;q_0)w(s)ds$$
$$- bz'(t-r_0)h_1 + \int_{-\infty}^t [D_qK(-s;q_0)]h_2z(t+s)ds$$
$$w(0) = 0, \quad w_0 = 0.$$

ACKNOWLEDGEMENT

This research was partially supported by the National Science Foundation under the Arkansas EPSCOR Grant.

REFERENCES

1. D. W. Brewer, The differentiability with respect to a parameter of the solution of a linear abstract Cauchy problem, *SIAM J. Math. Anal.*, to appear.

2. J. A. Burns and T. L. Herdman, Adjoint semigroup theory for a class of functional differential equations, *SIAM. J. Math. Anal.* 7 (1976), 729-745.

3. J. S. Gibson and L. G. Clark, Sensitivity analysis for a class of evolution equations, *J. Math. Anal. Appl.* 58 (1977), 22-31.

4. E. Hille and R. S. Phillips, *Functional Analysis and Semigroups*, Amer. Math. Soc. Colloq. Publ., Vol. 31, Providence, R. I., 1957.

APPROXIMATION OF FUNCTIONAL DIFFERENTIAL EQUATIONS BY DIFFERENTIAL SYSTEMS

Stavros N. Busenberg

Department of Mathematics
Harvey Mudd College
Claremont, California

Curtis C. Travis

Health and Safety Research Division
Oak Ridge National Laboratory
Oak Ridge, Tennessee

Functional differential equations were introduced by Volterra near the beginning of the present century to analyze systems whose future behavior was dependent on past states. We quote from Volterra [6]: "However, when heredity plays a role in physics, differential equations and partial differential equations are no longer sufficient, for otherwise, the initial conditions would determine the future. In order to allow the continuum of past states to play a role, ...one had to resort to integral and integrodifferential equations." Since the introduction of such equations, a dispute has periodically erupted concerning the propriety of explicitly including hereditary influences in models of physical systems. It has been argued that explicit expression of delays in such models is only necessary because the complete set of variables controlling the dynamics of the physical system is unknown. If one had access to these variables, the functional differential equations could be replaced by an extended set of differential equations having exactly the same dynamics. In 1907, E. Picard [5] wrote, "Examples are numerous where the future of a system seems to depend upon former states. Here we have heredity. In some complex cases, one sees that it is necessary, perhaps, to abandon differential equations and consider functional equations in which there appear integrals taken from a distant time to the present, integrals which will be, in fact, this hereditary part. The proponents of classical mechanics, however, are able to pretend that heredity is only apparent and that it amounts merely to this: that we have fixed our attention upon too small a number of variables."

This divergence of viewpoints has resurfaced in modern times in a different context. A recent book of MacDonald [4] on time lags in biological models is devoted almost entirely to the analysis of models leading to functional differential equations which, because of their special form, are reducible to systems of ordinary differential equations. For example, consider the system,

$$\frac{dx^1(t)}{dt} = \int_0^\infty e^{-\lambda s} x^2(t-s)\,ds,$$

$$\frac{dx^2(t)}{dt} = \int_0^\infty e^{-\lambda s} x^1(t-s)\,ds.$$

If we define

$$x^3(t) = \int_0^\infty e^{-\lambda s} x^2(t-s)\,ds,$$

$$x^4(t) = \int_0^\infty e^{-\lambda s} x^1(t-s)\,ds,$$

then the above hereditary system reduces to the system of ordinary differential equations,

$$\frac{dx^1(t)}{dt} = x^3(t), \quad \frac{dx^2(t)}{dt} = x^4(t),$$

$$\frac{dx^3(t)}{dt} = x^2(t) - \lambda x^3(t), \quad \frac{dx^4(t)}{dt} = x^1(t) - \lambda x^4(t).$$

A question arises as one reads MacDonald's book, as well as other biomathematical papers that employ models with reducible equations, as to the generality of the class of reducible functional differential equations. One would like to know if an arbitrary functional differential equation can be replaced by an equivalent system of ordinary differential equations, and, if this is possible, how does one effectively construct the equivalent system? Our first result establishes that given any retarded functional differential equation, one can construct a reducible system or, equivalently, a system of ordinary differential equations, whose solutions approximate those of the delay equation to any prescribed degree of accuracy.

Let $C = C([-T,0],\mathbb{R}^n)$ denote the space of continuous functions on $[-T,0]$ to $\mathbb{R}^n$ with the uniform norm topology, and let L be a continuous linear function mapping C into $\mathbb{R}^n$. For $\phi: \mathbb{R} \to \mathbb{R}^n$, define $\phi_t: [-T,0] \to \mathbb{R}^n$ by $\phi_t(s) = \phi(t-s)$, $s \in [0,T]$, and consider the linear retarded functional differential equation,

$$\frac{dX(t)}{dt} = L(X_t) = \int_0^\infty [d\eta(s)]X_t(s), \tag{1}$$

where $\eta(s)$, $0 \le s \le \infty$, is an $n \times n$ matrix whose elements are of bounded variation.

We formally define a reducible system to be a retarded functional differential equation of the form

$$\frac{dX(t)}{dt} = \int_0^\infty [d\eta(s)]X_t(s), \tag{2}$$

whose kernel $\eta(s)$ satisfies

$$d\eta_{ij}(s) = \sum_{k,\ell=1}^{M,N} \gamma_{ij}^{k\ell} \frac{s^{k-1}}{(k-1)!} e^{-\lambda_\ell s} ds, \qquad i,j=1,\dots,n,$$

where η_{ij} is the i,j-th element of η. By introducing new variables in a manner similar to that of the example treated above, it is possible to obtain an ordinary differential system which is equivalent to the reducible functional differential equation.

THEOREM 1. Given any linear retarded functional differential equation of the form (1) and any positive constant α, there exists a reducible functional differential equation of the form (2) whose equivalent system of ordinary differential equations has dimension m, a positive constant K, and continuous linear maps V and S, $V: \mathbb{R}^m \to \mathbb{R}^n$, $S: C \to \mathbb{R}^m$, such that for any $\phi \in C$,

$$|X_t(\phi) - VY_t(S\phi)| < K|\phi|e^{-\alpha t}, \quad t \geq 0,$$

where $X_t(\phi)$ solves (1) with $X_0(\phi) = \phi$ and $Y_t(S\phi)(s) = Y(t+s,S\phi)$, where $Y(t,S\phi)$ is the solution of the differential system with $Y(0,S\phi) = S\phi$.

The proof of Theorem 1 relies on the fact that given any linear retarded functional differential equation and a finite subset of its spectrum, it is possible to find a reducible functional differential equation whose spectrum exactly coincides with this set. Specifically, given $\alpha > 0$, let $\Lambda = (\lambda_1, \ldots, \lambda_p)$, where λ_i are roots to the characteristic equation of (1) obeying $Re\lambda_i > -\alpha$, and let $m = m_1 + m_2 + \ldots + m_p$, where m_i is the dimension of the generalized eigenspace of the root λ_i. Then for $i = 1,2,\ldots,p$, the system

$$\frac{dy^1(t)}{dt} = \lambda_i y^1(t),$$

$$\frac{dy^\ell(t)}{dt} = \lambda_i y^\ell(t) + y^{\ell-1}(t), \quad \ell = 2,3,\ldots,m_i,$$

is the differential equation form of the reducible system

$$\frac{du(t)}{dt} = \lambda_i u(t),$$

$$\frac{dv(t)}{dt} = \begin{cases} \lambda_i v(t) + \int_{-\infty}^{t} g_{m_i-2}(t-s)u(s)ds, & m_i > 2, \\ \lambda_i v(t) + u(t), & m_i = 2, \end{cases}$$

obtained by defining $y^1 = u$, $y^{m_i} = v$, and $y^{\ell+1}(t) = \int_{-\infty}^{t} g_\ell(t-s)y^1(s)ds$ for $\ell = 1,2,\ldots,m_i-2$, and $i = 1,2,\ldots,p$, where $g_\ell(t) = t^{\ell-1}e^{\lambda_\ell t}/(\ell-1)!$. The details of the proof are given in [1].

The method used in this proof for constructing the reducible functional differential equation requires accurate computation of part of the spectrum of the original functional differential equation as well as knowledge of the generalized eigenspaces which correspond to this part of the spectrum. This is usually a very difficult task. It is, hence, useful to have easily computable criteria which assure that a differential equation system will approximate the delay equation to a desired accuracy. The next two results address part of this issue by providing conditions which can be used to determine when the degree of approximation is sufficient for the stability properties of the original functional differential equation to be reflected in the approximating ordinary differential system. We will also give sufficient conditions for stability of certain families of delay differential equations which are obtained by varying the delays that enter in these equations.

We rewrite the system of delay functional differential equations (1) in component form

$$\frac{dx^i(t)}{dt} = \sum_{j=1}^{n} a_{ij}\int_0^\infty x^j(t-s)d\eta_{ij}(s), \tag{3}$$

where η_{ij}: $[0,\infty) \to \mathbb{R}$ are functions of bounded variation and the a_{ij} are real numbers. We assume that the η_{ij} are normalized by $\int_0^\infty |d\eta_{ij}(s)| = 1$ and that $\int_0^\infty s|d\eta_{ii}(s)| < \infty$. The characteristic equation of (3) is

$$F_\eta(\lambda) = \det\left\{\lambda I - [a_{ij}\int_0^\infty e^{-\lambda s}d\eta_{ij}(s)]\right\} = 0, \tag{4}$$

and we say that $F_\eta(\lambda)$ is stable if all the roots of (4) have real parts less than zero. We introduce the approximating reducible systems by setting for $i,j = 1,2,\dots,n$,

$$b_{ij}r_{ij}(s) + C_{ij}\frac{d\mu_{ij}(s)}{ds} = a_{ij}\frac{d\eta_{ij}(s)}{ds},$$

where the $r_{ij}(s)$ are reducible kernels of the form,

$$r_{ij}(s) = \sum_{k,\ell}\gamma_{ij}^{k\ell}\frac{s^{k-1}}{(k-1)!}e^{-\lambda_{ij}^{k\ell}},$$

where the b_{ij}, and C_{ij} and $\gamma_{ij}^{k\ell}$ are real numbers and the $\lambda_{ij}^{k\ell}$ are complex numbers. Also, it is assumed that μ_{ij} and r_{ij} are normalized by $\int_0^\infty |d\mu_{ij}(s)| = \int_0^\infty |r_{ij}(s)|ds = 1$ and that $\int_0^\infty s|d\mu_{ij}(s)| < \infty$. The characteristic equation (4) then takes the form,

$$F_{r+\mu}(\lambda) \equiv \det\left\{\lambda I - \left[b_{ij}\int_0^\infty e^{-\lambda s} r_{ij}(s)ds + C_{ij}\int_0^\infty e^{-\lambda s} d\mu_{ij}(s)\right]\right\} \tag{5}$$
$$= 0.$$

We can now give a sufficient condition for stability of (4).

THEOREM 2. The characteristic equation (4) is stable if, for each $i = 1,2,\ldots,n$, either one of the following two conditions holds.

(A). $a_{ii}\int_0^\infty d\eta_{ii}(s) < 0, \quad -|a_{ii}|\int_0^\infty s|d\eta_{ii}(s)| + 1 > 0,$

and there exist $d_i > 0$ such that

$$|a_{ii}\int_0^\infty \cos(sv)d\eta_{ii}(s)| - d_i^{-1}\sum_{j\neq i} d_j|a_{ij}| > 0,$$

for all real v satisfying

$$|v| \leq [d_i^{-1}\sum_{j\neq i} d_j|a_{ij}|]/[1 - |a_{ii}|\int_0^\infty s|d\eta_{ii}(s)|];$$

(B). η_{ii} is atomic at zero with $a_{ii}\int_0^{0^+} d\eta_{ii}(s) < 0$, and there exist $d_i > 0$, such that

$$|a_{ii}\int_0^{0+} d\eta_{ii}(s)| - d_i^{-1}\left\{\sum_{j\neq i} d_j|a_{ij}\int_0^\infty d\eta_{ij}(s)| + d_i|a_{ii}\int_{0+}^\infty d\eta_{ii}(s)|\right\} > 0.$$

An outline of the proof of this result will be given below. In stating the result, our aim was to provide a computable criterion for stability rather than one of the greatest possible generality. In the next result, we rephrase Theorem 2 in order to apply it to the form (5) of the characteristic equation where the reducible approximation is explicitly shown.

THEOREM 3. The characteristic equation (5) is stable if

$$b_{ii}\int_0^\infty r_{ii}(s)ds < -C_{ii}\int_0^\infty d\mu_{ii}(s),$$

$$-|b_{ii}|\int_0^\infty s|r_{ii}(s)|ds + 1 > |C_{ii}|\int_0^\infty s|d\mu_{ii}(s)|,$$

and there exist $d_i > 0$ such that

$$|b_{ii}\int_0^\infty \cos(sv)r_{ii}(s)ds| - d_i^{-1}\sum_{j\neq i} d_j|a_{ij}|$$

$$> |C_{ii}\int_0^\infty \cos(sv)d\mu_{ii}(s)|,$$

for all real v satisfying

$$|v|\ [1 - |b_{ii}|\int_0^\infty s|r_{ii}(s)|ds - |C_{ii}|\int_0^\infty s|d\mu_{ii}(s)|]$$

$$\le [d_i^{-1}\sum_{j\neq i} d_j|b_{ij}|].$$

This result follows from the triangle inequality applied to Condition (*A*) of Theorem 2.

The method of proof of Theorem 2 can be used to obtain computable stability criteria for certain classes of delay equations that apply for all values of the delays. We give below such a condition applying to certain classes of differential-difference equations and based on some results of Datko [3]. To this end, consider the following equation:

$$\frac{d}{dt}x^i(t) = \sum_{\ell,j} a_{ij}^{\ell} x^j(t - \alpha h_\ell), \quad \alpha > 0,$$

$\ell = 0,1,\ldots,m$, $0 = h_0 < h_1 < \ldots < h_m$ fixed constants. The characteristic equation of this differential difference equation is

$$G_\alpha(\lambda) = \det[\lambda I - \sum_\ell A_\ell e^{-\alpha\lambda h_\ell}] = 0 \tag{6}$$

where $A_\ell = (a_{ij}^\ell)$ are the matrices of coefficients in the differential equation. We can now give the following sufficient condition for stability of (6).

THEOREM 4. Suppose that there exist positive constants $d_j > 0$ with $a_{ii}^0 < 0$ and

$$|a_{ii}^0| > \sum_{\ell\neq 0} |a_{ii}^\ell| + d_i^{-1} \sum_\ell \sum_{j\neq i} d_j |a_{ij}^\ell|.$$

Then the characteristic equation (6) is stable for all $\alpha \in [0,\infty)$.

The proof of Theorem 2 is based on the following reasoning. If λ is a root of the characteristic equation (4), then, by the arguments leading to the Gerschgorin Theorem, we have for at least one value of $i = 1,2,\ldots,n$, the following relation

$$\lambda - a_{ii}\int_0^\infty e^{-\lambda s} d\eta_{ii}(s) = d_i^{-1} \sum_{j\neq i} d_j K_j a_{ij} \int_0^\infty e^{-\lambda s} d\eta_{ij}(s),$$

where $|K_j| = 1$, and $d_j > 0$. Now, conditions (*A*) or (*B*) exclude the possibility of a purely imaginary root, $\lambda = i\nu$. Also, if $a_{ij} = 0$, when $i \neq j$, either of these conditions implies that the real part of λ is negative. If we construct a homotopy joining the system which has $a_{ij} = 0$ for $i \neq j$ to the one in (4), we note that none of the roots of these homotopic equations can be pure imaginary, while they all have negative real parts when $a_{ij} = 0$ for $i \neq j$. This fact, and continuity of the roots of the homotopic equations with respect to the parameters a_{ij}, establishes the stability of (4). The detailed argument of this proof is given in [1].

As we stated before, Theorem 3 is a rephrasing of Theorem 2. The proof of Theorem 4 is based on a result of Datko [3] which establishes continuity of the roots of (6) as functions of the parameter $\alpha \in [0,\infty)$. The condition in Theorem 4 implies that $G_0(\lambda)$ is stable and that $G_\alpha(\lambda)$ has no pure imaginary roots. Using the homotopy argument employed in the proof of Theorem 2, we conclude the $G_\alpha(\lambda)$ is stable for all $\alpha \in [0,\infty)$.

ACKNOWLEDGEMENTS

We wish to thank Theodore A. Burton for bringing the quotation of E. Picard [5] to our attention.

The work of Professor Busenberg was partially supported by the National Science Foundation under Grant MCS7903497.

The work of Dr. Travis was partially supported by the National Science Foundation's Ecosystem Studies Program under Interagency Agreement No. DEB 80-21024 with the U. S. Department of Energy under contract W-7405-eng-26 with the Union Carbide Corporation.

REFERENCES

1. S. N. Busenberg and C. C. Travis, On the use of reducible functional differential equations in biological models, *J. Math. Anal. Appl.*, to appear.

2. D. S. Cohen, E. Coutsias, and H. Neu, Stable oscillations in single species growth models with hereditary effects, *Math. Biosciences* 44 (1979), 255-267.

3. R. Datko, A procedure for determination of the exponential stability of certain differential-difference equations, *Quarterly of Appl. Math.* 36 (1978), 279-292.

4. N. MacDonald, *Time Lags in Biological Models*, Lecture Notes in Biomathematics 27 (1978), Springer Verlag, Berlin.

5. E. Picard, La mécanique classique et ses approximations successives, *Revista di Scienza* 1 (1907), p. 15.

6. V. Volterra, *Leçons sur la Théorie Mathématique de la Lutte Pour la Vie*, Cahiers Scientifiques 7, Gauthier-Villars, Paris (1931), p. 142.

INTEGRABLE RESOLVENT OPERATORS FOR AN ABSTRACT VOLTERRA EQUATION IN HILBERT SPACE

R. W. Carr

Department of Mathematics
St. Cloud State University
St. Cloud, Minnesota

We study the asymptotic behavior of the solution to the initial value problem

$$\underset{\sim}{y}'(t) + \int_0^t [d + a(t - s)]\underset{\sim}{L}\underset{\sim}{y}(s)ds = \underset{\sim}{f}(t), \quad t > 0 \tag{1}$$

$$\underset{\sim}{y}(0) = \underset{\sim}{y}_0 \in H,$$

($' = d/dt$) where $\underset{\sim}{L}$ is a positive self-adjoint linear operator defined on a dense subspace $\mathcal{D}$ of the Hilbert space H. The kernel $d + a(t)$ satisfies

$$\begin{cases} a \in L^1_{loc}(\mathbb{R}^+, \overline{\mathbb{R}}^+), \quad (\mathbb{R}^+ = (0,\infty), \ \overline{\mathbb{R}}^+ = [0,\infty)); \\ a \text{ is nonincreasing and convex with } a(\infty) = 0 < a(0^+) \le \infty, \\ \text{and } d \ge 0, \end{cases} \tag{2}$$

and $\underset{\sim}{f}$ belongs to $\mathcal{B}^1_{loc}(\overline{\mathbb{R}}^+, H)$, the class of locally Bochner integrable functions from $\overline{\mathbb{R}}^+$ to H. The problem

$$y_t(x,t) = \int_0^t [d + a(t - s)]y_{xx}(x,s)ds + f(x,t), \quad t \geq 0, \quad 0 < x < \pi$$

$$y(x,0) = y_0(x), \quad 0 \leq x \leq \pi \tag{3}$$

$$y(0,t) = y(\pi,t) = 0, \quad t \geq 0,$$

which arises in one-dimensional viscoelasticity, provides an example for the abstract equation (1). Here $\underset{\sim}{L} = -(\partial/\partial x)^2$, $H = L^2(0,\pi)$, and $\mathcal{D} = H^2 \cap H_0^1(0,\pi)$.

Let $u(t,\lambda)$ denote the solution of the scalar equation

$$u'(t) + \lambda\int_0^t [d + a(t - s)]u(s)ds = 0, \quad t \geq 0, \ (\lambda > 0),$$

$$u(0) = 1.$$

Define $v(t,\lambda) = \frac{\partial}{\partial t} u(t,\lambda)$ and the resolvent operators

$$\underset{\sim}{U}(t) = \int_0^\infty u(t,\lambda)d\underset{\sim}{E}_\lambda, \quad \underset{\sim}{V}(t) = \int_0^\infty v(t,\lambda)d\underset{\sim}{E}_\lambda,$$

where $\{\underset{\sim}{E}_\lambda\}_{\lambda>0}$ is the spectral family associated with $\underset{\sim}{L}$. (We assume that the spectrum of $\underset{\sim}{L}$ is contained in an interval $[\Lambda,\infty)$ with $\Lambda > 0$. Without loss of generality, we take $\Lambda = 1$.)

In [1] we studied the resolvent formula

$$\underset{\sim}{y}(t) = \underset{\sim}{U}(t)\underset{\sim}{y}_0 + \int_0^t \underset{\sim}{U}(t - s)\underset{\sim}{f}(s)ds$$

for the solution of (1) and gave sufficient conditions for

$$\text{(a)} \quad \int_0^\infty \|\underset{\sim}{U}(t)\|dt \leq \int_0^\infty \sup_{1\leq\lambda<\infty} |u(t,\lambda)|dt < \infty$$

$$\text{(b)} \quad \|\underset{\sim}{U}(t)\| \leq \sup_{1\leq\lambda<\infty} |u(t,\lambda)| = \mathcal{O}(t^{-1}) \ (t \to \infty). \tag{4}$$

Now we are principally concerned with the alternate resolvent formula

$$\underset{\sim}{y}(t) = \underset{\sim}{F}(t) + \int_0^t \underset{\sim}{V}(t - s)\underset{\sim}{F}(s)ds,$$

where

$$\underset{\sim}{F}(t) = \int_0^t \underset{\sim}{f}(s)ds + \underset{\sim}{y}_0, \quad A(t) = \int_0^t a(s)ds,$$

for the integrated version of (1), i.e.,

$$\underset{\sim}{y}(t) + \int_0^t [(t - s)d + A(t - s)]\underset{\sim}{L}\underset{\sim}{y}(s)ds = \underset{\sim}{F}(t), \quad t \geq 0.$$

We desire results analogous to (4) for the alternate resolvent equation. Although

$$||\underset{\sim}{V}(t)|| \leq \sup_{1\leq\lambda<\infty} |v(t,\lambda)|$$

is valid, an estimate like

$$\int_0^\infty \sup_{1\leq\lambda<\infty} |v(t,\lambda)|dt < \infty$$

is always false. Instead, we use the fact that

$$||\underset{\sim}{V}(t)\underset{\sim}{L}^{-\gamma}|| \leq \sup_{1\leq\lambda<\infty} |v(t,\lambda)|\lambda^{-\gamma} \equiv v_\gamma(t)$$

to establish a family of inequalities

$$\text{(a)} \quad \int_0^\infty ||\underset{\sim}{V}(t)\underset{\sim}{L}^{-\gamma}||dt \leq \int_0^\infty v_\gamma(t)dt < \infty,$$

$$\text{(b)} \quad ||\underset{\sim}{V}(t)\underset{\sim}{L}^{-\gamma}|| = O(t^{-1}) \text{ as } t \to \infty, \tag{5}$$

where $\gamma > 0$ depends upon properties of the kernel $d + a(t)$. In particular, (5) holds with $\gamma = 1/2$ whenever $-a'$ is convex.

We construct a family of piecewise linear kernels a_n $(n = 1,2,3,\ldots)$ for which the second inequality in (5a) holds whenever $\gamma > n/2$ but fails when $\gamma \leq n/2$.

Writing the alternate resolvent formula in the form

$$\underset{\sim}{y}(t) = \underset{\sim}{F}(t) + \int_0^t (\underset{\sim}{V}(t - s)\underset{\sim}{L}^{-\gamma})(\underset{\sim}{L}^{\gamma}\underset{\sim}{F}(s))ds,$$

we see that stability in $\underset{\sim}{y}$ depends upon the degree of smoothness we impose on $\underset{\sim}{F}$ to compensate for the appropriate value of γ which applies to a given kernel.

In order to state our results precisely, we define the Fourier transform

$$\hat{a}(\tau) = \lim_{R\to\infty} \int_0^R e^{-i\tau t}a(t)dt = \phi(\tau) - i\tau\theta(\tau), \quad \tau > 0,$$

with both ϕ and θ real; note that $\hat{a}$ is continuous. Let $A(x) = \int_0^x a(t)dt$ and $A_1(x) = \int_0^x ta(t)dt$.

The following notation will be useful. If $f, g \in C(D,\mathbb{R}^+)$, $D \subseteq \overline{\mathbb{R}}^+$, and if there exist positive constants M_i $(i = 1,2)$ such that

$$M_1 g(t) \leq f(t) \leq M_2 g(t), \quad t \in D, \tag{6}$$

then write $f \overset{E}{=} g$. If the first inequality of (6) holds but the second does not, then write $f \overset{E}{>} g$ or $g \overset{E}{<} f$. Finally, if $f \overset{E}{=} g$ or $f \overset{E}{<} g$, then write $f \overset{E}{\leq} g$ or $g \overset{E}{\geq} f$. Clearly, $\overset{E}{=}$ is an equivalence relation and the others are transitive.

By estimates of Hannsgen [3], Shea and Wainger [4], and Hannsgen and Carr [1], we have

$$\begin{aligned} &|\hat{a}(\tau)| \overset{E}{=} A(\tau^{-1}), \quad \theta(\tau) \overset{E}{=} A_1(\tau^{-1}), \\ &|\hat{a}'(\tau)| \overset{E}{\leq} A_1(\tau^{-1}), \quad -\theta'(\tau) \overset{E}{\geq} \tau \int_0^{1/\tau} t^3 a(t)dt, \quad \tau > 0. \end{aligned} \tag{7}$$

Formally taking the Fourier transform of the equation defining u and using the definition of v leads to

$$\hat{v}(\tau,\lambda) = \frac{-D(\tau)}{D(\tau,\lambda)}, \tag{8}$$

where $D(\tau) = \hat{a}(\tau) - id\tau^{-1}$, $D(\tau,\lambda) = D(\tau) + i\tau\lambda^{-1}$, and thus Re $D(\tau,\lambda) = \phi(\tau)$, Im $D(\tau,\lambda) = \tau(\lambda^{-1} - \theta(\tau) - d\tau^{-2})$.

Equation (2) implies $\phi(\tau) \geq 0$. Furthermore, Hannsgen has shown that $\phi \in C(\mathbb{R}^+,\mathbb{R}^+)$ unless a(t) is piecewise linear with changes of slope only at integral multiples of a fixed positive number t_0.

Throughout this paper we restrict ourselves to the case $\phi(\tau) > 0$, thus ensuring that $D(\tau,\lambda)$ never vanishes. In particular, the hypotheses of [4, Theorem 2] hold and we have

$$\int_0^\infty |v(t,\lambda)|dt < \infty, \quad (\lambda > 0),$$

and (8) is valid.

Our estimates on the size of $v_\gamma(t)$ will depend crucially on the behavior of two implicitly defined functions of λ. Let $\omega = \omega(\lambda)$ be the solution of the equation

$$\theta(\omega) + \frac{d}{\omega^2} = \frac{1}{\lambda},$$

i.e., Im $D(\omega(\lambda),\lambda) = 0$. Conditions (2) and (7), along with the implicit function theorem ensure that ω is well-defined, monotonically increasing and continuous on an interval $[\lambda_0,\infty)$ for some $\lambda_0 > 0$. If necessary, extend ω to the interval $[1,\lambda_0)$ by defining $\omega(\lambda) = \omega(\lambda_0)$ for $1 \leq \lambda < \lambda_0$.

Let $\sigma = \sigma(\lambda)$ be the solution to the equation

$$\frac{1}{\sigma} A\left(\frac{1}{\sigma}\right) = \frac{1}{\lambda}.$$

Thus σ is also a continuous increasing function on $(0,\infty)$. Furthermore (7) implies

$$\theta(\omega) \overset{E}{=} A_1(\omega^{-1}) \overset{E}{=} \lambda^{-1} \overset{E}{=} \sigma^{-1}A(\sigma^{-1}), \tag{9}$$

from which one easily obtains

$$\sqrt{\lambda} \overset{E}{\leq} \omega \overset{E}{\leq} \sigma \overset{E}{<} \lambda, \quad (\lambda \geq 1).$$

If $a(0^+) < \infty$, then (9) implies

$$\sqrt{\lambda} \overset{E}{=} \omega \overset{E}{=} \sigma,$$

and if $a(0^+) = \infty$, then (9) gives

$$\sqrt{\lambda} \overset{E}{<} \omega \overset{E}{\leq} \sigma.$$

As examples, consider the following three kernels all satisfying (2).

(i) $d + a(t) = e^{-t}$. Then $\omega(\lambda) \overset{E}{=} \sigma(\lambda) \overset{E}{=} \sqrt{\lambda}$, $(\lambda \geq 1)$.
(ii) $d + a(t) = t^{-1/2}$. Then $\omega(\lambda) \overset{E}{=} \sigma(\lambda) \overset{E}{=} \lambda^{2/3}$, $(\lambda \geq 1)$.
(iii) $d + a(t) = 1/t((\log^+(1/t))^{3/2} + 10t)$. Then $\omega(\lambda) \overset{E}{=} \lambda(\log(\lambda + 1))^{-1}$ and $\sigma(\lambda) \overset{E}{=} \lambda(\log(\lambda + 1))^{-1/2}$, $(\lambda \geq 1)$.

The third example shows that σ and ω need not grow at the same rate.

The importance of a good lower bound on $\phi(\tau)$ can be seen in the following estimate, which uses (8) and (9).

$$\lambda^{-\gamma} \int_0^\infty |v(t,\lambda)|dt \overset{E}{\geq} \lambda^{-\gamma}|\hat{v}(\omega(\lambda),\lambda)|$$

$$\overset{E}{=} [\theta(\omega)]^\gamma \frac{|D(\omega)|}{\phi(\omega)} \overset{E}{\geq} \frac{\omega(\theta(\omega))^{1+\gamma}}{\phi(\omega)}.$$

Thus, we have the following necessary condition:

THEOREM 1. If $\int_0^\infty v_\gamma(t)dt < \infty$, then

$$\sup_{\tau>1} \frac{\tau\theta(\tau)^{1+\gamma}}{\phi(\tau)} < \infty. \tag{F1}$$

We would like the converse to be true as well. However, we were unable to prove this, and for $\gamma = 0$ we have a counterexample.

THEOREM 2. If (2) holds, then

$$|v(t,\lambda)| \overset{E}{\leq} \sigma(\lambda), \quad t \geq 0.$$

Furthermore,

$$\sup_{t>0} |v(t,\lambda)| \overset{E}{=} \sigma(\lambda),$$

and this occurs when $t = t(\lambda) = 1/c\sigma(\lambda)$ for a suitably chosen constant $c > 0$.

Thus, on some interval $0 < t < \varepsilon$,

$$|v(t,\lambda)| \overset{E}{=} \sigma(\lambda) \overset{E}{=} \frac{1}{t},$$

so that $v_0 \notin L^1(0,\varepsilon)$. Choosing $d + a(t) = t^{-1/2}$ gives $\sup_{\tau>1} \tau\theta(\tau)/\phi(\tau) < \infty$, which is (F1) with $\gamma = 0$, hence providing this weak counterexample to the converse of Theorem 1.

In order to state further positive results, we need additional hypotheses:

$$\left\{\begin{array}{l} a(t) = b(t) + c(t), \text{ where } b \text{ and } c \text{ both satisfy (2) except} \\ \text{that } b \equiv 0 \text{ or } c \equiv 0 \text{ is permitted. Moreover, } \int_1^\infty \frac{b(t)}{t}\,dt < \infty, \\ \text{and } -c' \text{ is convex.} \end{array}\right. \tag{10}$$

THEOREM 3. (i) If (2) holds and $\phi \in C(\mathbb{R}^+,\mathbb{R}^+)$, and if (F1) holds, then $v_\gamma(t) \overset{E}{\leq} t^{-1}$ for $t > 0$. (ii) If (10) holds and $\phi \in C(\mathbb{R}^+,\mathbb{R}^+)$ and if

$$\sup_{\tau>1} \frac{\tau\theta(\tau)^{1+\gamma-\varepsilon}}{\phi(\tau)} < \infty, \quad (0 < \varepsilon < \gamma), \tag{F2}$$

or if

$$\sup_{\tau>1} \frac{\tau\theta(\tau)^{2+\gamma}}{\phi(\tau)^2} < \infty, \quad (\gamma \geq 1), \tag{F3}$$

then $\int_0^\infty v_\gamma(t)dt < \infty$. (iii) If (10) holds and $\phi \in C(\mathbb{R}^+, \mathbb{R}^+)$, and if

$$\sup_{\tau>1} \frac{\tau^2\theta^2(\tau)}{\phi(\tau)} < \infty, \tag{F4}$$

then $\int_0^\infty v_{1/2}(t)dt < \infty$. Note that (F4) holds whenever $-a'$ is convex [1].

For comparison, in [1] we showed

THEOREM A. If (10) holds and if $\phi \in C(\mathbb{R}^+, \mathbb{R}^+)$ then

$$\int_0^\infty \sup_{1\le\lambda<\infty} |u(t,\lambda)|dt < \infty$$

if and only if

$$\sup_{\tau>1} \frac{\theta(\tau)}{\phi(\tau)} < \infty. \tag{F5}$$

The relations among these conditions are as follows: (F4) implies (F1) ($\gamma = 1/2$) implies (F5) implies (F3) ($\gamma \ge 1$). Furthermore if $a(0^+) < \infty$, then (F5) if and only if (F4) if and only if (F1) ($\gamma = 1/2$), and each of these conditions is equivalent to the condition $\inf_{\tau>0} (1 + \tau^2)\phi(\tau) > 0$, i.e., a is strongly positive. When $a(0^+) = \infty$, these conditions are distinct.

Finally, we note that our results apply also to the nonlinear problem

$$\underset{\sim}{y}'(t) + \int_0^t [d + a(t - s)][\underset{\sim}{L}\underset{\sim}{y}(s) + \underset{\sim}{N}\underset{\sim}{y}(s)]ds = \underset{\sim}{f}(t), \quad t \ge 0,$$

$$\underset{\sim}{y}(0) = \underset{\sim}{y}_0,$$

with the formal solution

$$\underset{\sim}{y}(t) = \underset{\sim}{U}(t)\underset{\sim}{y}_0 + \int_0^t \underset{\sim}{U}(t - s)\underset{\sim}{f}(s)ds + \int_0^t \underset{\sim}{V}(t - s)\underset{\sim}{L}^{-1}\underset{\sim}{N}\underset{\sim}{y}(s)ds,$$

provided $\underset{\sim}{N}$ is a possibly nonlinear operator with

$$\begin{cases} \underset{\sim}{N}(0) = 0 \\ \sup_{\|\underset{\sim}{x}_1\|_{\mathcal{D}}, \|\underset{\sim}{x}_2\|_{\mathcal{D}} \le \Delta} \|\underset{\sim}{N}\underset{\sim}{x}_1 - \underset{\sim}{N}\underset{\sim}{x}_2\|_{\mathcal{D}_1} \le \varepsilon(\Delta)\|\underset{\sim}{x}_1 - \underset{\sim}{x}_2\|_{\mathcal{D}} \\ \text{where } \varepsilon\colon (0,\alpha) \to \mathbb{R}^+ \text{ and } \varepsilon \to 0 \text{ as } \Delta \to 0, \end{cases} \tag{11}$$

and

$$\|\underset{\sim}{x}\|_{\mathcal{D}}^2 \equiv \|\underset{\sim}{x}\|^2 + \|\underset{\sim}{L}\underset{\sim}{x}\|^2; \quad \|\underset{\sim}{x}\|_{\mathcal{D}_1}^2 \equiv \|\underset{\sim}{x}\|^2 + \|\underset{\sim}{L}^{1/2}\underset{\sim}{x}\|^2.$$

One possible nonlinear perturbation of (3) which satisfies (11) is $\underset{\sim}{N}\underset{\sim}{y} = yy_x$. However, other nonlinearities, such as $(y_x)^2$ and $(h(y_x))_x$ are not included in our results. Complete statements and proofs of our results are contained in [2].

ACKNOWLEDGEMENT

This paper is based on joint work with K. B. Hannsgen [2].

REFERENCES

1. R. W. Carr and K. B. Hannsgen, A nonhomogeneous integrodifferential equation in Hilbert space, *SIAM J. Math. Anal.*, 10 (1979), 961-984.

2. R. W. Carr and K. B. Hannsgen, Resolvent formulas for a Volterra equation in Hilbert space, *SIAM J. Math. Anal.*, 13 (1982), 459-483.

3. K. B. Hannsgen, Indirect abelian theorems and a linear Volterra equation, *Trans. Amer. Math. Soc.*, 142 (1969), 539-555.

4. D. F. Shea and S. Wainger, Variants of the Wiener-Lévy theorem, with applications to stability problems for some Volterra integral equations, *Amer. J. Math.*, 97 (1975), 312-343.

ASYMPTOTIC EXPANSIONS OF A PENALTY METHOD FOR COMPUTING A REGULATOR PROBLEM GOVERNED BY VOLTERRA EQUATIONS

Goong Chen

Department of Mathematics
Pennsylvania State University
University Park, Pennsylvania

Ronald Grimmer

Department of Mathematics
Southern Illinois University
Carbondale, Illinois

1. INTRODUCTION

In this study we give a formal justification of a special penalty approach to solving an optimal regulator problem governed by a Volterra integro-differential equation. This approach was motivated by some numerical considerations.

Consider the following quadratic cost optimal control problem

$$\min J(u) \equiv \tfrac{1}{2}\int_0^T \left[\langle y(t), My(t)\rangle_{H_1} + \langle u(t), Nu(t)\rangle_{H_2} \right] dt \tag{1.1}$$

subject to

$$\begin{aligned} &\frac{d}{dt} y(t) = Ay(t) + \int_0^t k(t-s)y(s)ds + B(t)u(t) + f(t), \quad 0 \le t \le T \\ &y(0) = y_0 \in H_1, \end{aligned} \tag{1.2}$$

where

(i) H_1, H_2 are two Hilbert spaces, with inner products $\langle\cdot,\cdot\rangle_{H_1}$, $\langle\cdot,\cdot\rangle_{H_2}$

(ii) y(t): the state of the system at time t, εH_1

(iii) u(t): control at time t, εH_2

(iv) y_0: initial state

(v) A: a densely defined, closed operator with domain $D(A) \subseteq H_1$, which generates a C_0-semigroup in H_1

(vi) B(t): a bounded operator mapping H_2 into H_1, smooth in t, such that B maps $L^2(0,\infty;H_2)$ into $L^2(0,\infty;H_1)$

(vii) k(t): the convolution kernel which is defined at least on D(A); k induces a mapping from $L^2(0,T;D(A))$ into $L^2(0,T;H_1)$ with certain smoothness properties

(viii) M, N: symmetric, strictly positive definite, bounded (smooth) operators on H_1 and H_2, respectively, satisfying

$$\langle y,My\rangle_{H_1} \geq \mu \|y\|^2_{H_1}, \quad \langle u,Nu\rangle_{H_2} \geq \nu \|u\|^2_{H_2}, \quad \mu,\nu > 0$$

for all $y \varepsilon H_1$, $u \varepsilon H_2$.

There are several standard methods for computing the unique optimal control and the corresponding optimal state (denoted by $\hat{u}$ and $\hat{y}$, respectively) which we briefly describe below.

EXAMPLE 1. *The duality method.* Let $H_1 = \mathbb{R}^n$ and $H_2 = \mathbb{R}^m$. This is the simplest case when (1.2) is a controlled Volterra ODE. The usual dual algorithm for computing the optimal control using *finite elements* involves introducing the adjoint state $\hat{p}(t)$ satisfying

$$\frac{d}{dt}\hat{p}(t) = -A^*\hat{p}(t) - \int_t^T k^*(s - t)\hat{p}(s)ds - M\hat{y}(t), \quad 0 \leq t \leq T$$
$$\hat{p}(T) = 0. \tag{1.3}$$

Using

$$\hat{u}(t) = -N^{-1}B^*(t)\hat{p}(t) \tag{1.4}$$

and, from (1.3),

$$\hat{y}(t) = M^{-1}[\dot{\hat{p}}(t) + A^*\hat{p}(t) + \int_t^T k^*(s - t)\hat{p}(s)ds] \tag{1.5}$$

(where A^*, B^* and k^* are appropriate transpose matrices of A, B and k, respectively) as well as using the min-max principle, one reduces the problem of finding the *minimizing* solution $\hat{u}$ of (1.1) to that of finding the *maximizing* solution $\hat{p}$ of

$$I(p) \equiv -\tfrac{1}{2} \int_0^T \{<\dot{p}(t) + A^*p(t) + \int_t^T k^*(s - t)p(s)ds,\ M^{-1}[\dot{p}(t) + A^*p(t)$$
$$+ \int_t^T k^*(s - t)p(s)ds]> + <B^*p(t), N^{-1}B^*p(t)> + <p(t), f(t)>\}dt - <p(0), y_0> \tag{1.6}$$

subject to $p(T) = 0$, $p \in [H^1(0,T)]^n$. Approximations to $\hat{p}$ can be obtained by the finite element method. One then uses (1.4), (1.5) to find approximations to the optimal state and control $\hat{y}$, $\hat{u}$.

EXAMPLE 2. *The direct penalty method.* This method was first introduced by Balakrishnan [1] and Lions [5]. We consider minimizing

$$J_\varepsilon(y,u) \equiv \int_0^T [<y(t), My(t)> + <u(t), Nu(t)>]dt + \frac{1}{\varepsilon}\int_0^T \|\dot{y}(t) - Ay(t)$$
$$- \int_0^t k(t - s)y(s)ds - B(t)u(t) - f(t)\|^2_{H_1} dt, \qquad \varepsilon > 0, \tag{1.7}$$

with respect to y, u, subject to $y(0) = y_0$ only. Let the optimal solution to (1.7) be $(\hat{y}_\varepsilon, \hat{u}_\varepsilon)$. Using some standard arguments, one can show that $(\hat{y}_\varepsilon, \hat{u}_\varepsilon)$ tends to the optimal state and control $(\hat{y}, \hat{u})$ as ε tends to 0. In practice, (1.7) is minimized and approximated over finite element spaces $V_h^1 \times V_h^2$ containing (y,u).

Both approaches mentioned above produce excellent computational results when there is no delay, i.e., when $k \equiv 0$. However, if $k \not\equiv 0$, although these methods remain valid, computationally they become very inefficient for the finite element method because it is found that for each *locally supported* spline basis function $\phi_h(t)$, the convolution integrals

$$\int_t^T k^*(s - t)\phi_h(s)ds \quad \text{(in (1.6))}, \qquad \int_0^t k(t - s)\phi_h(s)ds \quad \text{(in (1.7))}$$

are no longer locally supported. Thus one will get *full stiffness matrices* whose computations are slow and expensive. It seems very desirable to look for other methods for (1.1), (1.2) which would involve only *sparse matrices*. This is the main motivation for this study.

In the following discussion, we will present a *modified penalty approach* which meets this requirement. It is based upon a *semigroup formulation of Volterra equations* studied earlier by R. K. Miller [6], R. K. Miller and R. L. Wheeler [7], and G. Chen and R. Grimmer [2]. In this article, we will only study the asymptotic expansion with respect to the penalty parameter ε of this special approach. Its numerical feasibility and error estimates are being investigated and will appear elsewhere.

2. AN ASSOCIATED SEMIGROUP, PENALTY AND ITS COMPUTATIONAL PERSPECTIVE

We assume that the operator A in (1.1) generates a contractive C_0-semigroup in H_1 and that the convolution kernel k is just the product of a sufficiently smooth scalar function on [0,T] with the identity operator in H_1. We can always extend k to be a smooth function on $[0,\infty)$ with compact support. From now on, without further mention, we will assume that such an extension for k has been done whenever necessary. Similarly, we can extend B and f outside [0,T] in the same manner, if necessary. We also assume that A^* generates a contractive semigroup.

For any function F: $[0,\infty) \times [0,\infty)$ (or $[0,T] \times [0,T]) \to H_1$, we define the Dirac delta function δ_0 by

$$\delta_0 F(t,s) = F(t,0),$$

i.e., $\delta_0 F$ is the trace of F on the t-axis.

LEMMA 1, [2]. The operator

$$C = \begin{bmatrix} A & \delta_0 \\ k & \frac{\partial}{\partial s} \end{bmatrix} \tag{2.1}$$

generates a C_0-semigroup in $H_1 \times L^2(0,\infty;H_1)$. The solution (z,F) of the differential equation

$$\frac{d}{dt}\begin{bmatrix} z(t) \\ F(t,\cdot) \end{bmatrix} = \begin{bmatrix} A & \delta_0 \\ k & \frac{\partial}{\partial s} \end{bmatrix} \begin{bmatrix} z(t) \\ F(t,\cdot) \end{bmatrix}$$
$$\begin{bmatrix} z(0) \\ F(0,\cdot) \end{bmatrix} = \begin{bmatrix} y_0 \\ (Bu)(\cdot) + f(\cdot) \end{bmatrix} \tag{2.2}$$

with $y_0 \in D(A)$, $Bu + f \in H^1(0,\infty;H_1)$ is equivalent to that of the Volterra equation

$$\frac{d}{dt} y(t) = Ay(t) + \int_0^t k(t-s)y(s)ds + B(t)u(t) + f(t), \quad 0 \le t < \infty$$
$$y(0) = y_0$$

through the identification of z(t) and y(t).

REMARK. The kernel k is allowed to be an unbounded operator of the form $k_1(t)A^{1/2}$ for some smooth scalar function $k_1(t)$; in that case, the corresponding generator C of the semigroup becomes slightly more complicated than (2.2). We refer to [2] for details.

In (2.2), we see that the control u becomes part of the initial condition. From this point of view, we can regard the optimal control problem governed by Volterra equations as a *control problem of the initial condition.*

DEFINITION 2. We define two Hilbert spaces H_3 and H_4 by

$$H_3 \equiv \{z \mid z \in L^2(0,T;H_1),\ \dot{z} - Az \in L^2(0,T;H_1)\},$$

$$\|z\|_{H_3}^2 \equiv \|z\|_{L^2(0,T;H_1)}^2 + \|\dot{z} - Az\|_{L^2(0,T;H_1)}^2,$$

$$H_4 \equiv \{G | G \in L^2([0,T] \times [0,T];H_1), \ \frac{\partial}{\partial t} G - \frac{\partial}{\partial s} G \in L^2([0,T] \times [0,T];H_1)\},$$

$$\|G\|^2_{H_4} \equiv \|G\|^2_{L^2([0,T]\times[0,T];H_1)} + \|\frac{\partial}{\partial t} G - \frac{\partial}{\partial s} G\|^2_{L^2([0,T]\times[0,T];H_1)}.$$

We are now in a position to consider the following penalty problem:

$$\begin{aligned} \min J_\varepsilon(y,u,F) \equiv & \int_0^T [\langle y(t),My(t)\rangle_{H_1} + \langle u(t),Nu(t)\rangle_{H_2}]dt \\ & + \frac{1}{\varepsilon} \{\int_0^T \|\dot{y}(t) - Ay(t) - F(t,0)\|^2_{H_1} dt \\ & + \int_0^T\int_0^T \|\frac{\partial}{\partial t} F(t,s) - \frac{\partial}{\partial s} F(t,s) - k(s)y(t)\|^2_{H_1} dt\, ds\}, \quad \varepsilon > 0, \end{aligned} \tag{2.3}$$

subject to

$$\begin{aligned} & y(0) = y_0, \quad y \in H_3, \quad u \in H^1(0,T;H_2), \\ & F \in H_4, \quad F(0,s) = B(s)u(s) + f(s), \quad F(t,T) = 0. \end{aligned} \tag{2.4}$$

We note that in (2.3), the term which is penalized is just

$$\frac{1}{\varepsilon}\int_0^T \|\frac{d}{dt}\begin{bmatrix} z(t) \\ F(t,\cdot)\end{bmatrix} - C\begin{bmatrix} z(t) \\ F(t,\cdot)\end{bmatrix}\|^2_{H_1 \times L^2(0,T;H_1)} dt,$$

that is, the expression associated with the *equivalent differential equation* (2.2).

We need the following important assumptions:

(A1) $y_0 \in D(A)$.

(A2) The optimal control $\hat{u}$ solving (1.1), (1.2) lies in $H^1(0,T;H_2)$.

It follows from (A1), (A2) and [2, Theorem 2.5] that the optimal state $\hat{y}$ belongs to $C^1([0,T];H_1) \cap C^0([0,T];D(A))$.

For control systems governed by partial differential equations, the question of *regularity* of the optimal control $\hat{u}$ is always difficult. Here, we make the following strong assumption:

(A3) The following coupled system

$$\begin{cases} \frac{\partial}{\partial t} q(t,s) - \frac{\partial}{\partial s} q(t,s) = 0, \quad (t,s) \in [0,T] \times [0,T] \\ q(t,0) = p(t),\ q(T,s) = 0, \quad t,s \in [0,T], \end{cases} \tag{2.5}$$

$$\begin{cases} \frac{d}{dt} p(t) + A^* p(t) + \int_t^T k^*(s - t)p(s)ds + My(t) = 0 \\ p(T) = 0, \end{cases} \tag{2.6}$$

$$\begin{cases} \frac{\partial}{\partial t} F(t,s) - \frac{\partial}{\partial s} F(t,s) - k(s)y(t) = -q(t,s) \\ F(0,s) = -B(s)N^{-1}B^*(s)q(0,s) + f(s), \quad F(t,T) = 0, \quad f \in L^2(0,T;H_1), \end{cases} \tag{2.7}$$

$$\begin{cases} \frac{d}{dt} y(t) - Ay(t) - F(t,0) = -p(t) \\ y(0) = y_0 \in H_1, \end{cases} \tag{2.8}$$

has a weak solution (q,p,y,F). Furthermore, for each $y_0 \in D(A)$ and $f \in H^1(0,T;H_1)$, the solutions (q,p,y,F) to (2.5) - (2.8) are "classical solutions". (F is allowed to have discontinuities along $\{(s,T-s) \mid s \in [0,T]\}$.)

REMARK. If (1.2) is an ODE, the above assumption can be verified indirectly through ODE methods (or perhaps through using the Riccati decoupling and the method of characteristics techniques as in [8]). It is not clear, however, if (A3) is a valid assumption when A is a partial differential operator.

THEOREM 3. Under assumptions (A1) - (A3), the penalty problem (2.3) has a unique solution $(\hat{y}_\varepsilon,\hat{u}_\varepsilon,\hat{F}_\varepsilon)$ minimizing (2.3) with

$$\hat{u}_\varepsilon \in H^1(0,T;H_2),\ \hat{y}_\varepsilon \in C^1([0,T];H_1) \cap C^0([0,T];D(A)),\ F_\varepsilon \in H_4.$$

As $\varepsilon \downarrow 0$, $(\hat{u}_\varepsilon,\hat{y}_\varepsilon)$ converges to the optimal solution $(\hat{u},\hat{y})$ solving (1.1), (1.2).

Now we briefly describe the computational aspect of (2.3). In practice, due to the constraints (2.4), it is more convenient to introduce two additional penalty parameters $\varepsilon_1, \varepsilon_2 > 0$ and consider instead

$$\inf \bar{J}_{\varepsilon,\varepsilon_1,\varepsilon_2}(y,u,F) \equiv J_\varepsilon(y,u,F) + \frac{1}{\varepsilon_1}\|y(0) - y_0\|^2_{H_1} + \frac{1}{\varepsilon_2}\int_0^T \|F(0,s) - B(s)u(s) + f(s)\|^2_{H_1} ds$$

subject to $y \in H_3$, $u \in H^1(0,T;H_2)$, $F \in H_4$, $F(t,T) = 0$. One then proceeds to introduce three finite dimensional spline spaces V_h^1, V_h^2, and V_h^3 such that

$$V_h^1 \subseteq H_3, \quad V_h^2 \subseteq H^1(0,T;H_2), \quad V_h^3 \subseteq H_4$$

$$\lim_{h\to 0} V_h^1 = H_3, \quad \lim_{h\to 0} V_h^2 = H^1(0,T;H_2), \quad \lim_{h\to 0} V_h^3 = H_4.$$

Write

$$y_h = \sum_{i_1=1}^{n_1} \alpha_{i_1}\phi_{hi_1}^{(1)}, \quad u_h = \sum_{i_2=1}^{n_2} \alpha_{i_2}\phi_{hi_2}^{(2)}, \quad F_h = \sum_{i_3=1}^{n_3} \alpha_{i_3}\phi_{hi_3}^{(3)}.$$

The coefficients $\alpha = (\alpha_{i_1}, \alpha_{i_2}, \alpha_{i_3})$ $(1 \le i_1 \le n_1,\ 1 \le i_2 \le n_2,\ 1 \le i_3 \le n_3)$ can be determined by solving a finite dimensional variational equation of the form

$$A_{\varepsilon,\varepsilon_1,\varepsilon_2}\left(\begin{bmatrix} y_h \\ u_h \\ F_h \end{bmatrix}, \phi_i\right) = \theta_{\varepsilon,\varepsilon_1,\varepsilon_2}(\phi_i) \quad \text{for all} \quad \phi_i = \begin{bmatrix} \phi_{hi_1}^{(1)} \\ \phi_{hi_2}^{(2)} \\ \phi_{hi_3}^{(3)} \end{bmatrix}$$

where $A_{\varepsilon,\varepsilon_1,\varepsilon_2}$ is a semi-positive definite quadratic form and $\theta_{\varepsilon,\varepsilon_1,\varepsilon_2}$ is a linear form on $H_3 \times H^1(0,T;H_2) \times H_4$. It can be easily seen that the stiffness matrix

$$A_{\varepsilon,\varepsilon_1,\varepsilon_2}\left(\begin{bmatrix} y_h \\ u_h \\ F_h \end{bmatrix}, \phi_i\right)$$

is *sparse*.

Such computations have been found to be extremely accurate when $k = 0$, and some error estimates with respect to ε, h have been obtained. See [4].

3. ASYMPTOTIC EXPANSIONS

In this section, we compute the following asymptotic expansions

$$\hat{y}_\varepsilon = \hat{y} + \varepsilon y_1 + \varepsilon^2 y_2 + \ldots + \varepsilon^n y_n + \ldots \tag{3.1}$$

$$\hat{u}_\varepsilon = \hat{u} + \varepsilon u_1 + \varepsilon^2 u_2 + \ldots + \varepsilon^n u_n + \ldots \tag{3.2}$$

$$\hat{F}_\varepsilon = \hat{F} + \varepsilon F_1 + \varepsilon^2 F_2 + \ldots + \varepsilon^n F_n + \ldots \tag{3.3}$$

for the minimizing solution $(\hat{y}_\varepsilon, \hat{u}_\varepsilon, \hat{F}_\varepsilon)$ of (2.3).

We make our last assumption:

(A4) The optimal regulator problem (1.1), (1.2) is *well-posed* with respect to (y_0, f), that is to say if $\hat{u}^1$, $\hat{u}^2$ are the optimal controls corresponding to (y_0^1, f^1), (y_0^2, f^2), respectively, in (1.1) and (1.2), then

$$\|\hat{u}^1 - \hat{u}^2\|_{L^2(0,T;H_2)} \le K[\|y_0^1 - y_0^2\|_{H_1} + \|f^1 - f^2\|_{L^2(0,T;H_1)}]$$

holds for some constant K independent of (y_0, f).

We let $(\hat{u}, \hat{y})$ be the optimal control and state as in Sections 1 and 2, and let $\hat{F}$ be the solution of

$$\begin{cases} \dfrac{\partial}{\partial t}\hat{F}(t,s) - \dfrac{\partial}{\partial s}\hat{F}(t,s) - k(s)\hat{y}(t) = 0, & (t,s) \in [0,T] \times [0,T], \\ \hat{F}(0,s) = B(s)\hat{u}(s) + f(s), \quad \hat{F}(t,T) = 0. & \end{cases}$$

Then $\hat{F} \in H_4$. We have

LEMMA 4. Under assumptions (A1) - (A4), there exists a constant $\bar{C}$ depending on (y_0, f) only, such that

$$\|[\dot{\hat{y}}_\varepsilon - A\hat{y}_\varepsilon - \hat{F}_\varepsilon(\cdot,0)] - [\dot{\hat{y}} - A\hat{y} - \hat{F}(\cdot,0)]\|_{L^2(0,T;H_1)}$$

$$+ \|[\frac{\partial}{\partial t}\hat{F}_\varepsilon - \frac{\partial}{\partial s}\hat{F}_\varepsilon - k\hat{y}_\varepsilon] - [\frac{\partial}{\partial t}\hat{F} - \frac{\partial}{\partial s}\hat{F} - k\hat{y}]\|_{L^2([0,T]\times[0,T];H_1)} \le \bar{C}\varepsilon.$$

The proof of this lemma is a straightforward generalization of Theorem 2.4 in [3].

An immediate consequence of Lemma 4 is that there exists (p_0,q_0) such that

$$\underset{\varepsilon_n\downarrow 0}{\text{w.-lim}}\ \frac{1}{\varepsilon_n}\{[\dot{\hat{y}}_{\varepsilon_n} - A\hat{y}_{\varepsilon_n} - \hat{F}_{\varepsilon_n}(\cdot,0)] - [\dot{\hat{y}} - A\hat{y} - \hat{F}(\cdot,0)]\} \equiv -p_0 \quad (3.4)$$

in $L^2(0,T;H_1)$

$$\underset{\varepsilon_n\downarrow 0}{\text{w.-lim}}\ \frac{1}{\varepsilon_n}\{[\frac{\partial}{\partial t}\hat{F}_{\varepsilon_n} - \frac{\partial}{\partial s}\hat{F}_{\varepsilon_n} - k\hat{y}_{\varepsilon_n}] - [\frac{\partial}{\partial t}\hat{F} - \frac{\partial}{\partial s}\hat{F} - k\hat{y}]\} = -q_0 \quad (3.5)$$

in $L^2([0,T]\times[0,T];H_1)$

for a subsequence $\{\varepsilon_n\}$ tending to 0^+. Therefore (p_0,q_0) satisfies the variational equation

$$\langle\hat{y}, Mz\rangle_{L^2(0,T;H_1)} + \langle\hat{u}, Nv\rangle_{L^2(0,T;H_2)} - \langle p_0, \dot{z} - Az - G(\cdot,0)\rangle_{L^2(0,T;H_1)}$$
$$- \langle q, \frac{\partial}{\partial t}G - \frac{\partial}{\partial s}G - kz\rangle_{L^2([0,T]\times[0,T];H_1)} = 0 \quad (3.6)$$

for all z, v and G satisfying

$$v \in H^1(0,T;H_2), \quad z \in H_3,$$
$$G \in H_4, \quad G(0,s) = B(s)v(s), \quad G(t,T) = 0. \quad (3.7)$$

LEMMA 5. Let p,q,u be any three functions satisfying

$$\begin{cases} \dot{p}(t) + A^*p(t) + \int_t^T k^*(s - t)p(s)ds + My(t) = 0, \quad t \in [0,T] \\ p(t) = 0 \\ \frac{\partial}{\partial t} q(t,s) - \frac{\partial}{\partial s} q(t,s) = 0 \\ q(t,0) = p(t), \quad q(T,s) = 0 \\ u(t) = -N^{-1}B(t)^*q(0,t). \end{cases}$$

Then p,q and u satisfy the variational equation

$$\langle y, Mz\rangle_{L^2(0,T;H_1)} + \langle u, Nv\rangle_{L^2(0,T;H_2)} - \langle p, \dot{z} - Az - G(\cdot,0)\rangle_{L^2(0,T;H_2)}$$
$$- \langle q, \frac{\partial}{\partial t} G - \frac{\partial}{\partial s} G - kz\rangle_{L^2([0,T]\times[0,T];H_1)} = 0$$

for all z,v and G satisfying (3.7).

Proof. Integration by parts.

We note from (3.4) that since $(-p_0,-q_0)$ are weak limits, there is no guarantee about their regularity. However, from (1.3), we see that $\hat{p}$ satisfies

$$\begin{cases} \dot{\hat{p}}(t) + A^*\hat{p}(t) + \int_t^T k^*(s - t)\hat{p}(s)ds + M\hat{y}(t) = 0 \\ \hat{p}(T) = 0. \end{cases} \tag{3.8}$$

We let $\hat{q}$ satisfy

$$\begin{cases} \frac{\partial}{\partial t} \hat{q} - \frac{\partial}{\partial s} \hat{q} = 0 \quad \text{on } [0,T]\times[0,T], \\ \hat{q}(t,0) = \hat{p}(t), \quad \hat{q}(T,s) = 0. \end{cases} \tag{3.9}$$

Then

$$\hat{q}(t,s) = \begin{cases} \hat{p}(t+s) & \text{if } t + s \leq T \\ 0 & \text{if } t + s \geq T. \end{cases} \tag{3.10}$$

From the dual theory of optimal control, we have (1.4). Therefore

$$\hat{u}(t) = -N^{-1}B^*(t)\hat{p}(t) = -N^{-1}B^*(t)\hat{q}(0,t). \tag{3.11}$$

Combining (3.8), (3.9) and (3.11) and applying Lemma 5, we see that the variational equation

$$\begin{aligned} &\langle \hat{y}, Mz\rangle_{L^2(0,T;H_1)} + \langle \hat{u}, Nv\rangle_{L^2(0,T;H_2)} \\ &\quad - \langle \hat{p}_0, \dot{z} - Az - G(\cdot,0)\rangle_{L^2(0,T;H_1)} \\ &\quad - \langle \hat{q}_0, \frac{\partial}{\partial t} G - \frac{\partial}{\partial s} G - kz\rangle_{L^2([0,T]\times[0,T];H_1)} = 0 \end{aligned} \tag{3.12}$$

is satisfied for all z,v,G satisfying (3.7). Now comparing (3.6) with (3.12) and using a denseness argument, we conclude that $p_0 = \hat{p}$ and $q_0 = \hat{q}$. Therefore, $p_0 \in H_3$ and $q_0 \in H_4$. The above argument also shows *the connection between the duality and the penalty method, namely, the weak limit* p_0 *in* (3.4) *is just the adjoint state* $\hat{p}$.

Now, for $n = 1,2,3,\ldots$, we define, recursively, functions q_n, p_n, u_n and F_n satisfying

$$\begin{cases} \dot{y}_n(t) - Ay_n(t) - F_n(t,0) = -p_{n-1}(t) \\ y_n(0) = 0 \end{cases} \tag{3.13}$$

$$\begin{cases} \frac{\partial}{\partial t} F_n(t,s) - \frac{\partial}{\partial s} F_n(t,s) - k(s)y_n(t) = -q_{n-1}(t,s) \\ F_n(0,s) = -B(s)u_{n-1}(s), \quad F_n(t,T) = 0 \end{cases} \tag{3.14}$$

$$\begin{cases} \dot{p}_n(t) + A^*p_n(t) + \int_t^T k^*(s - t)p_n(s)ds + My_n(t) = 0 \\ p_n(T) = 0 \end{cases} \tag{3.15}$$

$$\begin{cases} \frac{\partial}{\partial t} q_n(t,s) - \frac{\partial}{\partial s} q_n(t,s) = 0 \\ q_n(t,0) = p_n(t), \quad q_n(T,s) = 0 \end{cases} \tag{3.16}$$

for $t,s \in [0,T]$, and let $y_0 \equiv \hat{y}$, $u_0 \equiv \hat{u}$ and $F_0 \equiv \hat{F}$. By assumption (A3), all these functions q_n, p_n, y_n, u_n and F_n have sufficient regularity properties.

We let

$$\hat{y}^n_\varepsilon \equiv \sum_{j=0}^n \varepsilon^j y_j, \quad \hat{p}^n_\varepsilon \equiv \sum_{j=0}^n \varepsilon^j p_j,$$

$$\hat{u}^n_\varepsilon \equiv \sum_{j=0}^n \varepsilon^j u_j, \quad \hat{q}^n_\varepsilon \equiv \sum_{j=0}^n \varepsilon^j q_j,$$

$$\hat{F}^n_\varepsilon \equiv \sum_{j=0}^n \varepsilon^j F_j.$$

Multiplying (3.13) - (3.16) by ε^j and summing over j from 1 to n, then applying Lemma 5 and adding (3.6), we get

$$\begin{aligned} &<\hat{y}^n_\varepsilon, Mz> + <\hat{u}^n_\varepsilon, Nv> + \frac{1}{\varepsilon}\{<\dot{\hat{y}}^n_\varepsilon - A\hat{y}^n_\varepsilon - \hat{F}^n_\varepsilon(\cdot,0), \dot{z} - Az - G(\cdot,0)> \\ &\quad + <\frac{\partial}{\partial t}\hat{F}^n_\varepsilon - \frac{\partial}{\partial s}\hat{F}^n_\varepsilon - k\hat{y}^n_\varepsilon, \frac{\partial}{\partial t}G - \frac{\partial}{\partial s}G - kz>\} \\ &= \varepsilon^n[<p_n, \dot{z} - Az - G(\cdot,0)> + <q_n, \frac{\partial}{\partial t}G - \frac{\partial}{\partial s}G - kz>] \end{aligned} \tag{3.17}$$

for all z, v and G satisfying (3.7), where the inner products are taken in appropriate spaces.

On the other hand, from (2.3) and Theorem 3, we know that $\hat{y}_\varepsilon$, $\hat{u}_\varepsilon$ and $\hat{F}_\varepsilon$ satisfy the variational equation

$$<\hat{y}_\varepsilon, Mz> + <\hat{u}_\varepsilon, Nv> + \frac{1}{\varepsilon}\{<\dot{\hat{y}}_\varepsilon - A\hat{y}_\varepsilon - \hat{F}_\varepsilon(\cdot,0), \dot{z} - Az - G(\cdot,0)> \\ + <\frac{\partial}{\partial t}\hat{F}_\varepsilon - \frac{\partial}{\partial s}\hat{F}_\varepsilon - k\hat{y}_\varepsilon, \frac{\partial}{\partial t}G - \frac{\partial}{\partial s}G - kz>\} = 0. \tag{3.18}$$

Subtracting (3.18) from (3.17) and setting $z = \hat{y}_\varepsilon - \hat{y}^n_\varepsilon$, $v = \hat{u}_\varepsilon - \hat{u}^n_\varepsilon$ and $G = \hat{F}_\varepsilon - \hat{F}^n_\varepsilon$, we get

$$<\hat{y}_\varepsilon - \hat{y}^n_\varepsilon, M(\hat{y}_\varepsilon - \hat{y}^n_\varepsilon)> + <\hat{u}_\varepsilon - \hat{u}^n_\varepsilon, N(\hat{u}_\varepsilon - \hat{u}^n_\varepsilon)>$$

$$+ \frac{1}{\varepsilon}\{\|(\dot{\hat{y}}_\varepsilon - \dot{\hat{y}}^n_\varepsilon) - A(\hat{y}_\varepsilon - \hat{y}^n_\varepsilon) - (\hat{F}_\varepsilon(\cdot,0) - \hat{F}^n_\varepsilon(\cdot,0))\|^2$$

$$+ \|\frac{\partial}{\partial t}(\hat{F}_\varepsilon - \hat{F}^n_\varepsilon) - \frac{\partial}{\partial s}(\hat{F}_\varepsilon - \hat{F}^n_\varepsilon) - k(\hat{y}_\varepsilon - \hat{y}^n_\varepsilon)\|^2\}$$

$$= \varepsilon^n[<p_n, (\dot{\hat{y}}_\varepsilon - \dot{\hat{y}}^n_\varepsilon) - A(\hat{y}_\varepsilon - \hat{y}^n_\varepsilon) - (\hat{F}_\varepsilon(\cdot,0) - \hat{F}^n_\varepsilon(\cdot,0))>$$

$$+ <q_n, \frac{\partial}{\partial t}(\hat{F}_\varepsilon - \hat{F}^n_\varepsilon) - \frac{\partial}{\partial s}(\hat{F}_\varepsilon - \hat{F}^n_\varepsilon) - k(\hat{y}_\varepsilon - \hat{y}^n_\varepsilon)>],$$

implying that

$$\mu\|\hat{y}_\varepsilon - \hat{y}^n_\varepsilon\|^2 + \nu\|\hat{u}_\varepsilon - \hat{u}^n_\varepsilon\|^2 + \frac{1}{\varepsilon}\{\|(\dot{\hat{y}}_\varepsilon - \dot{\hat{y}}^n_\varepsilon) - A(\hat{y}_\varepsilon - \hat{y}^n_\varepsilon)$$

$$- (\hat{F}_\varepsilon(\cdot,0) - \hat{F}^n_\varepsilon(\cdot,0))\|^2 + \|\frac{\partial}{\partial t}(\hat{F}_\varepsilon - \hat{F}^n_\varepsilon)$$

$$- \frac{\partial}{\partial s}(\hat{F}_\varepsilon - \hat{F}^n_\varepsilon) - k(\hat{y}_\varepsilon - \hat{y}^n_\varepsilon)\|^2\} \le \frac{1}{2\varepsilon}\|(\dot{\hat{y}}_\varepsilon - \dot{\hat{y}}^n_\varepsilon)$$

$$- A(\hat{y}_\varepsilon - \hat{y}^n_\varepsilon) - (\hat{F}_\varepsilon(\cdot,0) - \hat{F}^n_\varepsilon(\cdot,0))\|^2 + \frac{\varepsilon^{2n+1}}{2}\|p_n\|^2$$

$$+ \frac{1}{2\varepsilon}\|\frac{\partial}{\partial t}(\hat{F}_\varepsilon - \hat{F}^n_\varepsilon) - \frac{\partial}{\partial s}(\hat{F}_\varepsilon - \hat{F}^n_\varepsilon) - k(\hat{y}_\varepsilon - \hat{y}^n_\varepsilon)\|^2 + \frac{\varepsilon^{2n+1}}{2}\|q_n\|^2.$$

Hence, for each n,

$$\varepsilon^{-2n}[\mu\|\hat{y}_\varepsilon - \hat{y}^n_\varepsilon\|^2 + \nu\|\hat{u}_\varepsilon - \hat{u}^n_\varepsilon\|^2] \le \frac{\varepsilon}{2}[\|p_n\|^2 + \|q_n\|^2] \to 0$$

as $\varepsilon \to 0$, i.e.,

$$\|\hat{y}_\varepsilon - \hat{y}^n_\varepsilon\| = o(\varepsilon^n), \quad \|\hat{u}_\varepsilon - \hat{u}^n_\varepsilon\| = o(\varepsilon^n),$$

proving the validity of the asymptotic expansion (3.1) and (3.2).

ACKNOWLEDGEMENT

Professor Chen's research was supported in part by the National Science Foundation under Grant MCS 81-01892.

REFERENCES

1. A. V. Balakrishnan, On a new computing technique in optimal control, *SIAM J. Control* 6 (1968), 149-173.

2. G. Chen and R. Grimmer, Semigroups and integral equations, *J. Integral Equations* 2 (1980), 133-154.

3. G. Chen and W. H. Mills, Finite elements and terminal penalization for quadratic cost optimal controls governed by ordinary differential equations, *SIAM J. Control Opt.* 19 (1981), 744-764.

4. G. Chen, W. H. Mills, S. Sun and D. Yost, Sharp error estimates for a finite element-penalty approach to a class of regulator problems, *Math. Comp.* 37 (1983), to appear.

5. J. L. Lions, *Optimal control of systems governed by partial differential equations*, Springer-Verlag, New York, 1971.

6. R. K. Miller, Volterra integral equations in a Banach space, *Funkcialaj Ekvacioj* 18 (1975), 163-193.

7. R. K. Miller and R. L. Wheeler, Well-posedness and stability of linear Volterra integrodifferential equations in abstract spaces, *Funkcialaj Ekvacioj* 21 (1978), 279-305.

8. D. L. Russell, Quadratic performance criteria in boundary control of linear symmetric hyperbolic systems, *SIAM J. Control* 11 (1973), 475-509.

INTEGRODIFFERENTIAL EQUATIONS WITH ALMOST PERIODIC SOLUTIONS

C. Corduneanu

Department of Mathematics
University of Texas at Arlington
Arlington, Texas

1. INTRODUCTION

In a recent paper [8], we dealt with the existence of almost periodic solutions to certain classes of differential equations with infinite delay, which can be represented in the form

$$\dot{x}(t) = \int_0^\infty [dA(s)]x(t-s) + f(t). \tag{0}$$

A(s) designates a square matrix of order n whose entries are real-valued functions with bounded variation on the positive half axis, while x and f stand for vector-valued functions with n real coordinates.

The aim of this paper is to point out that similar results hold true for a class of integrodifferential equations containing the above equations with infinite delay as special cases, namely

$$\dot{x}(t) = \int_{-\infty}^\infty [dA(s)]x(t-s) + f(t), \tag{1}$$

where A(s) will have as entries real-valued functions with bounded variation on the whole real axis.

Moreover, a partial answer will be given to the following open problem formulated in [8]: Assuming that the integral operator generated by $dA(s)$ is such that its spectrum does not contain points of the imaginary axis, can we assert the existence of an almost periodic solution in Bohr's sense, for any $f(t)$ almost periodic in the same sense? It turns out that this property holds for at least those equations of the form (1) in which $A(s)$ does not have a singular part. More precisely, it holds for those equations which can be written as

$$\dot{x}(t) = \sum_{j=0}^{\infty} A_j x(t-t_j) + \int_{-\infty}^{\infty} B(s)x(t-s)ds + f(t), \tag{2}$$

with t_j, $j = 0,1,2,\ldots$, an arbitrary sequence of real numbers and A_j, $j = 0,1,2,\ldots$, $B(t)$ square matrices of order n, such that

$$\sum_{j=0}^{\infty} \|A_j\| < \infty, \quad \|B(t)\| \in L^1(\mathbb{R}) . \tag{3}$$

As will be shown in this paper, equation (2) has a unique solution in a variety of function spaces provided that $f(t)$ belongs to those spaces and the above quoted spectral condition holds true.

The final part of the paper is dedicated to the problem of the existence of almost periodic solutions for equations of the form (1) under conditions that do not imply the validity of the above mentioned spectral property.

Let us point out that many authors have dealt with the problem of almost periodicity, usually in Bohr's sense, for various classes of equations of the form (1), often constituting particular cases of the general one. For instance, S. Bochner [3] generalizes the classical result of Bohr and Neugebauer (see [5] or [13]) to the case of differential-difference equations providing conditions of existence when the characteristic equation has purely imaginary roots. R. Doss [12] investigates similar problems for integrodifferential equations of the form (2) in which $A_j = 0$ for $j > N$. In [6], the author dealt with integral equations of convolution type for which the kernel is integrable, obtaining existence of almost periodic solutions in Bohr's sense and in Stepanov's sense. In [9], the authors find almost periodicity of solutions for integral equations involving operators of the form appearing in the right hand side of the equation (2). In [19], J. J. Levin and D. F.

Shea investigate asymptotic behavior of solutions of some classes of integrodifferential equations, in general nonlinear, which contain as a particular case the equations of the form (1). Asymptotically almost periodic solutions occur in such circumstances, though the framework suggests a much broader picture for the asymptotic behavior. Some ideas in [19] are further developed by G. S. Jordan and R. L. Wheeler [16] and G. S. Jordan, W. R. Madych and R. L. Wheeler [17]. O. J. Staffans [21] obtains further results, considerably relaxing the spectral condition, but always requiring a Tauber type property for the solution. Related results are also contained in a recent paper by G. S. Jordan, O. J. Staffans, and R. L. Wheeler [18]. Relying mainly on Liapunov function techniques, Y. Hino [14], [15], and K. Sawano [20] find criteria for existence of almost periodic solutions to equations with infinite delay, starting from stability properties of the zero solution of the attached homogeneous system. A nonlinear result has been obtained by V. Alexiades [1], the almost periodicity concept being that due to A. S. Besicovitch [2]. J. M. Cushing [11] finds criteria for existence of periodic solutions to equations with infinite delay representable as integrodifferential equations. Further references can be found in the survey paper [10] by C. Corduneanu and V. Lakshmikantham. Finally, let us point out that a good number of almost periodicity results have accumulated for abstract differential equations (see S. Zaidman [22], which contains more references). Since delay equations can often be represented as differential equations in an abstract space, it would be interesting to approach the problem in this manner and make use of available results.

2. SPACES OF ALMOST PERIODIC FUNCTIONS

The spaces of almost periodic functions that will be used in this paper can be generated as follows: Let us denote by T the set of trigonometric polynomials of the form

$$T(t) = \sum_{k=1}^{m} a_k \exp(i\lambda_k t), \quad \lambda_k \neq \lambda_j \quad \text{for } k \neq j, \tag{4}$$

where λ_k are real numbers, and a_k are complex numbers or n-vectors with complex coordinates. For any p such that $1 \leq p \leq 2$, the map

$$T \to \|T\|_p = \left(\sum_{k=1}^{m} |a_k|^p\right)^{1/p} \tag{5}$$

is a norm on the linear space T. If one considers the completion of T with respect to the above norm, one obtains a Banach space which will be denoted by $AP_p(\mathbb{R}, \mathbb{C})$ when T is the space of scalar polynomials, or by $AP_p(\mathbb{R}, \mathbb{C}^n)$ when T is the space of trigonometric polynomials with vector coefficients. The meaning of such notations as $AP_p(\mathbb{R},\mathbb{R})$, or $AP_p(\mathbb{R},\mathbb{R}^n)$, is obvious.

As pointed out in [8], the case $p = 1$ corresponds to the almost periodic functions with absolutely (and uniformly) convergent Fourier series. The case $p = 2$ leads to the almost periodic functions in the Besicovitch sense, of order 2, in which the elements are classes of functions differing from each other by a function with mean square value equal to zero. This last space is usually denoted $B_2(\mathbb{R}, \mathbb{C})$ in the scalar case. In between the two spaces $AP_1(\mathbb{R}, \mathbb{C})$ and $AP_2(\mathbb{R}, \mathbb{C})$ there are several intermediate spaces of almost periodic functions. First of all, it can be easily seen that each space $AP_p(\mathbb{R}, \mathbb{C})$, $1 < p < 2$, satisfies

$$AP_1(\mathbb{R}, \mathbb{C}) \subset AP_p(\mathbb{R}, \mathbb{C}) \subset AP_2(\mathbb{R}, \mathbb{C}). \tag{6}$$

Actually, the double inclusion (6) is a special case of the following property of any pair of spaces $AP_p(\mathbb{R}, \mathbb{C})$, $AP_r(\mathbb{R}, \mathbb{C})$, with $1 \le p \le r \le 2$:

$$AP_p(\mathbb{R}, \mathbb{C}) \subset AP_r(\mathbb{R}, \mathbb{C}), \tag{7}$$

the immersion of the first space into the second being continuous.

While it is clear that any function in $AP_p(\mathbb{R}, \mathbb{C})$, $1 \le p < 2$, enjoys the almost periodicity properties of Besicovitch functions in $B_2(\mathbb{R}, \mathbb{C})$, it would be interesting to describe those properties that are characteristic for the almost periodic functions in $AP_p(\mathbb{R}, \mathbb{C})$. Of course, a first characteristic property is related to the Fourier series. Namely, the series $\sum_{k=1}^{\infty} |a_k|^p$ is convergent, where the a_k's are the Fourier coefficients of any function in the space $AP_p(\mathbb{R}, \mathbb{C})$.

Since any almost periodic function in Bohr's sense is also almost periodic in the sense of Besicovitch, one has that the space of Bohr's almost periodic functions satisfies the double inclusion

$$AP_1(\mathbb{R}, \mathbb{C}) \subset AP(\mathbb{R}, \mathbb{C}) \subset AP_2(\mathbb{R}, \mathbb{C}). \tag{8}$$

From (6), (7) and (8), one derives the following problem: Is there a $p < 2$, such that

$$AP(\mathbb{R}, \mathbb{C}) \subset AP_p(\mathbb{R}, \mathbb{C})? \tag{9}$$

Besides the spaces $AP_p(\mathbb{R}, \mathbb{C})$, $1 < p < 2$, and the space $AP(\mathbb{R}, \mathbb{C})$ of Bohr's almost periodic functions, there are other spaces of almost periodic functions (see [5], for instance) whose relationships with the spaces considered above is known. Let us mention the space $S_2(\mathbb{R}, \mathbb{C})$ of almost periodic functions in Stepanov's sense, for which

$$AP(\mathbb{R}, \mathbb{C}) \subset S_2(\mathbb{R}, \mathbb{C}) \subset AP_2(\mathbb{R}, \mathbb{C}). \tag{10}$$

Of course, it would be interesting to clarify the relationships between $S_2(\mathbb{R}, \mathbb{C})$, or Stepanov's spaces with index p, and the spaces $AP_p(\mathbb{R}, \mathbb{C})$.

3. EXISTENCE IN $AP_1(\mathbb{R},\mathbb{R}^n)$ OR $AP_1(\mathbb{R}, \mathbb{C}^n)$

We shall state now an existence result for solutions with absolutely convergent Fourier series which is similar to the result given in Theorem 1 in [8]. Let us first define the Fourier - Stieltjes transform of the matrix kernel A(t) occuring in equation (1):

$$\tilde{A}(is) = \int_{-\infty}^{\infty} \exp(-ist)dA(t), \quad s \in \mathbb{R}, \tag{11}$$

a formula which makes sense when the entries of A(t) are (real or complex) functions with bounded variation on the whole real axis.

THEOREM 1. Assume that the following conditions hold true for the integro-differential equation (1):

(a) The characteristic equation does not have roots on the imaginary axis, i.e.,

$$\det[isI - \tilde{A}(is)] \neq 0, \quad s \in \mathbb{R}; \tag{12}$$

(b) $f \in AP_1(\mathbb{R}, \mathbb{C}^n)$, or $f \in AP(\mathbb{R}, \mathbb{C}^n)$ and its Fourier exponents are such that

$$\sum_{k=1}^{\infty} (\lambda_k)^{-2} < \infty. \tag{13}$$

Then there exists a unique solution of (1) such that $x \in AP_1(\mathbb{R}, \mathbb{C}^n)$.

Since the proof of Theorem 1 does not differ essentially from the proof provided in [8] for the similar result, we shall not include it here. We only point out the fact that

$$\|[isI - \tilde{A}(is)]^{-1}\| = O(|s|^{-1}) \quad \text{as} \quad |s| \to \infty, \tag{14}$$

from which one can easily derive the existence of a constant $K > 0$, depending only on $A(t)$, such that

$$\|[isI - \tilde{A}(is)]^{-1}\| \leq K, \quad s \in \mathbb{R}. \tag{15}$$

Let us point out the fact that results of the same type as Theorem 1 above can be obtained for spaces like $AP_p(\mathbb{R}, \mathbb{C}^n)$, $1 < p \leq 2$. The case $p = 2$ has been considered in [8]. In such cases, one has to impose some restrictions on the solution being sought because the Stieltjes integral occurring in equation (1) does not necessarily have a meaning for any function in the space $AP_p(\mathbb{R}, \mathbb{C}^n)$, $1 < p$.

4. EXISTENCE RESULTS FOR THE EQUATION (2)

Let us consider now equation (2) under assumptions (3). The numbers t_j are arbitrary real numbers.

It can easily be seen that the Fourier - Stieltjes transform becomes

$$\tilde{A}(is) = \sum_{j=0}^{\infty} A_j \exp(-ist_j) + \int_{-\infty}^{\infty} B(t)\exp(-ist)dt. \tag{16}$$

The series in the right hand side of (16) is obviously the Fourier series of an almost periodic (matrix-valued) function which converges absolutely and uniformly. The second term in the right hand side of (16) is a continuous function which tends to zero at infinity. Hence, the Fourier - Stieltjes transform of the kernel is an asymptotically almost periodic function. This fact has been thoroughly exploited in the papers [16], [17], [18], and [21].

THEOREM 2. Consider the integrodifferential equation (2), under assumptions (3), and further assume that the characteristic equation does not have roots on the imaginary axis (i.e., (12) holds true with $\tilde{A}(is)$ given by (16)). If f(t) is in $AP(\mathbb{R}, \mathbb{C}^n)$, or in $S_p(\mathbb{R}, \mathbb{C}^n)$ with $p \geq 1$, then there exists a unique almost periodic solution of equation (2) belonging to the same space as f(t).

Proof. The method of proof is the same used in the paper [9], where equations of the form Ax = f have been considered, the operator A being the difference-integral operator occurring in the right hand side of (2). More specifically, it will be shown that the unique solution x of (2) can be represented by means of a difference-integral operator of the same form as the one appearing in the equation (2). The method relies on the same classical result due to Wiener and Pitt, but it is somewhat more difficult than in the case of integral equations discussed in [9].

It is known that matrix-valued functions of the form (16) with A_j, j = 0,1,2,..., B(t) satisfying (3), can be organized as an algebra. The invertibility condition in this algebra is

$$\inf|\det \tilde{A}(is)| > 0, \quad s \in \mathbb{R}. \tag{17}$$

Let us now take (formally) the Fourier - Stieltjes transform of both sides of equation (2). One obtains easily

$$\tilde{x}(is) = [isI - \tilde{A}(is)]^{-1}\tilde{f}(is) \tag{18}$$

because of condition (12). If we can show that $[isI - \tilde{A}(is)]^{-1}$ is in the above mentioned algebra, then the proof will follow immediately from rather simple considerations. Let us remark that

$$[isI - \tilde{A}(is)]^{-1} = [I - \frac{I}{is+1} - \frac{\tilde{A}(is)}{is+1}]^{-1} \cdot \frac{I}{is+1}, \quad s \in \mathbb{R}. \tag{19}$$

Since the matrix $\frac{I}{is+1}$ is the Fourier transform of the matrix valued function equal to $I \cdot \exp(-t)$ on the positive half-axis and equal to the zero matrix on the negative half-axis, there results that the function matrix

$$F(is) = I - \frac{I}{is+1} - \frac{\tilde{A}(is)}{is+1}, \quad s \in \mathbb{R}, \tag{20}$$

belongs to the algebra described above. To be sure that its inverse is also in the same algebra (we know that this inverse exists pointwise), we need to check a condition of the form (17). Because $\tilde{A}(is)$ is bounded on the real axis, one obtains from (20) that $F(is) \to I$ as $|s| \to \infty$. Therefore, one can find a positive number S, such that

$$|\det F(is)| > \frac{1}{2}, \quad \text{for } |s| > S. \tag{21}$$

On the other hand, taking into account (12) and (19), one obtains a positive number m such that

$$|\det F(is)| > m, \quad \text{for } |s| \le S. \tag{22}$$

From the inequalities (21) and (22) one derives

$$|\det F(is)| > \min \{\frac{1}{2}, m\}, \quad s \in \mathbb{R}, \tag{23}$$

which proves that the element F(is) is invertible in the algebra of those matrix functions which are representable in the form (16), with A_j and B(t) satisfying (3).

It has been shown in [9] that the operator A, formally given by

$$(Ax)(t) = \sum_{j=0}^{\infty} A_j x(t-t_j) + \int_{-\infty}^{\infty} B(s)x(t-s)ds, \quad t \in \mathbb{R}, \tag{24}$$

under conditions (3), has most usual function spaces as invariant spaces. On each such space, the invertibility of the operator A is an equivalent problem to the invertibility of its "symbol", i.e., of the function matrix $\tilde{A}(is)$ defined by formula (16).

Under the assumptions of Theorem 2, the matrix function $[isI - \tilde{A}(is)]^{-1}$ is in the algebra described above, and therefore, any time f(t) belongs to an invariant space with respect to the operators of the form (24), the solution of equation (2) will be given by a formula of the form

$$x(t) = \sum_{j=0}^{\infty} \overline{A}_j f(t-\overline{t}_j) + \int_{-\infty}^{\infty} \overline{B}(s)f(t-s)ds, \quad s \in \mathbb{R}, \tag{25}$$

where $\bar{A}_j$, $j = 0,1,2,\ldots$, and $\bar{B}(t)$ satisfy conditions similar to (3). To be more specific, we should represent the operators of the form (24) as convolution products of a distribution by a function, as done in [9]. In particular, one can easily check that the spaces $AP(\mathbb{R}, \mathbb{C}^n)$, and $S_p(\mathbb{R}, \mathbb{C}^n)$, $p \geq 1$, are invariant spaces for the operators of the form (24) under condition (3).

The discussion above ends the proof of Theorem 2.

REMARK 1. There are many more function spaces which are left invariant by the operators of the form (24) under conditions (3). In other words, the condition of absence of purely imaginary characteristic roots enables us to conclude that if f(t) in (2) belongs to one of those spaces, there exists a unique solution which also belongs to that space. As further examples of invariant spaces for operators of the form (24), under condition (3), one can mention the following (see [9] for more details): The L_p - spaces $L_p(\mathbb{R}, \mathbb{C}^n)$, the space $M(\mathbb{R}, \mathbb{C}^n)$ of locally integrable vector-valued functions which are bounded in the mean (this space contains all L_p-spaces, $p \geq 1$), the space of continuous and bounded maps from $\mathbb{R}$ into $\mathbb{C}^n$ or $\mathbb{R}^n$, the space of locally integrable functions which are periodic of period ω, say $A_\omega(\mathbb{R}, \mathbb{C}^n)$, and other spaces.

REMARK 2. If, instead of equation (2), one deals with the special case which corresponds also to (0), namely

$$\dot{x}(t) = \sum_{j=0}^{\infty} A_j x(t-t_j) + \int_0^{\infty} B(s)x(t-s)ds + f(t), \tag{26}$$

where $t_j \geq 0$, $j = 0,1,2,3,\ldots$, and A_j, B satisfy conditions (3), then the operator occurring in the right hand side of (25) is a nonanticipative operator. This feature is very helpful in dealing with continuation problems (in the future) for infinite delay systems.

REMARK 3. If equation (1) is considered, then one additional hypothesis is necessary besides the absence of purely imaginary roots for the characteristic equation. (This remark refers to Bohr's almost periodic functions only.) Such a result can be found in [21]. The extra condition imposed by the author is the Tauber type condition $\lim[x(t+s)-x(t)] = 0$, as $t \to \infty$, $s \to 0$. Actually, the spectral condition can be considerably relaxed if the Tauber type condition is imposed.

REMARK 4. It is possible to obtain existence and uniqueness of the almost periodic solution for a nonlinear perturbation of the equation (2). Indeed, substituting an operator $(fx)(t)$ for the forcing term $f(t)$ in equation (2), one easily obtains the result by fixed-point techniques. Under more restrictive conditions, such results have been obtained in the author's paper [7].

5. FURTHER EXISTENCE RESULTS

Though we do not intend to consider in detail in this paper the case when the characteristic equation has purely imaginary roots, we would like to give a sample result. It is concerned with the existence of almost periodic solutions to equation (1), and it can be viewed as a generalization of Theorem 1 in this paper.

Instead of (12), which excludes the presence of purely imaginary characteristic roots, we will assume that the characteristic equation

$$\det[isI - \tilde{A}(is)] = 0, \tag{27}$$

has only a finite set $S = \{is_1, is_2, \ldots, is_p\}$ of roots on the imaginary axis. In general, equation (27) might have a countable set of roots on the imaginary axis.

THEOREM 3. Consider equation (1) under the above stated hypothesis regarding the characteristic roots, and let $f(t) \in AP_1(\mathbb{R}, \mathbb{C}^n)$,

$$f(t) \sim \sum_{k=0}^{\infty} b_k \exp(i\lambda_k t), \tag{28}$$

be such that $\Lambda = \{\lambda_k | k = 0,1,2,\ldots\}$ satisfies

$$\text{dist}(\Lambda, S) > 0. \tag{29}$$

Then, there exists a unique solution $x(t) \in AP_1(\mathbb{R}, \mathbb{C}^n)$ of equation (1), and

$$x(t) \sim \sum_{k=0}^{\infty} [i\lambda_k I - \tilde{A}(i\lambda_k)]^{-1} b_k \exp(i\lambda_k t). \tag{30}$$

The same conclusion holds true in case $f(t) \in AP(\mathbb{R}, \mathbb{C}^n)$, and the Fourier exponents of $f(t)$ satisfy condition (13).

Proof. The formal Fourier series of the solution is easily obtainable (see [8] for details). In order to prove that $x(t) \in AP_1(\mathbb{R}, \mathbb{C}^n)$, it suffices to show that

$$\| [i\lambda_k I - \tilde{A}(i\lambda_k)]^{-1} \| \leq M, \quad k = 0,1,2,\ldots, \tag{31}$$

where M is a positive constant. In order to obtain (31), let us remark that the entries of the inverse matrix $(isI - \tilde{A}(is))^{-1}$, $s \notin S$, are rational functions of s, with bounded coefficients on the entire real axis. Moreover, the degree of the denominator is n, while the degree of the numerator is at most n - 1. Since the polynomial in the denominator vanishes only for $s \in S$, (31) follows immediately from (29). When $f(t) \in AP(\mathbb{R}, \mathbb{C}^n)$, and the Fourier exponents satisfy (13), it suffices to point out that an estimate of the form (14) holds true. The proof then proceeds as in case of Theorem 1 from [8].

REMARK. The hypothesis regarding the existence of purely imaginary roots to the characteristic equation could be relaxed such that the case of a countable set of roots be treated.

Theorem 3 is analogous to a result due to W. A. Coppel [4]. See also [13].

REFERENCES

1. V. Alexiades, Almost periodic solutions of an integrodifferential system with infinite delay, *Nonlinear Analysis - TMA* 5 (1981), 401-421.

2. A. S. Besicovitch, *Almost Periodic Functions*, Dover, New York, 1954.

3. S. Bochner, Über gewisse Differential-und allgemeinere Gleichungen deren Lösungen fastperiodisch sind, I, II, III, *Math. Annalen* 102 (1929), 489-504; 103 (1929), 588-597; 104 (1931), 579-587.

4. W. A. Coppel, Almost periodic properties of ordinary differential equations, *Ann. Mat. Pura Appl.* 76 (1967), 27-50.

5. C. Corduneanu, *Almost Periodic Functions*, John Wiley & Sons, New York, 1968.

6. C. Corduneanu, Periodic and almost periodic solutions of some convolution equations, *Trudy (Proceedings) Fifth Int. Symp. Nonlinear Oscillations*, Vol. I, 311-321, Kiev, 1970.

7. C. Corduneanu, Recent contributions to the theory of differential systems with infinite delay, Rapport No. 95, Institut Mathématique Université de Louvain, Vander, Louvain, 1976.

8. C. Corduneanu, Almost periodic solutions for infinite delay systems, *Spectral Theory of Differential Operators*, North Holland Publ. Co., Amsterdam, 1981, 99-106.

9. C. Corduneanu and S. I. Grossman, On Wiener-Hopf equations, *Revue Roumaine Math. Pures Appl.* 18 (1973), 1547-1554.

10. C. Corduneanu and V. Lakshmikantham, Equations with infinite delay: a survey, *Nonlinear Analysis - TMA* 4 (1980), 831-877.

11. J. M. Cushing, *Integro-differential Equations and Delay Models in Population Dynamics*, Lecture Notes in Biomathematics No. 20, Springer-Verlag, Berlin, 1977.

12. R. Doss, On the almost periodic solutions of a class of integro-differential-difference equations, *Annals of Math.* (2), 81 (1965). 117-123.

13. A. M. Fink, *Almost Periodic Differential Equations*, Springer-Verlag, Berlin, 1974.

14. Y. Hino, Almost periodic solutions of functional differential equations with infinite retardation, *Funkcialaj Ekvacioj* 21 (1978), 139-150.

15. Y. Hino, Almost periodic solutions of functional differential equations with infinite retardation, II, *Tôhoku Math. J.* 32 (1980), 525-530.

16. G. S. Jordan and R. L. Wheeler, Linear integral equations with asymptotically almost periodic solutions, *J. Math. Anal. Appl.* 52 (1975), 454-464.

17. G. S. Jordan, W. R. Madych and R. L. Wheeler, Linear convolution integral equations with asymptotically almost periodic solutions, *Proc. Amer. Math. Soc.* 78 (1980), 337-341.

18. G. S. Jordan, O. J. Staffans and R. L. Wheeler, Local analyticity in weighted L^1-spaces and applications to stability problems for Volterra equations, *Trans. Amer. Math. Soc.*, to appear.

19. J. J. Levin and D. F. Shea, On the asymptotic behavior of the bounded solutions of some integral equations, I, II, III. *J. Math. Anal. Appl.* 37 (1972), 42-82, 288-326, 537-575.

20. K. Sawano, Exponential asymptotic stability for functional differential equations with infinite retardations, *Tôhoku Math. J.* 31 (1979), 363-382.

21. O. J. Staffans, On asymptotically almost periodic solutions of a convolution equation, *Trans. Amer. Math. Soc.* 266 (1981), 603-616.

22. S. Zaidman, Solutions presque-périodiques des équations différentielles abstraites, *Ens. Mathématique* 24 (1978), 87-110.

WEIGHTED L^1-SPACES AND RESOLVENTS OF VOLTERRA EQUATIONS

G. Samuel Jordan

Department of Mathematics
University of Tennessee
Knoxville, Tennessee

1. INTRODUCTION

The resolvent kernels r_1 and r_2 associated with the scalar Volterra integral and integrodifferential equations

$$x(t) + \int_0^t x(t-s)a(s)ds = f(t), \quad t \in \mathbb{R}^+ \equiv [0,\infty), \tag{1.1}$$

$$x'(t) + \int_0^t x(t-s)d\mu(s) = f(t), \; t \in \mathbb{R}^+, \quad x(0) = x_0, \tag{1.2}$$

are defined by

$$r_1(t) + \int_0^t r_1(t-s)a(s)ds = a(t), \quad t \in \mathbb{R}^+, \tag{1.3}$$

$$r_2'(t) + \int_0^t r_2(t-s)d\mu(s) = 0, \quad t \in \mathbb{R}^+, \quad r_2(0) = 1, \tag{1.4}$$

respectively. Under mild assumptions, the equations (1.1) and (1.2) are solved by

$$x(t) = f(t) - \int_0^t f(t-s)r_1(s)ds, \quad t \in \mathbb{R}^+, \tag{1.5}$$

$$x(t) = x_0 r_2(t) + \int_0^t f(t-s)r_2(s)ds, \quad t \in \mathbb{R}^+, \tag{1.6}$$

respectively. Moreover, resolvents also occur in variation of constants formulas associated with certain nonlinear perturbed forms of (1.5) and (1.6) (see [9, Chapter 4] and [5]).

Here we are primarily concerned with finding conditions on a or μ which imply that r_1 or r_2 belongs to a weighted L^1-space on $\mathbb{R}^+$. It is clear from the representations (1.5) and (1.6) that such conclusions about r_1 and r_2 are useful in analyzing the behavior of solutions of (1.1) and (1.2). The results to be described, as well as many additional results, proofs and further discussion, may be found in the paper [8] coauthored with Olof J. Staffans and Robert L. Wheeler. Most of the theory in [8] applies equally well to Fredholm equations of convolution type; the only difference is that one replaces a weighted L^1-space on $\mathbb{R}^+$ with one on the whole line $\mathbb{R} \equiv (-\infty,\infty)$. For the most part, our definitions and results are phrased so that they include both cases.

2. WEIGHTED L^1-SPACES AND LOCALLY ANALYTIC FUNCTIONS

A weight function on $\mathbb{R}$ is any positive continuous function $\rho(t)$ such that

$$\rho(s+t) \le \rho(s)\rho(t) \tag{2.1}$$

for $s, t \in \mathbb{R}$. For example, we have

$$\rho_1(t) = (1 + |t|)^{\delta}, \quad t \in \mathbb{R}, \quad \delta \ge 0,$$

$$\rho_2(t) = (1 + \log(1+|t|))^{\gamma}\rho_1(t), \quad t \in \mathbb{R}, \quad \gamma \ge 0,$$

$$\rho_3(t) = \exp(|t|^{\alpha})\rho_2(t), \quad t \in \mathbb{R}, \quad 0 \le \alpha < 1,$$

$$\rho_4(t) = \exp(\beta t)\rho_3(t), \quad t \in \mathbb{R}, \quad \beta \in \mathbb{R}.$$

The weighted space $L^1(\mathbb{R}; \rho)$ consists of all complex measurable functions x on $\mathbb{R}$ for which

$$\int |x(t)|\rho(t)dt < \infty, \tag{2.2}$$

where the integration is over $\mathbb{R}$. The convolution of $x, y \in L^1(\mathbb{R}; \rho)$ is defined by

$$x * y(t) = \int x(t-s)y(s)ds, \tag{2.3}$$

where again the integration is over $\mathbb{R}$. With convolution as multiplication, $L^1(\mathbb{R}; \rho)$ is a Banach algebra. We let $V(\mathbb{R}; \rho)$ denote the Banach algebra one obtains from $L^1(\mathbb{R}; \rho)$ by adjoining a unit.

Define

$$\rho_* = - \inf_{t>0} \frac{\log \rho(t)}{t} = - \lim_{t\to\infty} \frac{\log \rho(t)}{t},$$

$$\rho^* = - \sup_{t<0} \frac{\log \rho(t)}{t} = - \lim_{t\to-\infty} \frac{\log \rho(t)}{t}.$$

Then $-\infty < \rho_* \le \rho^* < \infty$ and the maximal ideal space of $V(\mathbb{R}; \rho)$ can be identified with $\overline{\pi} = \pi \cup \{\infty\}$, where

$$\pi = \{z \in \mathbb{C} \mid \rho_* \le \mathrm{Re}(z) \le \rho^*\}$$

with the usual topology of the compactified plane [1, pp. 100, 113-115].

We shall also need the space $M(\mathbb{R}; \rho)$ of locally finite Borel measures μ on $\mathbb{R}$ satisfying

$$\int \rho(t)d|\mu|(t) < \infty, \tag{2.4}$$

where $|\mu|$ is the total variation measure of μ and the integration is over $\mathbb{R}$. If $a \in L^1(\mathbb{R}; \rho)$ and $\mu \in M(\mathbb{R}; \rho)$, then their (bilateral) Laplace transforms

$$\hat{a}(z) = \int e^{-zt}a(t)dt, \quad \hat{\mu}(z) = \int e^{-zt}d\mu(t) \tag{2.5}$$

converge absolutely for $\rho_* \le \mathrm{Re}(z) \le \rho^*$; of course, the integrations are over $\mathbb{R}$. Also, the convolution of a and μ, defined by

$$\mu*a(t) = a*\mu(t) = \int a(t-s)d\mu(s), \tag{2.6}$$

belongs to $L^1(\mathbb{R}; \rho)$.

Analogous definitions and observations may be made for the half-line $\mathbb{R}^+$. The inequality (2.1) must hold for s, t ε $\mathbb{R}^+$, the integrations in (2.2), (2.4), and (2.5) are now over $\mathbb{R}^+$, and those in (2.3) and (2.6) are over [0,t]. For weights on $\mathbb{R}^+$, only ρ_* is meaningful and it satisfies $-\infty < \rho_* \leq \infty$; to avoid trivialities, we assume $\rho_* < \infty$. Then the maximal ideal space of $V(\mathbb{R}^+;\rho)$ may be identified with $\overline{\pi} = \pi \cup \{\infty\}$ where now

$$\pi = \{z \in \mathbb{C} \mid \operatorname{Re}(z) \geq \rho_*\}.$$

We use the simpler notation $L^1(\rho)$, $V(\rho)$, and $M(\rho)$ for $L^1(X;\rho)$, $V(X;\rho)$, and $M(X;\rho)$ with $X = \mathbb{R}$ or $\mathbb{R}^+$ if our statements hold for both cases or if the appropriate space is clear from the context.

DEFINITION 1. We call a function ϕ locally analytic at a point $z_0 \in \pi$ if ϕ is defined in a neighborhood of z_0 and there exist measures $\mu_1,\ldots,\mu_k$ in $M(\rho)$ and a function $\psi(z, \xi_1,\ldots,\xi_k)$ analytic at $(z_0,\hat{\mu}_1(z_0),\ldots,\hat{\mu}_k(z_0))$ such that

$$\phi(z) = \psi(z,\hat{\mu}_1(z),\ldots,\hat{\mu}_k(z))$$

in a neighborhood of z_0. We say that ϕ is locally analytic at infinity if ϕ is defined in a neighborhood of infinity and there exist functions $a_1,\ldots,a_n$ in $L^1(\rho)$, measures $\mu_1,\ldots,\mu_k$ in $M(\rho)$, and a function $\psi(z,\eta_1,\ldots,\eta_n, \xi_1,\ldots,\xi_k)$ analytic at $(0,0,\ldots,0)$ such that

$$\phi(z) = \psi(z^{-1},\hat{a}_1(z),\ldots,\hat{a}_n(z),\hat{\mu}_1(z)/z,\ldots,\hat{\mu}_k(z)/z)$$

in a neighborhood of infinity. Finally, we say that ϕ is locally analytic if it is locally analytic at each point of $\overline{\pi}$.

Here "neighborhood" means an open subset of $\overline{\pi}$. Also, the numbers k and n and the function ψ may vary from point to point and, in fact, the sets $\{a_1,\ldots,a_n\}$ and $\{\mu_1,\ldots,\mu_k\}$ may be empty. Finally, since $L^1(\rho) \subset M(\rho)$, a measure μ_j may be replaced by a function a_j in the representations, but not conversely.

From Definition 1 and the fact that Laplace transforms are analytic in the interior of their domains of convergence, one sees easily that every locally analytic function is analytic in the interior of π and, conversely, every function analytic on $\overline{\pi}$ is locally analytic there.

Our basic result for locally analytic functions is

THEOREM 1. If ϕ is locally analytic, then ϕ is the Laplace transform of an element of $V(\rho)$. If also $\phi(\infty) = 0$, then ϕ is the Laplace transform of an element of $L^1(\rho)$.

This result is obtained by showing that ϕ is locally analytic in the sense of [1, p. 82] and then applying Theorem 1 in [1, p. 82]. The definition of local analyticity in [1] does not permit explicit dependence of the analytic function ψ on z and does not permit transforms of measures in the representations for ϕ.

As a first indication of how Theorem 1 can be used, we observe that it yields immediately an extension of the Wiener-Lévy Theorem due to D. F. Shea and S. Wainger [11, Theorem 2, Condition (10)]. They show that a function of the form $\phi(z) \equiv \psi(z,\hat{a}(z))$ is the Laplace transform of a function $r \in L^1(\mathbb{R}^+)$; here $a(t) = b + \beta(t)$ with b constant, $\beta \in L^1(\mathbb{R}^+)$ and ψ is analytic on $\{(z,\hat{a}(z)) | \mathrm{Re}(z) \geq 0\}$ (including $(0,\infty)$ if $b \neq 0$) and at $(\infty,0)$ with $\phi(\infty) = 0$. To obtain this result from Theorem 1, simply note that $\tilde{\psi}(z,w) \equiv \psi(z,b/z + w)$ is analytic on $\{(z,\hat{\beta}(z)) | \mathrm{Re}(z) \geq 0\}$ as well as at $(\infty,0)$, so that $\phi(z)$ is locally analytic. Since

$$\hat{r}_1(z) = \hat{a}(z)\ (1 + \hat{a}(z))^{-1}, \quad \hat{r}_2(z) = (z + \hat{\mu}(z))^{-1},$$

it is clear that taking $\psi(z,w) = \psi_1(z,w) \equiv w(1 + w)^{-1}$ or $\psi(z,w) = \psi_2(z,w) \equiv (z + w)^{-1}$ yields the integrability of the resolvent r_1 or r_2 associated with a function $a(t)$ of the above form or with a measure μ of the form $d\mu(t) = a(t)dt$ with $a(t)$ as above.

Before describing some new results in the next section, we remark that in [8] the authors define and discuss several additional concepts including locally analytic zeros and poles, smoothness of positive order of locally analytic functions, and the order of dependence of an analytic function $\psi(z, \xi_1,\ldots,\xi_k)$ on z with respect to one of its other variables. They then make several applications of Theorem 1 together with these concepts in order to obtain some new results about resolvents and to unify and extend a number of known results.

3. EXTENDED LOCALLY ANALYTIC FUNCTIONS

A classical result of Paley and Wiener [10] asserts that when $a \in L^1(\mathbb{R}^+)$, then $r_1 \in L^1(\mathbb{R}^+)$ if, and only if, $1 + \hat{a}(z) \neq 0$, $\mathrm{Re}(z) \geq 0$. When $\mu \in M(\mathbb{R}^+)$, a result of S. I. Grossman and R. K. Miller [6] asserts that $r_2 \in L^1(\mathbb{R}^+)$ if, and only if, $1 + \hat{\mu}(z) \neq 0$, $\mathrm{Re}(z) \geq 0$. Of course, if $r_2 \in L^1(\mathbb{R}^+)$, then it follows from (1.4) that also $r_2' \in L^1(\mathbb{R}^+)$. Analogues of these results for weighted spaces are also valid (see [1, p. 116] and [11, Theorem 3]).

We are interested in obtaining necessary and sufficient conditions for r_1 or r_2 to belong to $L^1(\mathbb{R}^+; \rho)$ when a or μ does *not* belong to $L^1(\mathbb{R}^+; \rho)$ or $M(\mathbb{R}^+; \rho)$, respectively. In our general setting, this involves allowing some of the functions and measures whose transforms occur in the representations for a function $\phi(z)$ to lie outside $L^1(\rho)$ and $M(\rho)$. This, in turn, means that transforms need not be finite at all points of $\bar{\pi}$. We use the standard arithmetical conventions: $1/\infty = 0$, $1/0 = \infty$, $a + \infty = \infty$ $(a \neq \infty)$, $a \times \infty = \infty$ $(a \neq 0)$. Also, we say that $\psi(\xi_1,\ldots,\xi_k)$ is analytic at $(\eta_1,\ldots,\eta_k)$ where some of the components η_j may be infinite provided $\psi(\xi_1,\ldots,\xi_k) = \tilde{\psi}(\xi_1^{\pm 1},\ldots,\xi_k^{\pm 1})$ in a neighborhood of $(\eta_1,\ldots,\eta_k)$ where $\tilde{\psi}$ is analytic at $(\eta_1^{\pm 1},\ldots,\eta_k^{\pm 1})$. Here -1 is used for precisely those components for which $\eta_j = \infty$.

DEFINITION 2. A function θ defined on $\bar{\pi}$ and possibly assuming the value infinity is extended locally analytic at $z_0 \in \bar{\pi}$ if θ is locally analytic at z_0 when $\theta(z_0) \neq \infty$ and $1/\theta$ is locally analytic at z_0 when $\theta(z_0) = \infty$. We call θ extended locally analytic when it is extended locally analytic at each point of $\bar{\pi}$.

Of course, $\theta(z) = z$ is a simple example of an extended locally analytic function. A sufficient condition for a function ϕ to be extended locally analytic is given in

THEOREM 2. Let ϕ be defined on $\bar{\pi}$ possibly assuming the value infinity, and suppose that at each point $z_0 \in \bar{\pi}$ there exist extended locally analytic functions $\theta_1,\ldots,\theta_k$ and a function $\psi(\xi_1,\ldots,\xi_k)$ analytic at $(\theta_1(z_0),\ldots,\theta_k(z_0))$ so that in a neighborhood of z_0

$$\phi(z) = \psi(\theta_1(z),\ldots,\theta_k(z)) \quad \text{if} \quad \phi(z_0) \neq \infty,$$

$$1/\phi(z) = \psi(\theta_1(z),\ldots,\theta_k(z)) \quad \text{if} \quad \phi(z_0) = \infty.$$

Then ϕ is extended locally analytic. If also ϕ is complex-valued, then ϕ is locally analytic on $\overline{\pi}$ and ϕ is the transform of an element of $V(\rho)$. If, in addition, $\phi(\infty) = 0$, then ϕ is the transform of a function in $L^1(\rho)$.

Again, the functions ψ and θ_j and the number k may vary from point to point.

For the resolvent r_1 associated with (1.1) we have

THEOREM 3. Let ρ be a weight on $\mathbb{R}^+$ and let a be locally integrable on $\mathbb{R}^+$, $e^{-st}a(t) \in L^1(\mathbb{R}^+)$ $(s > \rho_*)$, $\hat{a}(\rho_* + iy) \equiv \lim_{x \to \rho_*^+} \hat{a}(x + iy)$ exist in the extended complex plane for each $y \in \mathbb{R}$, $\hat{a}(z) \to 0$ as $z \to \infty$ in π, and $1 + \hat{a}(z) \neq 0$ $(z \in \pi)$. Then $r_1 \in L^1(\mathbb{R}^+; \rho)$ if, and only if, $\hat{a}(z)$ is extended locally analytic on $\overline{\pi}$.

Proof. Let $\phi_1(z)$ be the Laplace transform of r_1; that is, let $\phi_1(z) = \psi(\hat{a}(z)) = \hat{a}(z)(1 + \hat{a}(z))^{-1}$. If $\hat{a}(z)$ is extended locally analytic on $\overline{\pi}$, then by Theorem 2 $\phi_1(z)$ is extended locally analytic. In fact, since $\phi(z)$ is complex-valued and $\phi(\infty) = 0$, $\phi(z)$ is the transform of a function in $L^1(\mathbb{R}^+; \rho)$ and, hence, $r_1 \in L^1(\mathbb{R}^+; \rho)$ by the uniqueness of Laplace transforms. Conversely, if $r_1 \in L^1(\mathbb{R}^+; \rho)$, then $\phi_1(z_0) = 1$ for $z_0 \in \pi$ if, and only if, $\hat{a}(z_0) = \infty$. Thus, in a neighborhood of a point z_0 we may write $\hat{a}(z) = \phi_1(z)(1 - \phi_1(z))^{-1}$ if $\hat{a}(z_0) \neq \infty$, or $1/\hat{a}(z) = (1 - \phi_1(z))/\phi_1(z)$ if $\hat{a}(z_0) = \infty$, and then conclude by Theorem 2 that $\hat{a}(z)$ is extended locally analytic.

The result for the resolvent r_2 associated with (1.2) is somewhat different. Let μ be a locally finite Borel measure on $\mathbb{R}^+$, $e^{-st}d\mu(t) \in M(\mathbb{R}^+)$ $(s > \rho_*)$, $\hat{\mu}(\rho_* + iy) \equiv \lim_{x \to \rho_*^+} \hat{\mu}(x + iy)$ exist in the extended complex plane for each $y \in \mathbb{R}$, $\hat{\mu}(z) \to \hat{\mu}(\infty)$ (finite) as $z \to \infty$ in π, and $z + \hat{\mu}(z) \neq 0$ $(z \in \pi)$. It is true that if $\hat{\mu}(z)$ is extended locally analytic on $\overline{\pi}$, then by Theorem 2 $\phi_2(z) \equiv (z + \hat{\mu}(z))^{-1} = \hat{r}_2(z)$ and $\phi_3(z) \equiv -\hat{\mu}(z)(z + \hat{\mu}(z))^{-1} = \hat{r}_2'(z)$ are extended locally analytic and, in fact, are transforms of functions of $L^1(\mathbb{R}^+; \rho)$, so that $r_2, r_2' \in L^1(\mathbb{R}^+; \rho)$. On the other hand, if $r_2 \in L^1(\mathbb{R}^+; \rho)$, then $\hat{r}_2(z_0) = \phi_2(z_0) = 0$ for $z_0 \in \overline{\pi}$ if, and only if, $\hat{\mu}(z_0) = \infty$. Thus, in a neighborhood of a point $z_0 \in \pi$, we may write $\hat{\mu}(z) = (1 - z\hat{r}_2(z))/\hat{r}_2(z)$ if $\hat{\mu}(z_0) \neq \infty$ or $1/\hat{\mu}(z) = \hat{r}_2(z)(1 - z\hat{r}_2(z))^{-1}$ if $\hat{\mu}(z_0) = \infty$, and then conclude from Theorem 2 that $\hat{\mu}(z)$ is extended locally analytic on π. However, even if $r_2' \in L^1(\mathbb{R}^+; \rho)$, one cannot use Theorem 2 to deduce that $\hat{\mu}(z)$ is extended locally analytic at infinity. The condition $r_2' \in L^1(\mathbb{R}^+;\rho)$ does imply that $\phi_3(\infty) = 0$, from which it follows that $\hat{\mu}(z)/z = -\phi_3(z)(1+\phi_3(z))^{-1}$ is locally analytic at infinity. Moreover, if $\hat{\mu}(z)/z$ is locally analytic at infinity, then $\phi_2(z) = (1/z)(1+\hat{\mu}(z)/z)^{-1}$ and $\phi_3(z) = (\hat{\mu}(z)/z)(1+\hat{\mu}(z)/z)^{-1}$ also are locally analytic there. Thus, we have

THEOREM 4. Let μ be as above. Then $r_2, r_2' \in L^1(\mathbb{R}^+;\rho)$ if, and only if, $\hat{\mu}(z)$ is extended locally analytic on π and $\hat{\mu}(z)/z$ is locally analytic at infinity.

Clearly, Theorems 3 and 4 show that when using transform methods to determine whether r_1 or r_2 belongs to $L^1(\mathbb{R}^+;\rho)$, the extended local analyticity of $\hat{a}(z)$ or $\hat{\mu}(z)$ is the key property to be established if $a \notin L^1(\mathbb{R}^+;\rho)$ or $\mu \notin M(\mathbb{R}^+;\rho)$. Unfortunately, it may be difficult to determine whether a given function or measure has an extended locally analytic transform.

When $\rho(t) \equiv 1$, so that $L^1(\mathbb{R}^+;\rho) = L^1(\mathbb{R}^+)$ and $\pi = \{z \in \mathbb{C} | \operatorname{Re}(z) \geq 0\}$, there is the deep result of Shea and Wainger [11, Theorem 2, Condition (6)]. They assume that $a(t) = b(t) + \beta(t)$ where $b \in L^1_{loc}(\mathbb{R}^+)$, $b \notin L^1(\mathbb{R}^+)$ is nonnegative, nonincreasing and convex on $(0,\infty)$ and $\beta \in L^1(\mathbb{R}^+)$. They then consider functions $\phi(z) = \psi(z,\hat{a}(z))$, where ψ is analytic on $\{(z,\hat{a}(z)) | z \in \overline{\pi}\}$ with $\psi(\infty,0) = 0$, and they show that $\phi(z)$ is the transform of a function in $L^1(\mathbb{R}^+)$. Because $1 + \hat{b}(z) \neq 0$ $(z \in \pi)$, the special case of their result with $\psi(z,w) \equiv w(1+w)^{-1}$ and with $\beta(t) \equiv 0$ yields $r_1 \in L^1(\mathbb{R}^+)$ where $\hat{r}_1(z) = \hat{b}(z)(1+\hat{b}(z))^{-1}$. Thus, by Theorem 3 $\hat{b}(z)$ is extended locally analytic on $\overline{\pi}$ and, in particular, $1/\hat{b}(z)$ is locally analytic at $z = 0$. Now, of course, the general case of Shea and Wainger's result follows from Theorem 2 with $\tilde{\psi}(z,\hat{b}(z),\hat{\beta}(z)) \equiv \psi(z,\hat{b}(z) + \hat{\beta}(z))$.

Other conditions which imply that $1/\hat{b}(z)$ is locally analytic at $z = 0$ (with $\rho(t) \equiv 1$) when $b \notin L^1(\mathbb{R}^+)$ may be found in [3, Theorem 4].

Sharp results similar to that of Shea and Wainger are not available for weighted spaces at this time. Several recent papers [7], [12], [2] and [4] give $\mathcal{O}$ estimates for the rates of decay as $t \to \infty$ of the resolvents r_1 and r_2 of certain classes of nonintegrable kernels. These estimates can be used to show that the resolvents of such kernels belong to appropriate weighted L^1-spaces. For example, Gripenberg [4] shows that r_1 satisfies

$$r_1(t) = \mathcal{O}\left(t^{-2}\int_0^t \left[1 + \int_0^s a(u)\,du\right]^{-1} ds\right) \quad \text{as} \quad t \to \infty$$

whenever $a \in L^1_{loc}(\mathbb{R}^+)$, $a \notin L^1(\mathbb{R}^+)$ is nonnegative, nonincreasing and convex on $(0,\infty)$ with $-a'$ convex on $(0,\infty)$ and $a(t) \to 0$ as $t \to \infty$. When $a(t) = t^{m-1}$, $0 < m < 1$, this result gives the sharp estimate $r_1(t) = \mathcal{O}(t^{-(1+m)})$ as $t \to \infty$; actually $\lim_{t\to\infty} t^{1+m} r_1(t)$ exists and is nonzero. Thus, $r_1 \in L^1(\mathbb{R}^+;\rho)$ if, and only if, $\int_0^\infty (1+t)^{-(1+m)}\rho(t)dt < \infty$. On the other hand, Gripenberg's estimate becomes $r_1(t) = \mathcal{O}((t\log t)^{-1})$ as $t \to \infty$ when $a(t) = 1/(1+t)$, but this estimate is not sharp since one can show that $r_1(t) = \mathcal{O}(t^{-1}(\log t)^{-2})$ as $t \to \infty$ for this choice of the kernel $a(t)$.

ACKNOWLEDGEMENT

Professor Jordan's research was partially supported by National Science Foundation Grant MCS-7903358A01.

REFERENCES

1. I. M. Gelfand, D. A. Raikov and G. E. Shilov, *Commutative Normed Rings*, Chelsea, New York, 1964.

2. G. Gripenberg, On the asymptotic behavior of resolvents of Volterra equations, *SIAM J. Math. Anal.* 11 (1980), 654-682.

3. G. Gripenberg, Integrability of resolvents of systems of Volterra equations, *SIAM J. Math. Anal.* 12 (1981), 585-594.

4. G. Gripenberg, Decay estimates for resolvents of systems of Volterra equations, *J. Math. Anal. Appl.* 85 (1982), 473-487.

5. S. I. Grossman and R. K. Miller, Perturbation theory for Volterra integrodifferential systems, *J. Differential Equations* 8 (1970), 457-474.

6. S. I. Grossman and R. K. Miller, Nonlinear Volterra integrodifferential systems with L^1-kernels, *J. Differential Equations* 13 (1973), 551-566.

7. K. B. Hannsgen, A Volterra equation with completely monotonic convolution kernel, *J. Math. Anal. Appl.* 31 (1970), 459-471.

8. G. S. Jordan, O. J. Staffans and R. L. Wheeler, Local analyticity and applications to stability problems for Volterra equations in weighted L^1-spaces, *Trans. Amer. Math. Soc.*, to appear.

9. R. K. Miller, *Nonlinear Volterra Integral Equations*, W. A. Benjamin, Menlo Park, 1971.

10. R. E. A. C. Paley and N. Wiener, *Fourier Transforms in the Complex Domain*, Amer. Math. Soc. Colloq. Publ. 19, American Mathematical Society, Providence, 1934.

11. D. F. Shea and S. Wainger, Variants of the Wiener-Lévy theorem, with applications to stability problems for some Volterra integral equations, *Amer. J. Math.* 97 (1975), 312-343.

12. J. S. W. Wong and R. Wong, Asymptotic solutions of linear Volterra integral equations with singular kernels, *Trans. Amer. Math. Soc.* 189 (1974), 185-200.

A VOLTERRA EQUATION WITH A TIME DEPENDENT OPERATOR

T. R. Kiffe

Department of Mathematics
Texas A&M University
College Station, Texas

In this paper, we will be concerned with the existence and uniqueness of solutions of the Volterra integro-differential equation

$$\frac{\partial^2 u}{\partial t^2}(x,y,t) - \Delta u(x,y,t) - \int_0^t a(t-s)\Delta u(x,y,s)ds = f(x,y,t) \tag{1.1}$$

for x real, $y > 0$ and $t \geq 0$, with the conditions

$$\begin{aligned} &u(x,y,0) = u_0(x,y) \\ &\frac{\partial u}{\partial t}(x,y,0) = u_1(x,y) \\ &u(x,0,t) = 0 \quad \text{for} \quad x > p(t), \quad t > 0 \\ &\frac{\partial u}{\partial y}(x,0,t) + \int_0^t a(t-s)\frac{\partial u}{\partial y}(x,0,s)ds = g(x,t) \quad \text{for} \quad x \leq p(t), \end{aligned} \tag{1.2}$$

where a,f,g and p are known functions. We shall assume that p(t) is a positive, increasing, smooth function and that $a \in LAC[0,\infty)$ (i.e., locally absolutely continuous).

Equation (1.1) arises in the study of crack propagation in viscoelastic media [5]. In particular, for a body in anti-plane strain with a crack moving along the x-axis with speed $p'(t)$, the displacement $u(x,y,t)$ satisfies (1.1) and (1.2) with $g(x,t)$ representing the stress behind the crack tip.

We will consider (1.1) in the following setting. Let Ω denote the upper half plane, i.e., $\Omega = \{(x,y): y > 0\}$, and let $H = L^2(\Omega)$ and $W = H^1(\Omega)$, the usual Lebesgue and Sobolev spaces of real valued functions. Then W is a dense subspace of H and the injection from W into H is continuous. If we identify H with its dual H', we may write $W \subseteq H \subseteq W'$ with the understanding that if $w \in W$ and $h \in H$ then $\langle w, h\rangle = (w, h)$, where $\langle\cdot,\cdot\rangle$ is the inner product in H and $(\cdot,\cdot)$ is the pairing between W and W'.

We define a linear operator $A: W \to W'$ by

$$(v,Au) = \int\int_\Omega (\nabla v \cdot \nabla u)dxdy \quad \text{for all} \quad v, u \in W.$$

Then there are positive constants λ, α, K so that

$$(u, Au) + \lambda|u|_H^2 \geq \alpha\|u\|_W^2$$
$$\|Au\|_{W'} \leq K\|u\|_W. \tag{1.3}$$

If we assume that $g(\cdot,t) \in L^2(\mathbb{R}^1)$ for $t > 0$, we can define an element $f_1(t)$ of W' by

$$(v, f_1(t)) = \int_{-\infty}^{p(t)} v(x,0)g(x,t)dx \quad \text{for all} \quad v \in W, \tag{1.4}$$

since $v \in W$ implies $v(x,0) \in H^{1/2}(\mathbb{R}^1) \subseteq L^2(\mathbb{R}^1)$. If we assume further that $f(\cdot,\cdot,t) \in H$ for $t > 0$ and define an increasing family $\{V(t) \mid t \geq 0\}$ of closed subspaces of W by $V(t) = \{u \in W \mid u(x,0) = 0 \text{ for } x > p(t)\}$, then purely formal calculations show that (1.1) and (1.2) lead to the abstract equation

$$u''(t) + Au(t) + \int_0^t a(t-s)Au(s)ds = f(t) + f_1(t), \quad t > 0 \tag{1.5}$$

and

$$u(0) = u_0$$

$$u'(0) = u_1 \tag{1.6}$$

$$u(t) \in V(t) \quad \text{a.e.} \quad t \geq 0.$$

THEOREM. (Local Existence) Suppose $a \in LAC[0,\infty)$, $f \in L^2_{loc}[0,\infty; H]$, $f_1 \in W^{1,2}_{loc}[0,\infty; W']$, $u_0 \in V(0)$, $u_1 \in H$ and A satisfies (1.3). Then there exists a $T_0 > 0$ and a unique function $u \in L^2[0,T_0; W]$ with $u' \in L^2[0,T_0; H]$, $u'' \in L^2[0,T_0; V'(0)]$ satisfying (1.5) and (1.6) on $[0, T_0]$.

REMARK. $V'(0)$ is the dual space of $V(0)$ and all derivatives are to be taken in the sense of distributions. Since $u \in L^2[0,T_0; H]$ and $u' \in L^2[0,T_0; H]$, we have, by [1, Proposition A.7], that u is equal almost everywhere to a continuous function on H, so that $u(0) = u_0$ has a meaning. Since $u'' \in L^2[0,T_0;V'(0)]$, we have, by [4, p. 19], that u' is equal almost everywhere to a continuous function in $[H, V'(0)]_{1/2}$, an intermediate space of Lions; hence $u'(0) = u_1$ has a meaning.

Proof. Without loss of generality we may assume that $u_0 = 0$; for if not, simply replace $f_1(t)$ with

$$f_2(t) = f_1(t) - Au_0 - \int_0^t a(t-s)Au_0 ds.$$

Then f_2 satisfies the same conditions as f_1 and, if $\bar{u}(t)$ satisfies

$$\bar{u}''(t) + A\bar{u}(t) + \int_0^t a(t-s)A\bar{u}(s)ds = f_1(t) + f_2(t),$$

$\bar{u}(0) = 0$, $\bar{u}'(0) = u_1$, $\bar{u}(t) \in V(t)$, then $u(t) = \bar{u}(t) + u_0$ satisfies (1.5) and (1.6).

Our method of proof is an extension of the technique used by Dafermos [2]. For $T > 0$, consider the set

$$E_T = \{v \in C^\infty[0,T; W] \mid v(0) = 0, \quad v^{(i)}(t) \in V(t) \quad \text{for} \quad 0 \leq t \leq T, \quad i = 0,1,2,\dots\}$$

equipped with two inner products

$$[v,w]_1 = \int_0^T \{\langle v'(t), w'(t)\rangle_H + \langle v(t), w(t)\rangle_W\}dt + T\langle v'(0), w'(0)\rangle$$

and

$$[v,w]_2 = \int_0^T \{\langle v'(t), w'(t)\rangle_H + \langle v(t), w(t)\rangle_W\}dt,$$

which induce norms $\|\cdot\|_1$ and $\|\cdot\|_2$, respectively. By F_T, we denote the completion of E_T under the norm $\|\cdot\|_2$.

Multiply (1.5) by a test function of the form $(t-T)v'(t)$ and integrate over $[0,T]$. An integration by parts gives us

$$-\int_0^T \langle u'(t), v'(t)\rangle dt - \int_0^T (t-T)\{\langle u'(t), v''(t)\rangle$$

$$- (v'(t), Au(t)) - \int_0^t a(t-s)(v'(t), Au(s))ds\}dt \tag{1.7}$$

$$= \int_0^T (t-T)(f(t) + f_1(t), v'(t))dt - T\langle u_1, v'(0)\rangle.$$

Our first goal is to prove that there is a function u satisfying (1.7) and (1.6). To this end, choose a number $k > \sqrt{\lambda}$ (λ given by (1.3)) and define a bilinear form $B(w,z)$ on $F_T \times E_T$ by

$$B(w,z) = -\int_0^T \langle w'(t), z'(t)\rangle dt - k\int_0^T \langle w(t), z'(t)\rangle dt$$

$$- \int_0^T (t-T)\{\langle w'(t), z''(t)\rangle - k\langle w'(t), z'(t)\rangle$$

$$+ k\langle w(t), z''(t)\rangle - k^2\langle w(t), z'(t)\rangle - (Aw(t), z'(t))\}dt \tag{1.8}$$

$$+ \int_0^T (t-T) \int_0^t b(t-s)(Aw(s), z'(t))dsdt,$$

where $b(t) = a(t)e^{-kt}$. It is easy to see that, for each fixed $z \in E_T$, the form $w \to B(w,z)$ is continuous on F_T. Also, some simple estimates and a few integrations by parts show that there is a constant M which depends only on k and a so that

$$-B(z,z) \geq \frac{1}{2}\int_0^T |z'(t)|_H^2\,dt + \frac{1}{2}T|z'(0)|_H^2 + \frac{1}{2}\int_0^T \|z(t)\|_W^2\,dt - TM\int_0^T \|z(t)\|_W^2\,dt. \tag{1.9}$$

Thus, there exists a $T_0 > 0$, depending only on k and a, so that

$$-B(z,z) \geq \frac{1}{4}\|z\|_2^2 \quad \text{for all} \quad z \in E_{T_0}. \tag{1.10}$$

Now define a linear functional L on E_{T_0} by

$$L(z) = \int_0^T (t-T)(e^{-kt}f(t) + e^{-kt}f_1(t),\ z'(t))dt - T_0\langle u_1, z'(0)\rangle. \tag{1.11}$$

An integration by parts on the integral involving $(e^{-kt}f_1(t), z'(t))$ shows that there is a constant c so that

$$|L(z)| \leq c\|z\|_1 \quad \text{for all} \quad z \in E_{T_0}. \tag{1.12}$$

Hence by (1.10) and (1.12), we may apply the projection theorem of Lions [3, p. 37] to conclude that there is a function $w(t) \in F_{T_0}$ such that

$$B(w,z) = L(z) \quad \text{for all} \quad z \in E_{T_0}. \tag{1.13}$$

It is an easy exercise to see that a function w(t) satisfies (1.13) if and only if $u(t) = e^{kt}w(t)$ satisfies (1.7). Thus, if we set $u(t) = e^{kt}w(t)$, we have $u \in L^2[0,T_0;\ W]$, $u' \in L^2[0,T_0;\ H]$, $u(t) \in V(t)$ a.e. and u(t) satisfies (1.7) for all $v \in E_{T_0}$. Hence u(t) satisfies (1.5) and (1.6) on $[0,T_0]$.

To prove uniqueness on $[0,T_0]$, let $u_1(t)$ and $u_2(t)$ satisfy (1.5) and (1.6) on $[0,T_0]$. If we set $u(t) = u_1(t) - u_2(t)$, then $w(t) = \int_0^t u(s)ds$ satisfies

$$w''(t) + Aw(t) + \int_0^t a(t-s)Aw(s)ds = 0, \quad 0 < t < T_0 \tag{1.14}$$

and

$$w(0) = 0$$

$$w'(0) = 0 \tag{1.15}$$

$$w(t) \in V(t) \quad \text{a.e.} \quad 0 < t < T_0.$$

By the remark above, $\overline{w}(t) = e^{-kt}w(t)$ satisfies

$$B(\overline{w},z) = 0 \quad \text{for all} \quad z \in E_{T_0} \tag{1.16}$$

But (1.16) can be extended to all z in the completion of E_{T_0} under the norm

$$\left(\int_0^T \{|v''(t)|_H^2 + \|v'(t)\|_W^2\}dt\right)^{1/2}.$$

Since $\overline{w}(t)$ is in this space, we have

$$B(\overline{w},\overline{w}) = 0, \tag{1.17}$$

but (1.10) now implies that $\overline{w}(t) = 0$ a.e. on $[0,T_0]$, so that $u(t) = 0$ a.e. on $[0,T_0]$. This completes the proof of local existence.

REMARK. Since T_0 depends only on λ and a, the usual translation argument for Volterra equations will give us a function $u \in L^2_{loc}[0,\infty; W]$, $u' \in L^2_{loc}[0,\infty; H]$, $u'' \in L^2_{loc}[0,T; V'(0)]$ which satisfies (1.5) and (1.6).

REFERENCES

1. H. Brézis, *Opérateurs Maximaux Monotones et Semi-groupes de Contraction dans les Espaces de Hilbert*, North-Holland, Amsterdam, 1973.

2. C. Dafermos, An abstract Volterra equation with applications to linear viscoelasticity, *J. Differential Equations* 7 (1970), 554-569.

3. J. L. Lions, *Équations Différentielles Opérationelles et Problèmes aux Limites*, Springer-Verlag, Berlin, 1961.

4. J. L. Lions and E. Magenes, *Non-homogeneous Boundary Value Problems and Applications*, Springer-Verlag, Berlin, 1972.

5. J. R. Willis, Crack propagation in viscoelastic media, *J. Mech. Phys. Solids* 15 (1967), 229-240.

FORMATION OF SINGULARITIES FOR A CONSERVATION LAW WITH DAMPING TERM

Reza Malek-Madani*

Department of Mathematics
University of Wisconsin
Madison, Wisconsin

1. INTRODUCTION

In this paper, we consider the behavior of smooth solutions of two first order hyperbolic equations which have lower order dissipation. The equations are

$$\begin{aligned} &u_t + \phi(u)_x = a(x)u \\ &u(x,0) = u_0 \end{aligned} \tag{1.1}$$

and

$$\begin{aligned} &u_t + \phi(u)_x + \int_0^t a'(t-\tau)\psi(u)_x d\tau = 0 \\ &u(x,0) = u_0(x). \end{aligned} \tag{1.2}$$

Problem (1.1) is a very simple model for studying the flow of gas in a tube with varying cross section. Under proper hypotheses, the original equations of gas dynamics reduce to a one-dimensional system of balance laws with the

*Current affiliation: Department of Mathematics, Virginia Polytechnic Institute and State University, Blacksburg, Virginia.

nonhomogeneous term similar to the one in (1.1) (cf. [4]). The function a(x) defines the cross section of the tube. It would be interesting to study whether the spatial dependence of (1.1) plays a role, if any, in producing shocks. To be more precise, from physical considerations, it seems intuitively clear that if gas is pushed into a tube of narrowing cross section ($a(x) > 0$ and $a'(x) < 0$), a shock should be produced in finite time. A result of this nature is shown to be true in Section 3.

Equation (1.2) is a first order model for the equations of viscoelasticity when the viscosity is formulated in terms of the memory term $a' * \psi(u)_x$ (cf. [6]). As it is shown in [6], under appropriate assumptions on ϕ, ψ, a and u_0, this equation has the interesting property that for small enough initial data there exists a unique smooth and global solution to the initial value problem. That is, the convolution operator in (1.2) has the character of a weak dissipation term. In this paper we investigate the nature of smooth solutions for properly large initial data.

In Section 2, we formulate the problem of the formation of singularities and state a simple lemma which is helpful in determining when a shock will be produced. Sections 3 and 4 contain the applications of this lemma to equations (1.1) and (1.2).

2. PRELIMINARIES

The definition of a shock discussed in the theory of conservation laws lends itself naturally to both equations (1.1) and (1.2). Let a family of characteristics $x(t,\xi)$ be defined by

$$\frac{dx}{dt} = \phi'(u(x,t)), \quad x(0,\xi) = \xi. \tag{2.1}$$

We assume throughout this paper that equations (1.1) and (1.2) are hyperbolic and genuinely nonlinear, i.e., $\phi'(u) > 0$ and $\phi''(u) > 0$, respectively. Under these conditions, a singularity (shock) develops if two characteristics intersect. Let $x(t,\xi_1)$ and $x(t,\xi_2)$ be two characteristics intersecting at (T,x^*), that is, $x^* = x(T,\xi_1) = x(T,\xi_2)$. Further, suppose that the two characteristics actually cross each other and $\frac{dx}{dt}(T,\xi_1) \neq \frac{dx}{dt}(T,\xi_2)$. It follows from (2.1) that $\phi'(u(x(T^-,\xi_1),T^-)) \neq \phi'(u(x(T^-,\xi_2),T^-))$. Since ϕ' is assumed to be monotone, the solution $u(x,t)$ becomes multivalued at (T,x^*) and a shock develops. It should be noted that a shock formed due to the intersection of characteristics in forward time already has the entropy inequality embodied in it (cf. [3]).

An indication that the smooth solution $u(x,t)$ will be discontinuous at a point (x,t) is that $u_x(x,t)$ becomes unbounded in finite time. On the other hand, u_x can be evaluated along the characteristic $x(t,\xi)$ as

$$u_x(x(t,\xi),t) = \frac{u_\xi(x(t,\xi),t)}{x_\xi(t,\xi)} \ . \tag{2.2}$$

Thus u_x will become unbounded if $x_\xi(t,\xi)$ approaches zero. This presents an alternate way of establishing the formation of a shock (cf. [2]). The following lemma states that if $x_\xi(t,\xi)$ becomes zero in finite time, then two characteristics must intersect. Let $v(t,\xi) \equiv x_\xi(t,\xi)$. It follows from (2.1) that $v(0,\xi) = 1$.

LEMMA 2.1. Suppose that there exists $T < \infty$ and ξ such that $v(T,\xi) < 0$. Then there are distinct ξ_1 and ξ_2 with $x(T,\xi_1) = x(T,\xi_2)$. Moreover, if u satisfies an equation of the form $\frac{du}{dt} = g(x,u)$ along the characteristic, it follows that the solution develops a shock in finite time.

Proof. By way of contradiction, suppose that for all $\xi_1 \neq \xi_2$, $x(T,\xi_1) \neq x(T,\xi_2)$. This implies that the function $f(\xi)$ defined by $f(\xi) = x(T,\xi)$ is monotone. Therefore, $f'(\xi)$ will always be nonnegative, which contradicts the hypothesis. Hence there are two characteristics ξ_1 and ξ_2 which meet at $(\tilde{x},T)$ for some $\tilde{x}$. On the other hand, by the standard uniqueness theorem in ordinary differential equations, the above characteristics viewed in the (x,u) plane reach the line $x = \tilde{x}$ at two different values of u at time T. This completes the proof of the lemma.

3. FORMATION OF SHOCKS FOR A CONSERVATION LAW WITH LINEAR NONHOMOGENEITY

Consider the differential equation

$$\begin{aligned} &u_t + \phi(u)_x = a(x)u \\ &u(x,0) = u_0 \end{aligned} \tag{3.1}$$

where u_0 is a positive constant, $\phi'(u) > 0$, $\phi''(u) \geq k > 0$ for some k. The following theorem is a generalization of a result proved in [4]. It states that the global smoothness of the solution of (3.1) depends essentially on the sign of $a(x)$.

THEOREM 3.1. Suppose $a \in C^1[0,\infty)$.

1) If $a(x) \le 0$, then (3.1) has global smooth solutions.
2) If $a(x) > 0$ and $a'(x) > 0$, then (3.1) has global smooth solutions.
3) If $a(x) > 0$ and $a'(x) < 0$, then a shock develops in finite time.

Proof. We will prove part 3) only. The proofs for 1) and 2) are similar. Let $v(t,\xi) = x_\xi(t,\xi)$. As before, characteristics are defined by

$$\frac{dx}{dt} = \phi'(u), \quad x(0,\xi) = \xi$$
$$\frac{du}{dt} = a(x)u, \quad u(0,\xi) = u_0 \tag{3.2}$$

where $u(t,\xi) = u(x(t,\xi),t)$. Let $w(t,\xi) = \frac{\partial}{\partial\xi} u(t,\xi)$. Differentiating (3.2) with respect to ξ yields

$$\frac{dv}{dt} = \phi''(u)w, \quad v(0,\xi) = 1$$
$$\frac{dw}{dt} = a'(x)uv + a(x)w, \quad w(0,\xi) = 0. \tag{3.3}$$

Let

$$G(t,\xi) = \exp\{-\int_0^t a(x(s,\xi))ds\}. \tag{3.4}$$

Then it follows from the second equation in (3.3) that

$$w(t,\xi) = G^{-1}(t,\xi)\int_0^t G(s,\xi)a'(x(s,\xi))u(s,\xi)v(s,\xi)ds. \tag{3.5}$$

From the second equation in (3.2) we have $\frac{d}{dt}(Gu) = 0$, which implies that

$$u(t,\xi) = u_0 G^{-1}(t,\xi). \tag{3.6}$$

Equation (3.6) combined with (3.5) yields (the dependence on ξ will be omitted)

$$w(t) = G^{-1}(t)u_0\int_0^t a'(x(s))v(s)ds. \tag{3.7}$$

Thus, the analysis of showing the existence of a finite time t at which v becomes zero leads to the study of the equation

$$\frac{dv}{dt} = \phi''(u)G^{-1}(t)u_0\int_0^t a'(x(s))v(s)ds. \tag{3.8}$$

Let

$$V(t) \equiv \int_0^t a'(x(s))v(s)ds \tag{3.9}$$

so that $V(0) = 0$,

$$\dot{V}(t) = a'(x(t))v(t), \tag{3.10}$$

and $\dot{V}(0) = a'(\xi)$ from (3.3). Hence, the problem of finding the time such that $v(T) = 0$ becomes equivalent to showing $\dot{V}(T) = 0$. Differentiating (3.10) again, we obtain

$$\ddot{V} = \dot{a}'v + a'\dot{v}$$

which, combined with (3.8) and (3.10), yields

$$\ddot{V} = \frac{\dot{a}'}{a'}\dot{V} + a'\phi''G^{-1}u_0V$$

$$V(0) = 0 \tag{3.11}$$

$$\dot{V}(0) = a'(\xi).$$

We note that $G^{-1}(t) > 1$ since $a(x) > 0$. Without loss of generality, we can assume that $V(t) < 0$ for $t \in [0,T]$ since otherwise there exists $T^* > 0$ such that $V(T^*) = 0$ and, by Rolle's theorem, there exists a time T^{**} such that $\dot{V}(T^{**}) = 0$. With the same argument as above, we can assume that $V(t) < -\delta$ for $t > T_1$, for some $\delta > 0$ and $T_1 > 0$. Thus we obtain the inequality

$$\ddot{V} - \frac{\dot{a}'}{a'}\dot{V} - a'\phi''u_0V \geq 0. \tag{3.12}$$

Multiplying by the integrating factor $1/a'$ and integrating from 0 to t, we obtain

$$\frac{1}{a'}\dot{V} - \frac{1}{a'(\xi)} a'(\xi) \le u_0\int_0^t \phi''(u(s))V(s)ds, \qquad (3.13)$$

or

$$\dot{V} \ge a'(x(t))[1 + u_0\int_0^t \phi''(u(s))V(s)ds]. \qquad (3.14)$$

Finally, using the hypothesis on ϕ'' and the bound on $V(t)$, we arrive at the following estimate for $\dot{V}$,

$$\dot{V}(t) \ge a'(x(t))[1 - \delta u_0 k\ t]. \qquad (3.15)$$

We note that the term in brackets becomes negative in finite time which, combined with the sign of $a'(x)$, shows the existence of a finite time at which $\dot{V}(t)$ becomes zero. This completes the proof of the theorem.

4. FORMATION OF SHOCKS FOR A CONSERVATION LAW WITH MEMORY

We present a result similar to Theorem 3.1 for the equation

$$\begin{aligned} &u_t + \phi(u)_x + \int_0^t a'(t-\tau)\phi(u)_x d\tau = 0 \\ &u(x,0) = u_0(x) \end{aligned} \qquad (4.1)$$

with $a(t) = e^{-\alpha t}$, α positive. We note that (4.1) is a special case of (1.2) where $\phi(u) \equiv \psi(u)$. What distinguishes this special case, as is shown more explicitly in [6], is that it is possible to invert (4.1) using the kernel $k(t)$ defined by

$$k(t) + (a'*k)(t) = -a'(t) \qquad (4.2)$$

to obtain

$$u_t + \phi(u)_x + k*u_t = 0$$
$$u(x,0) = u(x). \tag{4.3}$$

An integration by parts on $k*u_t$ yields

$$u_t + \phi(u)_x + k(0)u + k'*u = k(t)u_0(x), \tag{4.4}$$

and thus the convolution term involves a lower order derivative of u in comparison with the balance law. A simple calculation shows that k(t) corresponding to our very special memory kernel $a(t) = e^{-\alpha t}$ is $k(t) \equiv \alpha$. Thus (4.4) takes the form

$$u_t + \phi(u)_x + \alpha u = \alpha u_0(x)$$
$$u(x,0) = u_0(x). \tag{4.5}$$

Although (4.1) together with the kernel $a(t) = e^{-\alpha t}$ does not fall into the category of the problems discussed in [6], it is not difficult to construe the same smoothness properties discovered in [6] for the solutions of (4.5). This program was also carried out, although in a somewhat different way, in [1]. The main feature of these results is that if the initial data $u_0(x)$ is smooth and "small enough", the dissipation term prohibits the formation of shocks and solutions remain globally smooth. We are interested in the limitation of this damping term. In other words, how large should the initial data be to insure breaking of waves? The following theorem answers this question.

THEOREM 4.1. Let ϕ' and ϕ'' be positive. Let $u_0(x)$ be a bounded $C^1(-\infty,\infty)$ function with $u_0(x) > 0$.

1) If $u_0'(x) > 0$ for all x, then the solution to (4.5) is globally smooth.

2) Suppose there exists a ξ such that $u_0'(\xi) < -\frac{\alpha}{k}$, where k depends on the bound on $u_0(x)$, and $u_0'(x) \leq 0$ for $x > \xi$. Then a shock develops in finite time.

Proof. As in the proof of Theorem 3.1, let $v = x_\xi$, $w = u_\xi$. Define characteristics by

$$\frac{dx}{dt} = \phi'(u), \quad x(0,\xi) = \xi$$
$$\frac{du}{dt} = -\alpha u + \alpha u_0(x), \quad u(0,\xi) = u_0(\xi). \tag{4.6}$$

Differentiating (4.6) with respect to ξ yields

$$\frac{dv}{dt} = \phi''(u)w, \quad v(0,\xi) = 1$$
$$\frac{dw}{dt} = -\alpha w + \alpha u_0'(x)v, \quad w(0,\xi) = u_0'(\xi). \tag{4.7}$$

The second equation in (4.7) can be integrated to give us

$$w(t) = e^{-\alpha t}u_0'(\xi) + \int_0^t e^{-\alpha(t-s)}u_0'(x(s))v(s)ds,$$

so that the first equation in (4.7) becomes

$$\frac{dv}{dt} = e^{-\alpha t}\phi''(u)u_0'(\xi) + \phi''(u)\int_0^t e^{-\alpha(t-s)}u_0'(x(s))v(s)ds$$
$$v(0) = 1. \tag{4.8}$$

The first part of the theorem becomes trivial since if u_0' is positive, dv/dt always remains positive and, therefore, the characteristics diverge from each other and the solution remains smooth. On the other hand, since $dx/dt = \phi'(u)$ and ϕ' is positive, we deduce that $x(t,\xi) > \xi$ for $t > 0$. By the hypothesis of the second part of the theorem, $u_0'(x) \le 0$ for $x > \xi$ so that the second term in (4.8) can be neglected to give us the inequality

$$\frac{dv}{dt} \le e^{-\alpha t}\phi''(u)u_0'(\xi). \tag{4.9}$$

Integrating (4.9) from 0 to t yields

$$v(t) \le 1 + u_0'(\xi)\int_0^t e^{-\alpha s}\phi''(u(s))ds. \tag{4.10}$$

The second equation in (4.6) gives us

$$u(t,\xi) = e^{-\alpha t}u_0(\xi) + \alpha\int_0^t e^{-\alpha(t-s)}u_0(x(s,\xi))ds \tag{4.11}$$

along the ξ-characteristic. Therefore

$$|u(t,\xi)| \le M + \alpha M\int_0^t e^{-\alpha(t-s)}ds, \tag{4.12}$$

where M is the pointwise bound on $u_0(x)$. Thus

$$|u(t,\xi)| \le M + M(1 - e^{-\alpha t}) \le 3M, \tag{4.13}$$

so that $u(t,\xi)$ is bounded along the ξ-characteristic. Since $\phi''(u)$ is assumed to be positive for all u, it follows that it has a positive minimum k on the interval [-3M,3M]. Inequality (4.10) then reduces to

$$v(t) \le 1 + ku_0'(\xi)\int_0^t e^{-\alpha s}ds$$

or

$$v(t) \le 1 + ku_0'(\xi)\left[\frac{e^{-\alpha t}}{-\alpha}\right] + \frac{k}{\alpha}u_0'(\xi). \tag{4.14}$$

Since $e^{-\alpha t}$ approaches zero as t tends to infinity, we see that v(t) becomes negative in finite time if $u_0'(\xi) < -\frac{\alpha}{k}$. We then apply Lemma 2.1. This completes the proof of the theorem.

We remark that the sufficient assumption on ϕ'' in Theorem 4.1 is its positivity, unlike the corresponding assumption for Theorem 3.1. The reason for the weaker assumption here is that it is possible to show for equations (4.5) that bounded initial data give rise to bounded solutions along each characteristic.

Theorem 4.1 was proved for the very special kernel $a(t) = e^{-\alpha t}$. A similar result, but with a more complicated proof, is shown to hold for a much more general kernel a(t). We state the result here; its proof will be presented in a forthcoming paper [5].

THEOREM 4.2. Consider the equation

$$\begin{aligned} &u_t + \phi(u)_x + a'*\phi(u)_x = 0 \\ &u(x,0) = u_0(x). \end{aligned} \tag{4.15}$$

Suppose that

1) $a(t)$ is $C^2[0,\infty)$ with $(-1)^i a^{(i)}(t) \quad 0$, $i = 0,1,2$, and

   $a''(t) \geq a'(0)a'(t)$.

2) $\phi'(u) > 0$ and $\phi''(u) > k > 0$.

3) u_0 is C^1 smooth with $u_0(x) > 0$.

Moreover, suppose that there exists a point ξ such that $u_0'(\xi)$ is sufficiently negative and $u_0'(x) \leq 0$ for all $x > \xi$. Then a shock develops in finite time.

ACKNOWLEDGEMENT

Professor Malek-Madani's research was sponsored by the United States Army under Contract No. DAAG29-80-C-0041.

REFERENCES

1. Dafermos, C. M., Can dissipation prevent the breaking of waves?, Transactions of the Twenty-Sixth Conference of Army Mathematicians, ARO Report 81-1, 187-198.

2. Klainerman, S. and Majda, A., Formation of singularities for wave equations including the nonlinear vibrating string, *Comm. Pure and Appl. Math.* 33 (1980), 241-263.

3. Lax, P. D., Hyperbolic systems of conservation laws II, *Comm. Pure and Appl. Math.* 10 (1957), 537-566.

4. Lin, S. S. and Malek-Madani, R., Solution to the Riemann problem for the equations of gas dynamics in a tube with varying cross section, Transactions of the Twenty-sixth Conference of Army Mathematicians, ARO Report 81-1, 341-354.

5. Malek-Madani, R. and Nohel, J. A., in preparation.

6. Nohel, J. A., A nonlinear conservation law with memory, *Volterra and Functional Differential Equations*, Marcel Dekker, New York, 1982 (this volume).

ON OBTAINING ULTIMATE BOUNDEDNESS FOR α-CONTRACTIONS

Paul Massatt

Department of Mathematics
University of Oklahoma
Norman, Oklahoma

The use of Lyapunov functions, maximum principles, and invariance principles for the study of the limiting behavior of evolution equations has generated interest in the study of dissipative systems. A dissipative system is a system where there is some bounded set which all trajectories enter into and remain in. In the study of dissipative systems we make hypotheses as general as possible -- hypotheses which are likely to occur in the applications; we seek to obtain as sharp as possible results on the limiting behavior of solutions. The results are generally concerned with autonomous, periodic, and certain types of nonautonomous flows (see [1], [2], [3], [4], and [21]).

Let X be a Banach space and let $T: X \to X$ be continuous. Let B and C be bounded sets in X. Let $B_r(y) = \{x \in X \mid \|x - y\| \le r\}$. We say B *dissipates* C if there is an $n_0 \ge 0$ such that $n \ge n_0$ implies $T^n B \subset C$. We say that B *attracts* C if for all $\varepsilon > 0$, $B + B_\varepsilon(0)$ dissipates C. B is *invariant* if $TB = B$. B is *positively invariant* if $TB \subset B$. B is *negatively invariant* if for all $x \in B$ there exists $y \in B$ with $Ty = x$. The *orbit* of B, $\gamma^+(B)$, is defined by $\gamma^+(B) = \bigcup_{n=0}^{\infty} T^n(B)$. The ω - *limit set* of B, $\omega(B)$, is defined by $\omega(B) = \bigcap_{m=0}^{\infty} \mathrm{Cl}\{\bigcup_{n=m}^{\infty} T^n B\}$. If B is negatively invariant, the *alpha limit set* of B, alpha(B), is defined by alpha(B) = $\{y \in X \mid$ there exists $\{y_j\} \subset B$,

$\{x_j\} \subset B$, and $\{k_j\} \to \infty$ with $T^{k_j} y_j = x_j$ and $\{y_j\} \to y\}$. B is *stable* if for every $\varepsilon > 0$ there is a $\delta > 0$ such that $\gamma^+(B + B_\delta(0)) \subset B + B_\varepsilon(0)$. B is *uniformly asymptotically stable* if it is stable and there is an $\varepsilon > 0$ such that B attracts $B + B_\varepsilon(0)$.

We say T is *point dissipative* if there is a bounded set which dissipates all points. T is *compact dissipative* if there is a bounded set which dissipates all compact sets. T is *local dissipative* if there is a bounded set which dissipates a neighborhood of any point. T is *local compact dissipative* if there is a bounded set which dissipates a neighborhood of any compact set. T is *bounded dissipative*, or *ultimately bounded*, if there is a bounded set which dissipates all bounded sets. If T is continuous (as we always assume), then compact dissipative, local dissipative, and local compact dissipative are all equivalent.

Normally, in order to obtain results on the limiting behavior of solutions, we need some type of compactness condition on T. Hale, LaSalle and Slemrod [12] have shown that if T is completely continuous and point dissipative, then there exists a maximal compact invariant set (MCI) which is uniformly asymptotically stable, attracts bounded sets, and has a fixed point. If T is *conditionally completely continuous*, i.e., B, TB ε $\mathcal{B}$ implies TB is precompact where $\mathcal{B}$ is the collection of all bounded sets in X, then the same conclusions hold except that the MCI attracts only those sets B ε $\mathcal{B}$ for which TB ε $\mathcal{B}$ also.

However, in applications the map T is often not conditionally completely continuous. Examples are stable neutral functional differential equations (SNFDE's), retarded functional differential equations (RFDE's) with infinite delay, strongly damped nonlinear wave equations, various parabolic and hyperbolic partial differential equations, etc. In many of these problems the Kuratowski measure of noncompactness, or α-measure, has proved useful. The *α-measure* is a map $\alpha: \mathcal{B} \to [0,\infty)$ defined by $\alpha(B) = \inf\{d \mid B$ can be covered by a finite collection of sets of diameter $d\}$. T is an *α-contraction* if there exists a $k \in [0,1)$ such that for all B ε $\mathcal{B}$ we have TB ε $\mathcal{B}$ and $\alpha(TB) \leq k\alpha(B)$. T is a *conditional α-contraction* if there exists a $k \in [0,1)$ such that for all B, TB ε $\mathcal{B}$ we have $\alpha(TB) \leq k\alpha(B)$.

Hale and Lopes ([13], [10]) have shown that if T is a conditional α-contraction and compact dissipative, then there exists an MCI which is uniformly asymptotically stable, attracts a neighborhood of any compact set, and has a fixed point. One can obtain similar results under more general assumptions on T (see [7], [11], [15], and [17]).

In this paper we show for certain α-contractions which allow for backward existence of solutions, and satisfy some uniform Lipschitz condition in the backward direction, that point dissipative implies the existence of a maximal compact invariant set which attracts bounded sets. We also assume that the equation is defined on two Banach spaces, with one compactly imbedded in the other. The assumptions will be shown to be natural enough to include stable neutral function differential equations of the form

$$\frac{d}{dt} Dx_t = f(t,x_t), \tag{1}$$

where f is periodic in t and uniformly Lipschitz in x_t, x_t is defined on $C[-r,0]$ and D is linear and atomic at both -r and 0. We will also apply the results to strongly damped nonlinear wave equations of the form

$$(1 - \varepsilon\Delta)u_{tt} - \lambda\Delta u_t - \Delta u = f(t,u,u_t), \tag{2}$$

where f is uniformly Lipschitz, $\varepsilon > 0$ and $\lambda \geq 0$. We also assume $u(t,x)$: $\mathbb{R}^+ \times \Lambda \to \mathbb{R}^n$ where $\Lambda \subset \mathbb{R}^m$ is a bounded domain with smooth boundary, and $u(t,x) = 0$ for all $x \in \partial\Lambda$.

Equation (2) arises in the modelling of longitudinal vibrations in a homogeneous bar in which there are inertial and viscous effects ([6], [8], [9], [22], and [23]). The term $\varepsilon\Delta u_{tt}$ takes into account the inertia of lateral motions in which the cross-sections are extended or contracted in their own planes. The term $\lambda\Delta u_t$ indicates that the stress is proportional not only to the strain, but also the strain rate, as in a linearized Kelvin material.

In two recent papers, I have shown that for most α-contractions which arise in the applications, point dissipative and compact dissipative are equivalent. In this paper we obtain the much stronger result of ultimate boundedness, and that the maximal compact invariant set attracts bounded sets. However, we do this at the sacrifice of making more restrictive assumptions. The class of equations which satisfy these hypotheses is sufficiently large to make the results interesting. I expect that similar results should be obtainable under more general assumptions. This is still an open area for further research (see [17], [18], and [19]). In this paper we are interested in proving the following theorem.

THEOREM. Let X_1 and X_2 be two Banach spaces with X_1 compactly imbedded into X_2, and let T, C and U: $X_i \to X_i$ be continuous and uniformly Lipschitz in each space. Let C be a linear contraction in both spaces and let U: $X_2 \to X_1$ be uniformly Lipschitz. We also assume that T^{-1}, C^{-1} and $\tilde{U}$: $X_i \to X_i$ exist and are continuous and uniformly Lipschitz, where $T^{-1} = C^{-1} + \tilde{U}$, and $\tilde{U}$: $X_2 \to X_1$ is uniformly Lipschitz. Under these assumptions, if T is point dissipative in X_2, then T is bounded dissipative in X_1 and X_2.

Before proving this theorem, it is important to point out some of its applications. As we mentioned before, this result will prove for equations (1) and (2) the equivalence of point dissipative and bounded dissipative. In equation (1) we may let $X_2 = C[-r,0]$ and $X_1 = W^{1,\infty}[-r,0]$ (see [17] for details). Other possibilities, such as $L^p \times \mathbb{R}^n$ and $W^{1,p} \times \mathbb{R}^n$ are also admissible. The hypothesis that $U,\tilde{U}$: $X_2 \to X_1$ are uniformly Lipschitz comes from the assumption that f is uniformly Lipschitz. The hypothesis on T^{-1}, C^{-1}, and $\tilde{U}$ is satisfied by the fact that D is atomic at -r. In equation (2), we may let $X_2 = L^p \times L^p$ and $X_1 = \mathring{W}^{1,p} \times \mathring{W}^{1,p}$. This equation fits into the theory of analytic semigroups (see, for example, [14]). The equation may be written as $u_{tt} - \alpha\Delta(1 - \varepsilon\Delta)^{-1}u_t - \Delta(1 - \varepsilon\Delta)^{-1}u = (1 - \varepsilon\Delta)^{-1}f(u)$. The homogeneous part $u_{tt} - \alpha\Delta(1 - \varepsilon\Delta)^{-1}u_t - \Delta(1 - \varepsilon\Delta)^{-1}u = 0$ has a solution map which is a contraction in some equivalent norm. Since $(1 - \varepsilon\Delta)^{-1}f(u)$ is compact, one can ascertain from the variation of parameters formula that the solution map for the entire equation is an α-contraction. The uniform Lipschitz condition on f insures that $U,\tilde{U}$: $X_2 \to X_1$ are uniformly Lipschitz. It is also easy to obtain the existence of T^{-1}, C^{-1} and $\tilde{U}$ here.

Proof of theorem. The fact that T is bounded dissipative in X_1 has already been proven in [17]. Also, in [18] we showed that point dissipative and compact dissipative are equivalent in X_2. Hence, there is a maximal compact invariant set K which attracts a neighborhood of any compact set. In fact, it attracts any bounded set whose orbit is bounded. We shall assume that T is not bounded dissipative and arrive at a contradiction. Let $R > 0$. There must exist some $M > 0$ such that there exist sequences $\{x_j\} \subset K + B_M(0)$ and $\{k_j\} \to \infty$ with $T^k x_j \notin K + B_R(0)$ for $1 \le k \le k_j$. Otherwise, it is easy to show that $\gamma^+(K + B_R(0))$ is bounded and dissipates all bounded sets.

In the following, for convenience and without loss of generality, we shall assume that if $x \in X_1$, then $\|x\|_2 \le \|x\|_1$.

Now let k be a contraction constant for C in both spaces and let ℓ be a uniform Lipschitz constant for T, T^{-1}, C^{-1}, U, and $\tilde{U}$: $X_i \to X_i$, $i = 1,2$, and for $U, \tilde{U}$: $X_2 \to X_1$. Since $T = C + U$, we have $T^2 = C^2 + CU + UT$, and $T^{n+1} = C^{n+1} + \sum_{j=0}^{n} C^j UT^{n-j}$. We shall define $U_{n+1} = \sum_{j=0}^{n} C^j UT^{n-j}$, $n \geq 0$, so that $T^{n+1} = C^{n+1} + U_{n+1}$. We notice that U_{n+1}: $X_2 \to X_1$ is Lipschitz and, hence, U_{n+1}: $X_2 \to X_2$ is completely continuous. Let $\{y_j^1\} \subset \{x_j\}$ be a subsequence with $\{U_1 y_j^1\}$ converging to some $u_1 \in X_2$. Let $\{y_j^2\} \subset \{y_j^1\}$ be a subsequence with $\{U_2 y_j^2\}$ converging to some u_2. Continuing this process, and then using a diagonalization process, we can find sequences $\{z_j\}$ and $\{u_n\}$ with $\{U_n z_j\} \to u_n$ in X_2 for all $n \geq 1$. From this, we get

$$\overline{\lim_{j\to\infty}} \|T^n z_j - u_n\|_2 \leq \overline{\lim_{j\to\infty}} \|C^n z_j\|_2 + \overline{\lim_{j\to\infty}} \|U_n z_j - u_n\|_2 \leq k^n M.$$

Using this and the fact that U: $X_2 \to X_1$ is uniformly Lipschitz gives us that

$$\lim_{q,p\to\infty} \|U_{n+1} z_q - U_{n+1} z_p\|_1 \leq \Big\| \sum_{j=0}^{n} C^j UT^{n-j} z_q - \sum_{j=0}^{n} C^j UT^{n-j} z_p \Big\|_1 \leq \ell(n + 1)k^n M.$$

We may also choose a sequence $\{\tilde{u}_n\} \subset X_1$ with

$$\overline{\lim_{q\to\infty}} \|U_{n+1} z_q - \tilde{u}_{n+1}\|_1 \leq 2(n + 1)\ell M k^n$$

and

$$\overline{\lim_{q\to\infty}} \|T^n y_q - \tilde{u}_n\|_2 \leq 2k^n M.$$

Now, let $\{h_n\}$ be a bounded nonnegative solution of $h_n = \ell h_{n+1} + 2(\ell^2 + \ell)(n + 2)Mk^{n-1}$. We will show that $T^{-1}B^1_{h_{n+1}}(\tilde{u}_{n+1}) \subset B^1_{h_n}(\tilde{u}_n)$. Remember that $B^i_r(a)$ is a ball of radius r, centered at a, in the space X_i.

Let $z \in B_{h_{n+1}}(\tilde{u}_{n+1})$. Then

$$\|T^{-1}z - \tilde{u}_n\|_1 \leq \|T^{-1}z - T^{-1}\tilde{u}_{n+1}\|_1 + \|T^{-1}\tilde{u}_{n+1} - \tilde{u}_n\|$$

$$\leq \ell \|z - \tilde{u}_{n+1}\|_1 + \overline{\lim_{q\to\infty}} \|T^{-1}\tilde{u}_{n+1} - T^{-1}U_{n+1}z_q\|_1$$

$$+ \overline{\lim_{q\to\infty}} \|T^{-1}U_{n+1}z_q - U_nz_q\|_1 + \overline{\lim_{q\to\infty}} \|U_nz_q - \tilde{u}_n\|_1$$

$$\le \ell \|z - \tilde{u}_{n+1}\|_1 + 2\ell^2(n + 1)Mk^n + \overline{\lim_{q\to\infty}} \|T^{-1}U_{n+1}z_q - U_nz_q\|_1$$

$$+ 2\ell nMk^{n-1}$$

$$\le \ell \|z - \tilde{u}_{n+1}\|_1 + 2(\ell^2 + \ell)(n + 1)Mk^{n-1}$$

$$+ \overline{\lim_{q\to\infty}} \|T^{-1}\sum_{j=0}^{n} C^jUT^{n-j}z_q - \sum_{j=0}^{n-1} C^jUT^{n-j-1}z_q\|_1$$

$$\le \ell \|z - \tilde{u}_{n+1}\|_1 + 2(\ell^2 + \ell)(n + 1)Mk^{n-1}$$

$$+ \overline{\lim_{q\to\infty}} \|\sum_{j=0}^{n-1} C^jUT^{n-j-1}z_q + C^{-1}UT^nz_q + \tilde{U}U_{n+1}z_q - \sum_{j=0}^{n} C^jUT^{n-j-1}z_q\|_1$$

$$\le \ell\|z - \tilde{u}_{n+1}\|_1 + 2(\ell^2 + \ell)(n + 1)Mk^{n-1} + \overline{\lim_{q\to\infty}} \|C^{-1}UT^nz_q + \tilde{U}U_{n+1}z_q\|_1$$

$$\le \ell \|z - \tilde{u}_{n+1}\|_1 + 2(\ell^2 + \ell)(n + 1)Mk^{n-1}$$

$$+ \overline{\lim_{q\to\infty}} \|(C^{-1}T^{n+1}z_q - T^nz_q) + (\tilde{U}T^{n+1}z_q - \tilde{U}C^{n+1}z_q)\|_1$$

$$\le \ell \|z - \tilde{u}_{n+1}\|_1 + 2(\ell^2 + \ell)(n + 1)Mk^{n-1} + \overline{\lim_{q\to\infty}} \|\tilde{U}C^{n+1}z_q\|_1$$

$$\le \ell \|z - \tilde{u}_{n+1}\|_1 + 2(\ell^2 + \ell)(n + 1)Mk^{n-1} + \ell Mk^{n+1}$$

$$\le \|z - \tilde{u}_{n+1}\|_1 + 2(\ell^2 + \ell)(n + 2)Mk^{n-1}.$$

Using this property, we see that $T^{-n}B^1_{h_{n+1}}(\tilde{u}_{n+1})$ are decreasing subsets in X_1 and $T^{-n}B^1_{h_{n+1}}(\tilde{u}_{n+1}) \subset B^1_{h_1}(\tilde{u}_1)$ for all $n \ge 0$. We denote the closure of a set B in the Banach space X_i by $Cl_i(B)$. Then, by forward and backward existence and continuous dependence, we have $T^{-n}Cl_2(B_{h_{n+1}}(\tilde{u}_{n+1})) = Cl_2(T^{-n}B_{h_{n+1}}(\tilde{u}_{n+1}))$. The sets $T^{-n}Cl_2(B_{h_{n+1}}(\tilde{u}_{n+1}))$ are nonempty, compact sets which are decreasing. Clearly, the intersection is nonempty. Let $z \in \bigcap_{n=0}^{\infty} T^{-n}Cl_2(B_{h_{n+1}}(\tilde{u}_{n+1}))$. Then, $T^n z \in Cl_2(B_{h_{n+1}}(\tilde{u}_{n+1}))$ for all $n \geq 0$ so that $\|T^n z - \tilde{u}_{n+1}\|_2 \to 0$. This implies that K doesn't attract the point z, which is clearly a contradiction. Notice that

$$\varliminf_{\substack{h\to\infty \\ x\in K}} \|\tilde{u}_n - x\|_2 \geq R > 0.$$

Since we obtain a contradiction, we get the result that T is bounded dissipative.

I feel it is important to remark here that we have used the property that T^{-1} also forms a discrete dynamical system. It is my feeling that this assumption is not necessary for this result to hold. However, for the proof given, it is necessary.

REFERENCES

1. Artstein, Z., Topological dynamics of an ordinary differential equation, *J. Differential Equations* 23 (1977), 216-223.

2. Artstein, Z., Topological dynamics of ordinary differential equations and Kurzweil equations, *J. Differential Equations* 23 (1977), 224-243.

3. Artstein, Z., The limiting equations of nonautonomous ordinary differential equations, *J. Differential Equations* 25 (1977), 184-202.

4. Artstein, Z., Uniform asymptotic stability via the limiting equations, *J. Differential Equations* 27 (1978), 172-189.

5. Billotti, J. E. and J. P. LaSalle, Periodic dissipative processes, *Bull. Amer. Math. Soc.* 6 (1971), 1082-1089.

6. Caroll, R. W. and R. E. Showalter, *Singular and Degenerate Cauchy Problems*, Academic Press, New York, 1976.

7. Cooperman, G. D., α-Condensing Maps and Dissipative Systems, Ph.D. Thesis, Brown University, June 1978.

8. Greenberg, J. M, R. C. MacCamy and V. J. Mizel, On the existence, uniqueness, and stability of solutions of the equation $\sigma'(u_x)u_{xx} + \lambda_0 u_{xtx} = \rho_0 u_{tt}$, *J. Math. Mech.* 17 (1968), 707-728.

9. Greenberg, J. M., On the existence, uniqueness, and stability of the equation $X_{tt} = E(X_x)X_{xx} + \lambda X_{xxt}$, *J. Math. Anal. Appl.* 25 (1969), 575-591.

10. Hale, J. K., α-contractions and differential equations, *Proc. Equations Differential Fon. Nonlin.* (Brussels 1975), Hermann, Paris, 15-42.

11. Hale, J. K., *Theory of Functional Differential Equations*, Appl. Math. Sci., Vol. 3, Springer-Verlag, New York, 1977.

12. Hale, J. K., J. P. LaSalle and M. Slemrod, Theory of a general class of dissipative processes, *J. Math. Anal. Appl.* 39 (1972), 177-191.

13. Hale, J. K. and O. Lopes, Fixed point theorems and dissipative processes, *J. Differential Equations* 13 (1973), 391-402.

14. Henry, D., *Geometric Theory of Semilinear Parabolic Equations*, Lecture Notes in Mathematics Vol. 840, Springer-Verlag, New York, 1981.

15. Massatt, P. D., Some properties of condensing maps, *Ann. di Mat. Pura ed Appl.* (IV) 125 (1981), 101-115.

16. Massatt, P. D., Stability and fixed points of point dissipative systems, *J. Differential Equations* 2 (1981), 217-231.

17. Massatt, P. D., Properties of Condensing Maps and Dissipative Systems, Ph.D. Thesis, Brown University, June 1980.

18. Massatt, P. D., Attractivity properties of α-contractions, *J. Differential Equations*, to appear.

19. Massatt, P. D., Limiting behavior for strongly damped nonlinear wave equations, *J. Differential Equations*, to appear.

20. Masuda, K., unpublished manuscript, Kyoto University.

21. Sell, G. R., *Lectures on Topological Dynamics and Differential Equations*, Van Nostrand-Reinhold, London, 1971.

22. Showalter, R. E., Regularization and approximation of second order evolution equations, *SIAM J. Math. Anal.* 7 (1976), 461-472.

23. White, L. W., Approximation of point controls of second order evolution equations of Sobolev type, *J. Math. Anal. Appl.*, to appear.

COSINE FAMILIES, PRODUCT SPACES AND INITIAL VALUE PROBLEMS WITH NONLOCAL BOUNDARY CONDITIONS

Samuel M. Rankin, III

Department of Mathematics
West Virginia University
Morgantown, West Virginia

1. INTRODUCTION

Consider the initial boundary value problem

$$
\begin{aligned}
&y_{tt}(x,t) = y_{xx}(x,t), \quad 0 < x < \pi, \quad t \in \mathbb{R} \\
&y_{tt}(0,t) = \int_0^\pi f(x)y(x,t)dx + u(t), \quad t \in \mathbb{R}, \quad u \in L^2_{loc}(0,\infty), \quad f \in L^2(0,\pi) \\
&y(\pi,t) = 0, \quad t \in \mathbb{R} \qquad\qquad (1) \\
&y(x,0) = y_0(x), \quad y_t(x,0) = y_1(x), \quad y_0, y_1 \in L^2(0,\pi) \\
&y(0,0) = z_0, \quad y_t(0,0) = z_1, \quad z_0, z_1 \in \mathbb{R}.
\end{aligned}
$$

Define the operators:

$$
\begin{aligned}
&A: D(A) \to L^2(0,\pi) \quad \text{by} \quad Ag = g'' \\
&D(A) = \{g \in H^2(0,\pi) \mid g(\pi) = 0\} \\
&B: D(A) \to \mathbb{R} \quad \text{by} \quad Bg = g(0)
\end{aligned}
$$

$$F: L^2(0,\pi) \to \mathbb{R} \quad \text{by} \quad Fg = \langle f,g\rangle \equiv \int_0^\pi f(x)g(x)dx$$

$$\mathcal{A}: D(\mathcal{A}) \to \mathbb{R} \times L^2(0,\pi) \quad \text{by}$$

$$\mathcal{A}(z,g) = (Fg, Ag)$$

$$D(\mathcal{A}) = \{(z,g) \in \mathbb{R} \times D(A) \mid Bg = z\}.$$

If one assumes that $y(x,t)$ is a solution of equation (1) and defines $z(t) = y(0,t)$, then

$$z''(t) = \langle f,y(x,t)\rangle + u(t) \quad \text{for} \quad t \in \mathbb{R}.$$

Thus equation (1) is related to the following evolution equation on $\mathbb{R} \times L^2(0,\pi)$:

$$\frac{d^2}{dt^2}(z(t),y(\cdot,t)) = \mathcal{A}(z(t),y(\cdot,t)) + (u(t),0)$$

$$(z(0),y(\cdot,0)) = (z_0,y_0) \tag{2}$$

$$\frac{d}{dt}(z(0),y(\cdot,0)) = (z_1,y_1).$$

The second component of the mild solution of equation (2) is interpreted as the weak solution of equation (1). If the operator $\mathcal{A}$ generates a strongly continuous cosine family $C(t)$, $t \in \mathbb{R}$, with associated sine family $S(t)$, then the mild solution of equation (2) can be written as

$$(z(t),y(\cdot,t)) = C(t)(z_0,y_0) + S(t)(z_1,y_1) + \int_0^t S(t-s)(u(s),0)ds, \quad t \in \mathbb{R}.$$

The purpose of this note is to give conditions on A, F, and B such that $\mathcal{A}$ will be the generator of a strongly continuous cosine family. The fact that $\mathcal{A}$ generates a strongly continuous cosine family is equivalent to equation (2), and therefore equation (1), being well-posed; that is, we have existence, uniqueness and continuous dependence of solutions.

2. STATEMENT AND OUTLINE OF PROOF OF RESULT

Let X and Y be Banach spaces, $A: D(A) \to X$, $D(A) \subset X$, $B: D(A) \to Y$, and $F: D(F) \to Y$, $D(A) \subset D(F) \subset X$, be linear operators. Let $\bar{A}$ be the linear operator A restricted to the kernel of B, that is, $D(\bar{A}) = D(A) \cap \ker B$. Throughout, we make the following assumptions on A, B and F:

i) F is $\bar{A}$ - bounded on $D(\bar{A})$, that is, there exist constants C_1, C_2 (with $C_2 < 1$) such that $\|F\psi\|_Y \le C_1\|\psi\|_X + C_2\|\bar{A}\psi\|_X$ for all $\psi \in D(\bar{A})$;

ii) there exists a right inverse $D \in L(Y,X)$ of B such that $AD: Y \to X$ and $FD: Y \to Y$ are bounded linear operators.

Define the operator $\mathcal{A}$ on $Y \times X$ by $\mathcal{A}(z,\psi) = (F\psi, A\psi)$ with

$$D(\mathcal{A}) = \{(z,\psi) \in Y \times D(A) \mid B\psi = z\}.$$

A one-parameter family $C(t)$, $t \in \mathbb{R}$, of bounded linear operators from a Banach space X into itself is called a strongly continuous cosine family if and only if

i) $C(0) = I$,
ii) $C(t + s) + C(t - s) = 2C(t)C(s)$, $s,t \in \mathbb{R}$,
iii) $C(t)x$ is continuous in t for each $x \in X$.

The associated sine family $S(t)$, $t \in \mathbb{R}$, of $C(t)$ is given by $S(t)x = \int_0^t C(r)x\,dr$ for all $x \in X$, and the infinitesimal generator G of $C(t)$ is given by $Gx = C''(0)x = \lim_{t\to 0} \frac{C(2t)x - x}{2t^2}$ for all $x \in D(G) \equiv \{x \in X \mid C(t)x$ is twice continuously differentiable$\}$. See the paper [1] for more detailed information on strongly continuous cosine families.

THEOREM. If $\bar{A} - DF$ with domain $D(\bar{A})$ generates a strongly continuous cosine family on X, then $\mathcal{A}$ generates a strongly continuous cosine family on $Y \times X$. Conversely, if $\mathcal{A}$ generates a strongly continuous cosine family on $Y \times X$, then $\bar{A} - DF$ has a closed linear extension which generates a strongly continuous cosine family on X. If $\|D\| < 1$, then $\bar{A} - DF$ itself generates a strongly continuous cosine family on X.

The basic idea of the proof is to show that

$$\mathcal{A} = \begin{bmatrix} 0 & F \\ 0 & A \end{bmatrix}$$

transforms into

$$\tilde{\mathcal{A}} = \begin{bmatrix} FD & F \\ AD - DFD & \bar{A} - DF \end{bmatrix}$$

under a change of coordinates, and then to show that $\mathcal{A}$ generates a strongly continuous cosine family if and only if $\tilde{\mathcal{A}}$ does. This technique is justified by the lemma below. See the paper [2] by R. Vinter in the case where $\mathcal{A}$ generates a strongly continuous semigroup.

LEMMA. Let X be a Banach space, $P: X \to X$ a linear homeomorphism, and $\tilde{G}$ a linear operator with $D(\tilde{G}) \subset X$. Then the operator $G = P\tilde{G}P^{-1}$ with $D(G) = PD(\tilde{G})$ generates a strongly continuous cosine family if and only if $\tilde{G}$ does.

Proof. The proof follows from the cosine generation theorem as found in [1].

Define P on $Y \times X$ into $Y \times X$ by $P(z,\psi) = (z,\psi + Dz)$ and P^{-1} by $P^{-1}(z,\psi) = (z,\psi - Dz)$. Thus we have

$$\begin{aligned} PD(\tilde{\mathcal{A}}) &= \{(z,\psi + Dz) \mid z \in Y, \ \psi \in D(\bar{A})\} \\ &= \{(B(\psi + Dz), \ \psi + Dz) \mid z \in Y, \ \psi \in D(\bar{A})\} \\ &= \{(Bw,w) \mid w \in D(\mathcal{A})\} \end{aligned}$$

and

$$P\tilde{\mathcal{A}}P^{-1} = \begin{bmatrix} 0 & F \\ 0 & A \end{bmatrix}.$$

3. OUTLINE OF PROOF OF THEOREM

First assume that $\overline{A} - DF$ generates a strongly continuous cosine family $C(t)$, $t \in \mathbb{R}$, on X with associated sine family $S(t)$. Rewrite $\tilde{A}$ as

$$\tilde{A} = \begin{bmatrix} 0 & F \\ 0 & \overline{A} - DF \end{bmatrix} + \begin{bmatrix} FD & 0 \\ AD - DFD & 0 \end{bmatrix} \equiv \tilde{\tilde{A}} + L.$$

Since D, FD and AD are bounded linear operators, so is L. By perturbation results for cosine families [1], it suffices to show that $\tilde{\tilde{A}}$ generates a strongly continuous cosine family. Without loss of generality assume that $||D|| < 1$, for if this is not the case, one could rescale the norm of X. Now we can show that since F is $\overline{A}$ - bounded, it is also $\overline{A}$-DF - bounded and from this deduce that $\tilde{\tilde{A}}$ generates the strongly continuous family

$$\tilde{\tilde{C}}(t)(z,\psi) = (z + F(\int_0^t S(r)\psi dr), C(t)\psi), \quad (z,\psi) \in Y \times X, \quad t \in \mathbb{R}.$$

Note that $\int_0^t S(r)\psi dr \in D(\overline{A} - DF) = D(\overline{A})$ for each $t \in \mathbb{R}$ and $\psi \in X$ (see [1]) and, therefore, since $D(A) \subset D(F)$, $\tilde{\tilde{C}}(t)$ is well defined on $Y \times X$. Using the lemma, we deduce that A generates a strongly continuous cosine family.

Suppose now that A generates a strongly continuous cosine family. By the lemma, $\tilde{A}$ does, and again by perturbation, $\tilde{\tilde{A}}$ does as well. Let $\tilde{\tilde{C}}(t)$, $t \in \mathbb{R}$, be the cosine family generated by $\tilde{\tilde{A}}$, and define the family of bounded linear operators $C(t)$, $t \in \mathbb{R}$, on X by

$$C(t)\psi = \pi_2 \tilde{\tilde{C}}(t)(0,\psi), \quad \psi \in X,$$

where π_2 denotes the projection onto the X component. One can show that $C(t)$, $t \in \mathbb{R}$, is a strongly continuous cosine family generated by a closed linear extension of $\overline{A}$ - DF. Furthermore, one can show that if $||D|| < 1$, then this extension coincides with $\overline{A}$ - DF.

REMARKS. In the example of the introduction, $D(\overline{A}) = H^2 \cap H_0^1$ and $\overline{A}$ generates the cosine family $C(t)h = \sum_{n=1}^{\infty} \cos nt(h,h_n)h_n$, where $h \in L^2(0,\pi)$ and $h_n(s) = \frac{\sqrt{2}}{\pi} \sin ns$, $s \in [0,\pi]$. Furthermore, if $f \in L^2(0,\pi)$, then F is a bounded linear operator and the operator $D: R \to D(A)$ defined by $Dz = \frac{z}{\pi^m}(\pi - x)^m$ is a bounded right inverse of B. Thus, by perturbation, $\overline{A}$ - DF generates a strongly continuous cosine family. If $f \in (H^2 \cap H_0^1)'$, then F is no longer bounded; however, F is $\overline{A}$ - bounded and, by Theorem 4.3 of [1], $\overline{A}$ - DF generates a strongly continuous cosine family.

The details of the proof of the Theorem and an expanded version of this paper will appear elsewhere.

ACKNOWLEDGEMENTS

The author would like to thank Professor John Burns for bringing this problem to his attention.

Professor Rankin's research was supported by the National Science Foundation under Grant ISP-8011453-15 and by the United States Army under Contract No. DAAG29-80-C-0041.

REFERENCES

1. C. Travis and G. Webb, Second order differential equations in Banach space, *Nonlinear Equations in Abstract Spaces*, edited by V. Lakshmikantham, Academic Press, New York, 1978, 331-361.

2. R. B. Vinter, On a problem of Zabczyk concerning semigroups generated by operators with nonlocal boundary conditions, Imperial College, Department of Computing and Control Report 77/8 (1977).

THEORY AND NUMERICS OF A QUASILINEAR PARABOLIC EQUATION IN RHEOLOGY

M. Renardy

Mathematics Research Center
University of Wisconsin
Madison, Wisconsin

1. INTRODUCTION

The following problem occurs in polymer processing: A thin filament of a viscoelastic liquid is stretched by pulling its ends as indicated in the diagram:

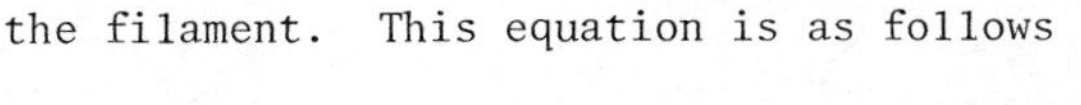

We investigate a model equation for the temporal evolution of the displacement, which is derived from the "rubberlike liquid" constitutive assumption for the stress-strain law [5] and a one-dimensional approximation based on the thinness of the filament. This equation is as follows (see [9]):

$$\rho \ddot{u} = 3\eta \frac{\partial^2}{\partial x \partial t} \left[- \frac{1}{u_x} \right] + \frac{\partial}{\partial x} \int_{-\infty}^{t} a(t - s) \left[\frac{u_x(t)}{u_x^2(s)} - \frac{u_x(s)}{u_x^2(t)} \right] ds \qquad (1.1)$$

Here u(x,t) is a real-valued function of $x \in [-1,1]$ and $t \in \mathbb{R}$. A subscript x denotes partial differentiation with respect to x and a "dot" denotes partial differentiation with respect to t. Equation (1.1) is supplemented by the nonlinear Neumann boundary condition

$$3\eta \frac{\partial}{\partial t}\left[-\frac{1}{u_x}\right] + \int_{-\infty}^{t} a(t-s)\left[\frac{u_x(t)}{u_x^2(s)} - \frac{u_x(s)}{u_x^2(t)}\right]ds = f(t) \qquad (1.2)$$

at the ends $x = \pm 1$.

Physically, $u(t,x)$ denotes the position of a fluid particle at time t, which occupies the position x in a certain reference state. This reference state is chosen such that the filament has uniform thickness. For the following, it will be assumed that in the limit $t \to -\infty$, the filament is in the reference state ($u = x$). The variable ρ denotes the density of the fluid, multiplied by the square of half the length of the filament (this scaling factor arises from the normalization of x to the interval $[-1,1]$), η is a Newtonian contribution to the viscosity, and f is the force acting on the ends of the filament divided by the cross-sectional area in the reference state. The memory kernel a is assumed to have the following properties:

(i) a has the representation

$$a(t) = \int e^{-\lambda t} d\mu(\lambda)$$

for some complex-valued Borel measure μ on the complex plane. The total variation $|\mu|$ of μ is finite, and there exist some $\phi < \frac{\pi}{2}$ and $\varepsilon > 0$, such that supp μ is contained in $\{\lambda \in \mathbb{C} \mid -\phi \le \arg \lambda \le \phi,\ |\lambda| \ge \varepsilon\}$. Since a is real, μ can always be chosen such that $d\mu(\bar{\lambda}) = \overline{d\mu(\lambda)}$.

(ii) $a(t) \ge 0$ for $t \in [0,\infty)$.

(iii) a is monotone decreasing.

Note that (i) implies, in particular, that a is analytic for $t > 0$, continuous at $t = 0$, and that a can be estimated by a decaying exponential. In physical theories derived from structural models of polymers (see [5]), a is a finite sum of decaying exponentials, which is clearly a special case for our assumptions (choose μ as a finite sum of Dirac measures). Other examples of functions satisfying (i) are e.g., $t^n e^{-\lambda t}$, $e^{-\lambda t^2}$. In the latter cases the form (i) can be established by using complex contour integrals.

The boundary condition (1.2) agrees with the equation governing the evolution of the length of the filament when inertial forces are neglected. Prior to the results that will be sketched in this paper (cf. [7], [9]), this problem was investigated by Lodge, McLeod and Nohel in [6]. They study problems where u_x is a given nondecreasing function for $t < 0$, and the force f vanishes for $t > 0$. Theorems concerning existence of solutions, asymptotic behavior and various monotonicity properties are proved. Whereas their arguments rely on monotonicity properties, the basic tools in [7], [9] are the implicit function theorem and Liapunov functions.

In Section 2 of this paper, I state the existence and uniqueness results for (1.1), (1.2) which I proved in [9]. Results concerning the boundary problem (1.2) are stated first. Then (1.1) is transformed to a system of equations that can be classified as "quasilinear parabolic" in the sense of Sobolevskii (see [2], [10]). This fact can be used to establish an existence theorem locally in time, and a global existence theorem for "small" forces f. For the details, the reader is referred to [9].

In Section 3, some numerical results are shown that are based on joint work with P. Markowich [7], [8]. We discretized (1.1), using an implicit Euler-type finite difference approximation. For the kernel a, we chose numbers given in [4] for a polyethylene melt at 150°C. The qualitative behavior of solutions is studied as the parameters ρ and η vary.

2. ANALYTICAL THEORY

We begin this section by stating some theorems concerning the solution of the boundary problem (1.2). For brevity, we omit the proofs, and the reader is referred to [9]. Before we can state the first theorem, we need the following definition.

DEFINITION 2.1. Let Z be a Banach space. Then $H^n(\mathbb{R}, Z)$ denotes the space of all functions $\mathbb{R} \to Z$ whose first n derivatives are square integrable in the sense of Bochner [3]. Moreover, let

$$\hat{X}_n^\sigma(Z) = \{v \in H^n(\mathbb{R}, Z) \mid e^{\sigma t}v, e^{-\sigma t}v \in H^n(\mathbb{R}, Z)\}$$

$$\hat{Y}_n^\sigma(Z) = \{v\colon \mathbb{R} \to Z \mid e^{-\sigma t}v \in H^n(\mathbb{R}, Z) \text{ and there exists a } v_\infty \in Z \text{ such that}$$

$$e^{+\sigma t}(v - v_\infty) \in H^n(\mathbb{R}, Z)\}.$$

The spaces $\hat{X}_n^\sigma(Z)$ and $\hat{Y}_n^\sigma(Z)$ have natural norms which make them Banach spaces. We quote the following two theorems from [9].

THEOREM 2.2. Let $\sigma > 0$ be sufficiently small. Then the following holds: If $f \in \hat{X}_n^\sigma(\mathbb{R})$ has sufficiently small norm, then (1.2) has a unique solution $u_x(t)$ which satisfies $u_x - 1 \in \hat{Y}_{n+1}^\sigma(\mathbb{R})$. The function u_x depends smoothly on f.

THEOREM 2.3. In addition to (a), assume that supp μ is contained in the real axis and μ is positive real. Let $f: \mathbb{R} \to \mathbb{R}$ be continuous and such that $\lim_{t\to-\infty} e^{-\sigma t} f(t) = 0$ for some $\sigma > 0$ and $f(t) = 0$ for $t \geq t_0$. For any such f, equation (1.2) has a unique solution satisfying $\lim_{t\to-\infty} u_x(t) = 1$. This solution exists globally in time; moreover, $\lim_{t\to\infty} u_x(t) = u_x(\infty)$ exists and $u_x(\infty) > 0$.

The proof for Theorem 2.2 is based on the implicit function theorem. In the situation of Theorem 2.3, the implicit function theorem yields the existence of a solution on some interval $(-\infty,-T]$. Estimates show that solutions cannot blow up in finite time, and the result about the limiting behavior at infinity follows from a Liapunov function argument (see [9]).

We now turn to the study of (1.1). According to the theorems above, we consider $u_x(t) = b(t) > 0$ as being given for $x = \pm 1$, where b is a smooth function of t. In [9], equation (1.1) was reformulated in such a way that it fits into the theory of quasilinear parabolic equations. This was achieved by the following substitutions:

$$p = u_x$$

$$q = u_{xx}$$

$$r = \dot{u}$$

$$g_1(\lambda) = \int_{-\infty}^{t} e^{-\lambda(t-s)}(u_x(s) - u_x(t))ds$$

$$g_2(\lambda) = \int_{-\infty}^{t} e^{-\lambda(t-s)}\left[\frac{u_{xx}(s)}{u_x^3(s)} - \frac{u_{xx}(t)u_x(s)}{u_x^4(t)}\right]ds$$

$$g_3(\lambda) = \int_{-\infty}^{t} e^{-\lambda(t-s)}(u_x^{-2}(s) - u_x^{-2}(t))ds$$

$$g_4(\lambda) = \int_{-\infty}^{t} e^{-\lambda(t-s)}[u_{xx}(s) - \frac{u_{xx}(t)u_x^2(t)}{u_x^2(s)}]ds.$$

Equation (1.1) now assumes the following form:

$$\dot{p} = r_x$$

$$\dot{q} = r_{xx}$$

$$\rho\dot{r} = 3\eta p^{-2}r_{xx} - 6\eta p^{-3}qr_x - 2p\int g_2(\lambda)d\mu(\lambda) - \frac{1}{p^2}\int g_4(\lambda)d\mu(\lambda)$$

$$\dot{g}_1(\lambda) = -\lambda g_1(\lambda) - \frac{r_x}{\lambda} \tag{2.1}$$

$$\dot{g}_2(\lambda) = -\lambda g_2(\lambda) - \frac{r_{xx}}{p^4}[g_1(\lambda) + \frac{p}{\lambda}] + \frac{4r_xq}{p^5}[g_1(\lambda) + \frac{p}{\lambda}]$$

$$\dot{g}_3(\lambda) = -\lambda g_3(\lambda) + \frac{2r_x}{\lambda p^3}$$

$$\dot{g}_4(\lambda) = -\lambda g_4(\lambda) - r_{xx}p^2[g_3(\lambda) + \frac{1}{\lambda p^2}] - 2r_xqp[g_3(\lambda) + \frac{1}{\lambda p^2}],$$

with boundary condition $p = b(t)$, $r_x = \dot{b}(t)$ at $x = \pm 1$. Provided it is satisfied initially, the first boundary condition follows from the second and the first equation of (2.1), and it can therefore be ignored.

Before we state our theorems, we need to define some function spaces. Let H^k denote the Sobolev space of those functions on [-1,1] which have k square integrable derivatives, and $L^s(\mu,H^k)$ is the space of H^k-valued functions defined on ₵ which are s-integrable with respect to $|\mu|$ in the Bochner sense (see [3]). Equation (2.1) is regarded as an evolution problem in the space $X_s = H^2 \times (H^1)^2 \times (L^s(\mu,H^1))^4$. Let us substitute $\hat{r} = r - \dot{b}(t)x$, and introduce the abbreviation $y = (p,q,r,g_1,g_2,g_3,g_4)$. We split (2.1) into a "quasilinear" and a "perturbing" part,

$$\dot{y} = A(y)y + f(y,t), \tag{2.2}$$

where A(y) is defined as the following linear operator

$$A(y)y' = (\hat{r}'_x, \hat{r}'_{xx}, \frac{1}{\rho}(3\eta p^{-2}\hat{r}'_{xx} - 6\eta p^{-3}q\hat{r}'_x), -\lambda g_1'(\lambda) - \frac{1}{\lambda}\hat{r}'_x,$$

$$-\lambda g_2'(\lambda) - (g_1(\lambda) + \frac{p}{\lambda})\frac{\hat{r}'_{xx}}{p^4} + \frac{4q}{p^5}(g_1(\lambda) + \frac{p}{\lambda})\hat{r}'_x, -\lambda g_3'(\lambda) + \frac{2}{\lambda p^3}\hat{r}'_x,$$

$$-\lambda g_4'(\lambda) - p^2(g_3(\lambda) + \frac{1}{\lambda p^2})\hat{r}'_{xx} - 2qp(g_3(\lambda) + \frac{1}{\lambda p^2})\hat{r}'_x);$$

A incorporates the boundary conditions $\hat{r}'_x = 0$ at $x = \pm 1$.

In [9], I proved the following theorem.

THEOREM 2.4. Let $1 \le s < \infty$ be arbitrary. Let $y_0 = (p_0, q_0, \hat{r}_0, g_{i,0}) \in X_s$ be given such that $\hat{r}_0 \in H^3$, $\hat{r}_{0,x} = 0$ at $x = \pm 1$, $\lambda g_{i,0} \in L^s(\mu, H^1)$ and $\min_{x\in[-1,1]} p_0(x) > 0$. Then, for some $T > 0$, equation (2.2) has a unique solution $y \in C^1([0,T], X_s)$ such that $y(0) = y_0$.

Idea of Proof. The result follows from the Sobolevskii theory [2], [10] after it is proved that A generates an analytic semigroup and that certain smoothness conditions hold. For the details, see [9].

It is also possible to establish an analogue of Theorem 2.2 for the equation (1.1), i.e., to prove existence of solutions globally in time for small forces f. The case f = 0 now corresponds to the boundary condition $u_x(t) = b(t) = 1$. Then (2.1) has the trivial solution $p = 1$, $q = 0$, $r = 0$, $g_i(\lambda) = 0$. We substitute $y = (p-1, q, \hat{r}, g_1, g_2, g_3, g_4)$, and split the right hand side of (1.1) into the linearization at the trivial solution and a nonlinear perturbation, leading to an equation of the form

$$\dot{y} = Ay + f(y,t). \qquad (2.3)$$

In [9], it is proved that the operator A generates an analytic semigroup, and that its spectrum consists of the semisimple eigenvalue zero and a remainder in the left half-plane. (At this point, assumptions (ii), (iii) on the kernel a enter essentially. In determining the spectrum, we use

$$\int \frac{1}{\lambda(\lambda+\alpha)}\, d\mu(\lambda) = \int_0^\infty a(t)\, \frac{1-e^{-\alpha t}}{\alpha}\, dt,$$

and (ii), (iii) yield that this has positive real part for Re $\alpha \geq 0$.) Using this fact and the implicit function theorem, I proved in [9]

THEOREM 2.5. Let $\sigma > 0$ be small enough. If $\dot{b} \in \hat{X}^{\sigma}_{n+1}(\mathbb{R})$ has sufficiently small norm, then equation (2.3) has a solution $y \in \hat{Y}^{\sigma}_{n+1}(N(A)) \oplus (\hat{X}^{\sigma}_{n+1}(R(A)) \cap \hat{X}^{\sigma}_{n}(D(A) \cap R(A)))$.

3. NUMERICAL RESULTS

The results presented in this section were obtained in joint work with P. Markowich [8]. We discretized (1.1), (1.2) using a finite difference method, which is second order in space and first order in time. The scheme is implicit, i.e., time derivatives are approximated by backward differences. In [8] we gave a convergence analysis for our scheme in the case that f is small. The stability analysis essentially copies the analysis used for the continuous problem with the spaces $\hat{X}^{\sigma}_{n}$, $\hat{Y}^{\sigma}_{n}$ replaced by their discrete analogues. I refer the reader to [8] for details about the numerical scheme, and I proceed straight to some results. In these computations, we chose the kernel $\sum_{i=1}^{8} K_i e^{-\lambda_i u}$ with the constants K_i, λ_i listed in Table 1.

TABLE 1

i	λ_i (sec^{-1})	K_i ($Nm^{-2}sec^{-1}$)
1	10^{-3}	1×10^{-3}
2	10^{-2}	1.8×10^{0}
3	10^{-1}	1.89×10^{2}
4	1	9.8×10^{3}
5	10	2.67×10^{5}
6	10^{2}	5.86×10^{6}
7	10^{3}	9.48×10^{7}
8	10^{4}	1.29×10^{9}

These numbers were obtained by Laun [4] from an experimental fit for a polyethylene melt at 150°C, which he calls "Melt 1".

The parameter η is physically identified as a Newtonian contribution to the viscosity. Experimental values are not available, and theoretically η is either a solvent viscosity (for polymer solutions), or it results from fractions of low molecular weight (for melts). The value of η should be compared to the viscosity resulting from the memory, which, for constant shear rate, is given by $\sum_{i=1}^{8} K_i \lambda_i^{-2} = 50{,}000\ Nm^{-2}sec$. One would expect η to influence the solution significantly only if it is at least comparable to this value. This heuristic argument was confirmed by our computations. The numbers given in the following are understood to be in the following units:

(i) η is given in $Nm^{-2}sec$.

(ii) f denotes the force acting on the ends of the filament divided by the cross-sectional area in the reference state ($u = x$). It is measured in Nm^{-2}.

(iii) ρ denotes the density of the filament multiplied by the square of half the length in the reference state (this latter scaling factor arises from the normalization of the variable x to the interval $[-1,1]$). ρ is measured in $kg\ m^{-1}$. Time is measured in seconds.

In the accompanying figures, we have chosen $f = 100{,}000\ e^{-t^2/25}$. In the first twelve figures, we have the initial condition $u(x,t = -\infty) = x$. The first three plots show u, u_x and u_{xx} for $\rho = 1$, $\eta = 1$. It can be seen that u_{xx} is negligibly small, and that u is almost linear in x. This means that the solution is determined by the evolution of the boundary condition, and inertial forces can be neglected. This changes if ρ is increased. Physically, this corresponds to changing the length of the filament. For realistic values of the density, $\rho = 1$ would correspond to an initial length of a few millimeters, $\rho = 1{,}000$ would correspond to an initial length of about 1m.

In figures 4 - 6, we have $\rho = 1{,}000$, $\eta = 1$. Three-dimensional plots of u, u_x and u_{xx} are shown. It can be seen that u_{xx} has increased by a factor of 1,000 compared to the previous plots. Otherwise the qualitative behavior remains roughly the same.

The next three plots (figures 7 - 9) were made for the same ρ and η = 100,000 (for $0 \leq \eta \leq 10{,}000$, the solutions changed very little). In comparison to $\eta = 1$, it was found here that the boundary value for u_x increases more slowly up to $t = -2$, and then increases rather suddenly around $t = 0$. In this region inertial forces become very important, a fact manifested in the plots by a rather pronounced spike in u_{xx}.

In Figure 10 we have $\rho = 1{,}000$, $\eta = 1{,}000{,}000$. In this case the behaviour becomes almost Newtonian, and there is hardly any elastic recovery (the maximal value for $u(x = 1,t)$ is 1.404, and the value at $t = 60$ is still 1.398). The dependence of u on x is again almost linear.

Several calculations were done for $\rho = 10{,}000$. Figures 11 and 12 show u and u_{xx} for $\eta = 100{,}000$. It can be seen that u_{xx} becomes rather large. When one looks carefully at the plot for u, one also finds that a little "overshoot" occurs in the relaxation, whereas in the previous plots u decreased monotonically after reaching its maximal value, this is no longer true here, as the following calculations show:

	$t = 6$	$t = 8$	$t = 10$	$t = 12$
$u(x = -1,t)$	-83.6	-22.7	-3.8	-4.3

If smaller values of η are chosen, this "overshoot" becomes even more pronounced. The mesh size becomes very crucial here, and computations with adaptive meshes would be more appropriate.

In the last plot (figure 13), a different initial state was chosen, namely $u(x,t = -\infty) = x + x^3$. This corresponds physically to a filament which is getting thinner towards the ends. We have $\rho = 1{,}000$, $\eta = 1$. It can be observed that u makes several oscillations during the elastic recovery.

I conclude this paper by mentioning an interesting open problem. The theory presented here relies essentially on the fact that $\eta \neq 0$. On the other hand, our computations indicate that even very big viscosities can still be considered small parameters (for comparison: η water $\approx 10^{-3}$, η glycerine ≈ 2). Besides the mathematical interest, there is thus also a very strong physical motivation for developing a theory which remains valid as $\eta \to 0$.

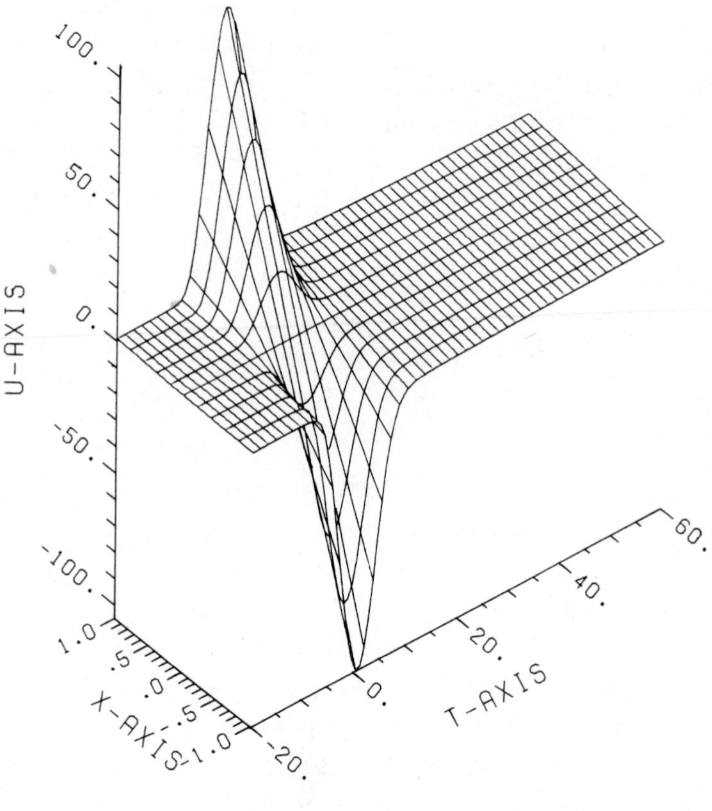

FIGURE 1

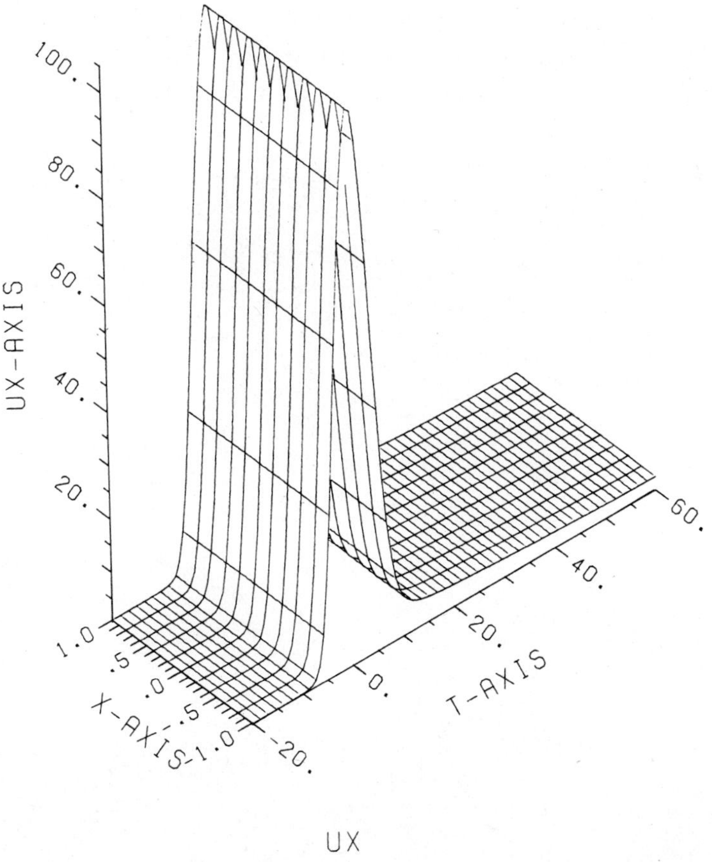

FIGURE 2

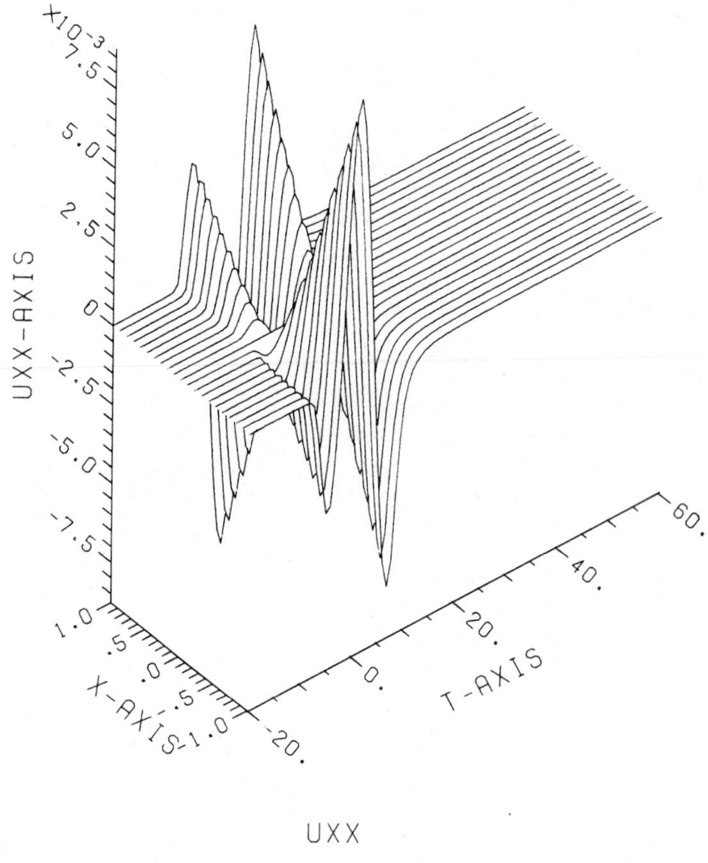
X10^-3
7.5
5.0
2.5
0
-2.5
-5.0
-7.5
UXX-AXIS
1.0
.5
0
-.5
-1.0
X-AXIS
-20.
0.
20.
40.
60.
T-AXIS
UXX

FIGURE 3

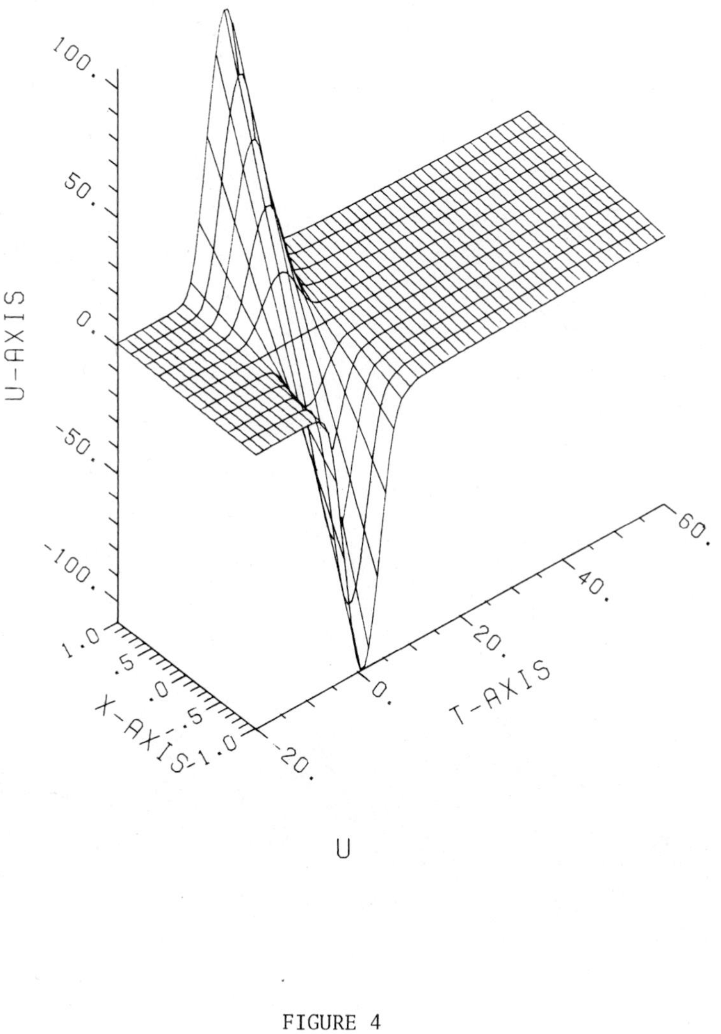

FIGURE 4

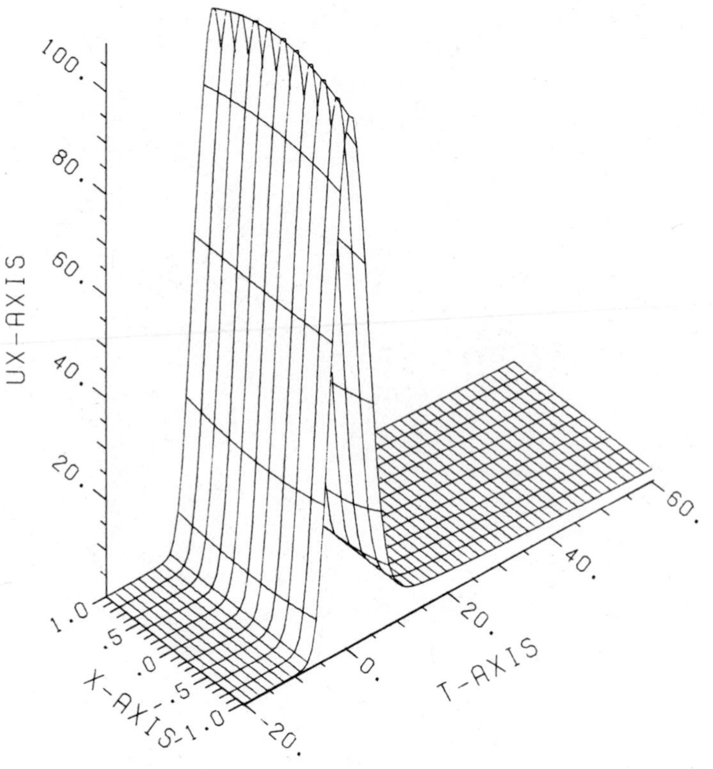

FIGURE 5

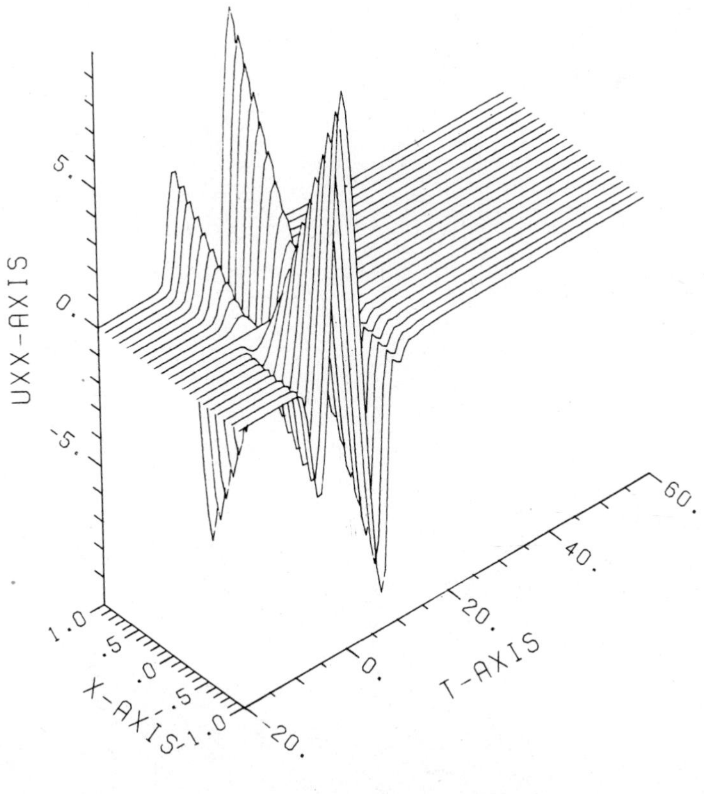

FIGURE 6

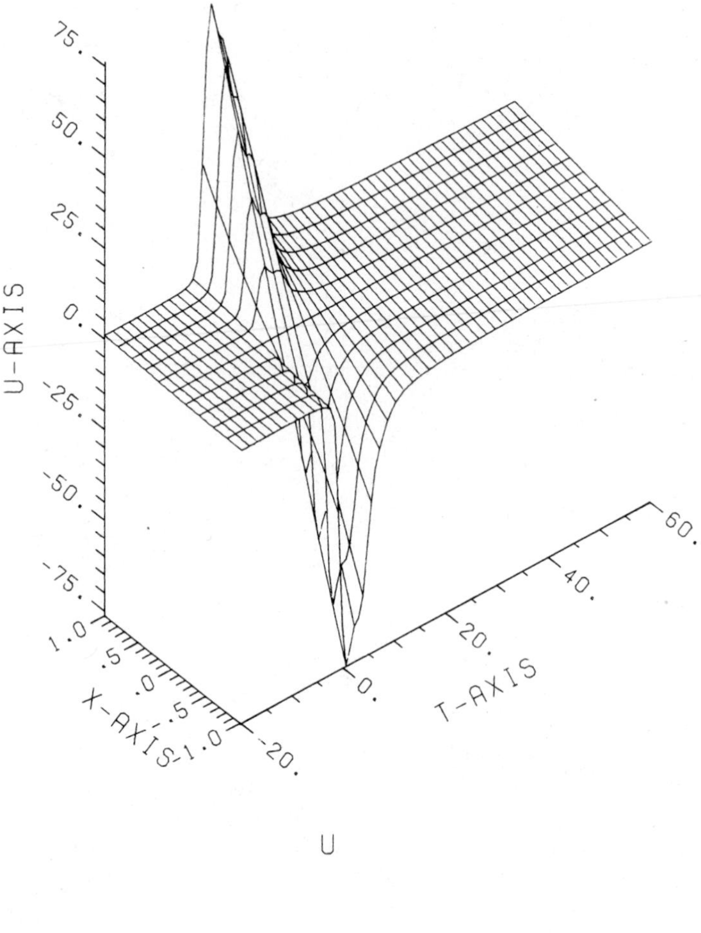

FIGURE 7

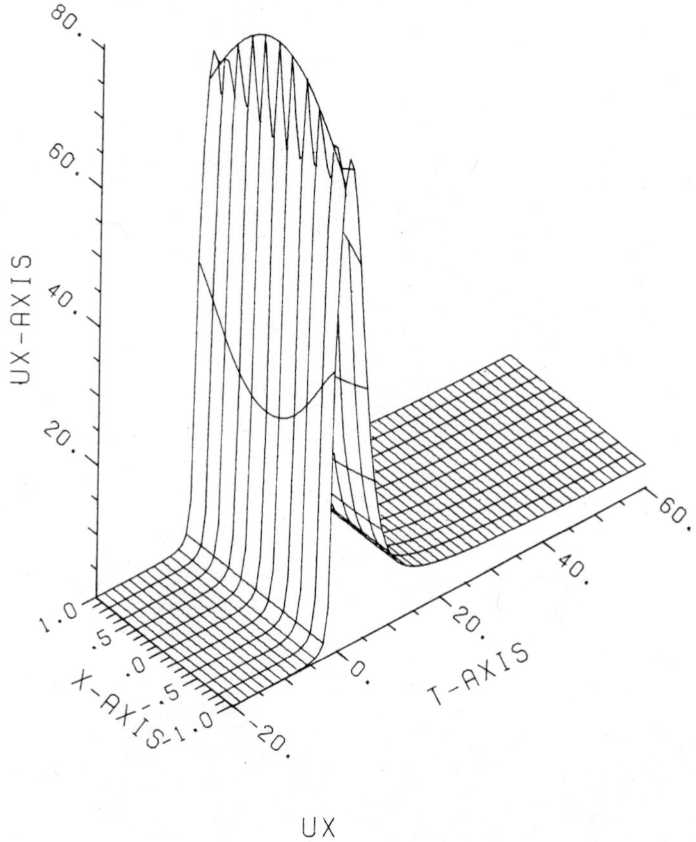

FIGURE 8

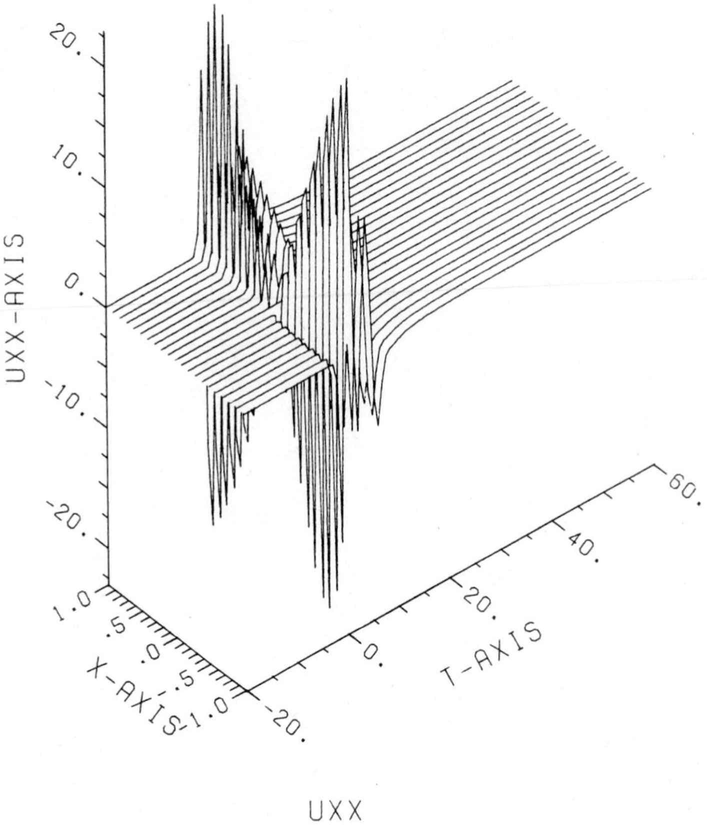

FIGURE 9

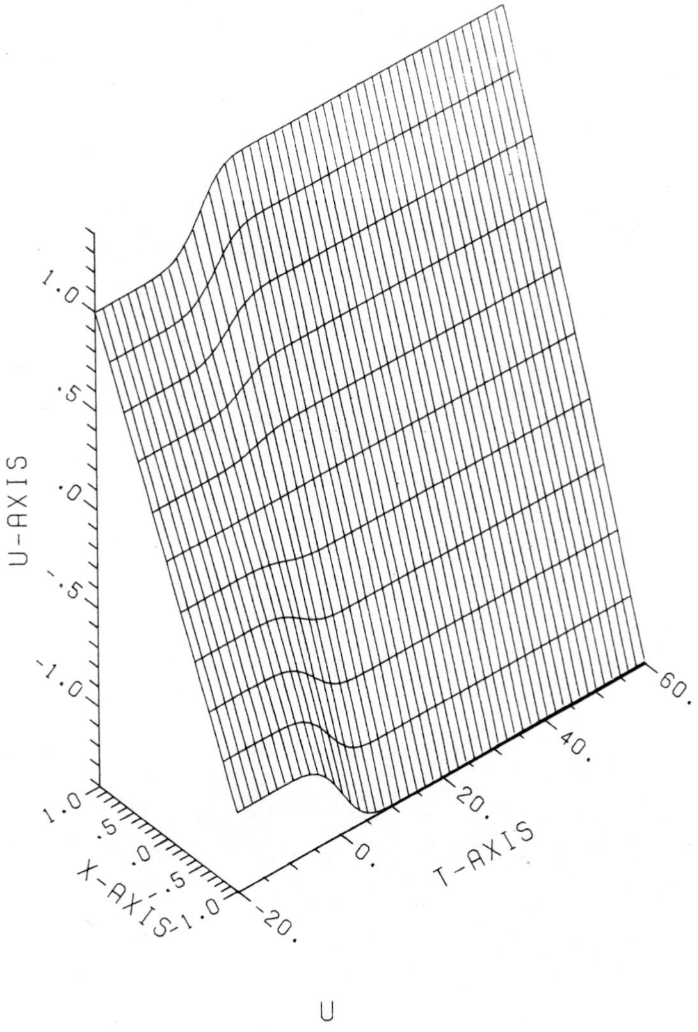

FIGURE 10

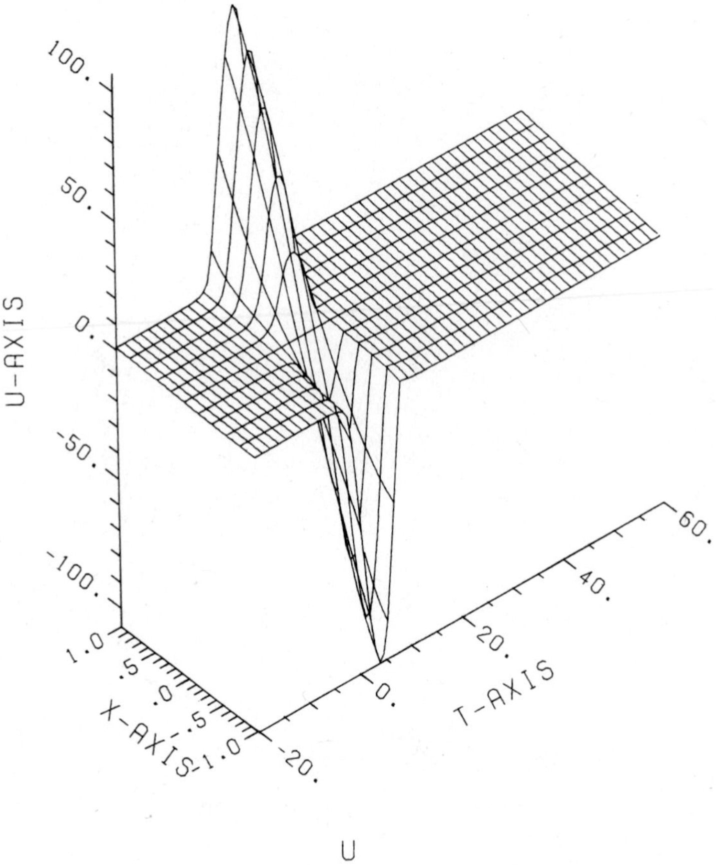

FIGURE 11

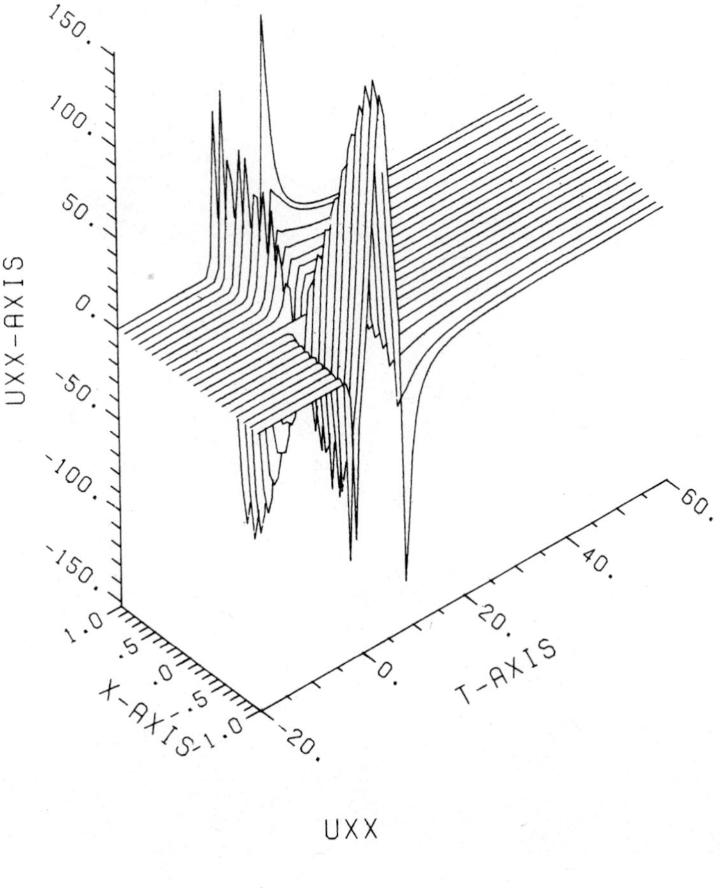

150.
100.
50.
0.
-50.
-100.
-150.
UXX-AXIS
1.0
.5
.0
-.5
-1.0
X-AXIS
-20.
0.
20.
40.
60.
T-AXIS
UXX

FIGURE 12

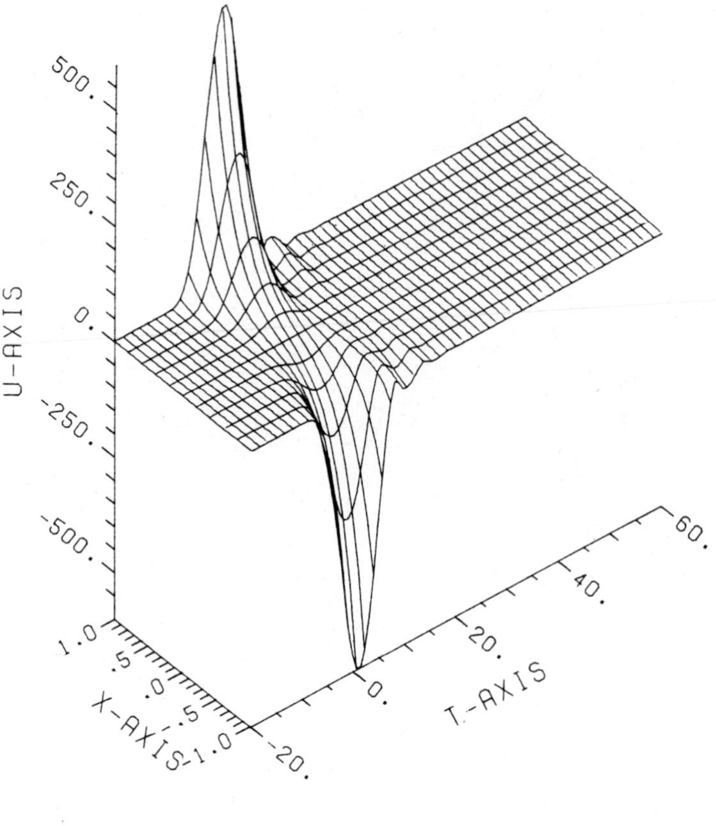

FIGURE 13

ACKNOWLEDGEMENTS

Professor Renardy's research was sponsored by the United States Army under Contract No. DAAG29-80-C-0041 and Deutsche Forschungsgemeinschaft.

REFERENCES

1. S. Agmon and L. Nirenberg, Properties of solutions of ordinary differential equations in Banach space, *Comm. Pure Appl. Math.* 16 (1963), 121-239.

2. A. Friedman, *Partial Differential Equations*, Holt, Rinehart & Winston, New York, 1969.

3. E. Hille and R. S. Phillips, *Functional Analysis and Semi-Groups*, American Mathematical Society, Providence, 1957.

4. H. M. Laun, Description of the non-linear shear behavior of a low density polyethylene melt by means of an experimentally determined strain dependent memory function, *Rheol. Acta* 17 (1978), 1-15.

5. A. S. Lodge, *Body Tensor Fields in Continuum Mechanics*, Academic Press, New York, 1974.

6. A. S. Lodge, J. B. McLeod and J. A. Nohel, A nonlinear singularly perturbed Volterra integrodifferential equation occurring in polymer rheology, *Proc. Roy. Soc. Edinburgh* 80A (1978), 99-137.

7. P. Markowich and M. Renardy, A nonlinear Volterra integrodifferential equation describing the stretching of polymeric liquids, *SIAM J. Math Anal.* (1983), to appear.

8. P. Markowich and M. Renardy, The numerical solution of a quasilinear parabolic equation arising in polymer rheology, MRC Technical Summary Report #2255, University of Wisconsin - Madison, 1981 (submitted to

9. M. Renardy, A quasilinear parabolic equation describing the elongation of thin filaments of polymeric liquids, *SIAM J. Math. Anal.* 13 (1982), 226-238.

10. P. E. Sobolevskii, Equations of parabolic type in a Banach space, *Amer. Math. Soc. Transl.* 49 (1966), 1-62.

ON PERIODIC SOLUTIONS IN SYSTEMS OF HIGH ORDER DIFFERENTIAL EQUATIONS

S. J. Skar

Department of Mathematics
Oklahoma State University
Stillwater, Oklahoma

R. K. Miller

Department of Mathematics
Iowa State University
Ames, Iowa

A. N. Michel

Department of Electrical Engineering
Iowa State University
Ames, Iowa

1. INTRODUCTION AND NOTATION

In this paper we establish conditions which ensure the existence or the nonexistence of periodic motions in complex autonomous dynamical systems. We consider those types of systems which can be viewed as an interconnection of several simpler subsytems. Our results use the describing function method (a Galerkin or harmonic balance procedure, see [3]) and answer the following question: Suppose that the describing function method predicts the existence (respectively, nonexistence) of periodic motions in the free subsystems. What conditions will ensure the existence (respectively, nonexistence) of limit cycles in the entire interconnected system? The hypotheses of our results may be validated by easily interpreted graphical criteria.

Perhaps the most widely accepted procedure for treating large scale dynamical systems consists of viewing complex systems as an interconnection of several less complex subsystems. The qualitative behavior of these isolated subsystems is analyzed and conditions on the interconnection terms are found which ensure that the behavior of the free subsystems is shared by the entire interconnected system. A summary of many of these results may be found in the books by Michel and Miller [6] and by Siljak [8].

The important problem of determining whether periodic motions exist in interconnected systems seems to have received little attention thus far. Miller and Michel [7] consider periodic motions in systems with periodic forcing. The results of the present paper are an improvement on the results in [9] which give criteria for the nonexistence of periodic motions in nonlinear autonomous systems.

Our results extend to interconnected systems some results of Mees and Bergen [5]. Other related work may be found, for example, in Cesari [2], Urabe [11], and Bergen and Franks [1].

Let $\mathbb{R}$ be the set of real numbers. To facilitate the study of periodic motions, we define the space $H(\omega)$ for each $\omega > 0$ as follows: The measurable function $x: \mathbb{R} \to \mathbb{R}$ is in the set $H(\omega)$ if and only if

(a) $x(t + \pi/\omega) = -x(t)$ for $t \in \mathbb{R}$ and

(b) $||x||_\omega^2 = \frac{\omega}{\pi} \int_0^{2\pi/\omega} |x(t)|^2 dt < \infty.$

Property (a) is called π-symmetry [5] or odd symmetry [4]; it implies that $H(\omega)$ contains only periodic functions of period $2\pi/\omega$ which have only odd harmonics. Thus, $H(\omega)$ is a Hilbert space with a canonical basis

$$\{\exp(in\omega t) \mid n = \pm 1, \pm 3, \pm 5, \ldots\}.$$

For $x \in H(\omega)$, define the nth Fourier coefficient to be

$$\hat{x}_n = \frac{\omega}{\pi} \int_0^{2\pi/\omega} e^{-in\omega t} x(t) dt.$$

Let $H_\ell(\omega)$ be the space of functions $x: \mathbb{R} \to \mathbb{R}^\ell$ such that $x_k \in H(\omega)$ for every component function x_k of x.

2. THE INTERCONNECTED SYSTEM

The purpose of this paper is to find sufficient conditions for the existence (or nonexistence) of a function x in $H_\ell(\omega)$ for some $\omega > 0$ with $x \neq 0$ such that the components x_k of x are solutions of the interconnected system

$$x_k + g_k T_k f_k x_k = \sum_{m=1}^{\ell} g_k b_{km} x_m, \qquad (\Sigma_k)$$

where the integral convolution operators g_k, the delay operators T_k, the nonlinear functions f_k, and the interconnection operators b_{km} are defined in the following assumptions: For $k = 1, \ldots, \ell$

(i) $g_k: H(\omega) \to H(\omega)$ is defined by

$$(g_k y)(t) = \int_{-\infty}^{\infty} \tilde{g}_k(t - s)y(s)ds$$

for some function $\tilde{g}_k \in L^1(\mathbb{R})$.

(ii) The function

$$G_k(i\omega) = \int_{-\infty}^{\infty} \tilde{g}_k(s)e^{-i\omega s}ds$$

has the properties $G_k(i\omega) = \overline{G}_k(-i\omega) \to 0$ as $\omega \to \infty$ and G_k is continuous for $\omega \in \mathbb{R}$.

(iii) $T_k: H(\omega) \to H(\omega)$ is defined by

$$(T_k y)(t) = \int_{-r}^{0} y(t + s)d\mu_k(s)$$

where μ_k is a function of bounded variation and $r > 0$.

(iv) If

$$\nu_k(\omega) = \int_{-r}^{0} e^{i\omega s}d\mu_k(s),$$

then

$$\zeta_k(\omega) = \inf_{n \text{ odd}, |n|>1} |\nu_k(n\omega)| > 0$$

(for values of interest).

Note that if $(T_k y)(t) = y(t - r_k)$ is a single delay, then this assumption is trivially satisfied since, in this case, $\zeta_k(\omega) \equiv 1$.

(v) $f_k: \mathbb{R} \to \mathbb{R}$, $f_k(-u) = -f_k(u)$ and $\alpha_k(u - v) \leq f_k(u) - f_k(v) \leq \beta_k(u - v)$ for some constants α_k and β_k and for $u \geq v$.

(vi) For $m = 1, 2, \ldots, \ell$, the operator b_{km} is a bounded linear operator on $H(\omega)$. For example, we may choose b_{km} to be a constant or

$$(b_{km}y)(t) = \int_{-\infty}^{\infty} \tilde{b}_{km}(t - s)y(s)ds$$

for some function $\tilde{b}_{km} \in L^1(\mathbb{R})$, or

$$(b_{km}y)(t) = \int_{-r}^{0} y(t + s)d\mu_{km}(s),$$

where μ_{km} is a function of bounded variation. We will use the symbol $|b_{km}|$ to represent the bounds of b_{km} on $H(\omega)$.

REMARK 1. The convolution operator g_k defined in (i) might be, for example, the inverse of the differential operator

$$L_k y = \sum_{m=0}^{n_k} a_m \frac{d^m y}{dt^m}$$

on $H(\omega)$. In this case, the function G_k defined in (ii) is

$$G_k(s) = \left(\sum_{m=0}^{n_k} a_m s^m \right)^{-1}.$$

REMARK 2. If $y \in H(\omega)$ and $z = g_k y$, then the nth Fourier coefficients are related by the equation $\hat{z}_n = G_k(in\omega)\hat{y}_n$ for $n = \pm 1, \pm 3, \ldots$. This expression may be taken to define the linear map g_k on $H(\omega)$.

REMARK 3. If $y \in H(\omega)$ and $z = T_k y$, then $\hat{z}_n = \nu_k(n\omega)\hat{y}_n$ or $\hat{y}_n = \frac{1}{\nu_k(n\omega)} \hat{z}_n$ for $n = \pm 1, \pm 3, \ldots$. The latter expression defines a linear map $y = T_k^{-1} z$ on $H(\omega)$.

REMARK 4. If $y(t) = a \cos(\omega t + b)$ with $a > 0$ and $z = f_k(y)$, then

$\hat{z}_1 = F_k(a)\, \hat{y}_1$ where

$$F_k(a) = \frac{1}{\pi a} \int_0^{2\pi} e^{-i\theta} f_k(a \cos \theta)d\theta$$

is the *describing function* for f_k. Under our assumptions, F_k is real and $\alpha_k \le F_k(a) \le \beta_k$ for $a \ge 0$ and $k = 1, \ldots, \ell$.

3. APPROXIMATIONS

We will use the method suggested in the introduction to determine whether (Σ_k) has a periodic solution $x \in H_\ell(\omega)$. Thus, we first approximate the interconnected system (Σ_k) by setting the interconnections $b_{km} = 0$. The resulting *free subsystem* has the form

$$x_k + g_k T_k f_k x_k = 0, \tag{S_k}$$

where g_k, T_k, and f_k satisfy assumptions (i) - (v) and Remarks 1 - 4 for $k = 1, \ldots, \ell$.

We now pose two questions. First, if none of the (S_k) has a solution $x_k \in H(\omega)$ with $x_k \neq 0$, then what additional conditions ensure that the interconnected system (Σ_k) has no solution $x \in H_\ell(\omega)$ with $x \neq 0$? Second, if exactly one (say (S_1)) of the (S_k) has a solution $x_1 \in H(\omega)$ with $x_1 \neq 0$, then what additional conditions ensure that the interconnected system (Σ_k) has a solution $x \in H_\ell(\omega)$ with $x \neq 0$?

To determine whether the (S_k) have periodic solutions, we begin by replacing (S_k) with its *describing function approximation*. To do so, we define a projection operator P on $H(\omega)$ by the equation

$$(Py)(t) = \frac{1}{2}\hat{y}_{-1}e^{-i\omega t} + \frac{1}{2}\hat{y}_1 e^{i\omega t}.$$

The describing function approximation for (S_k) is

$$\bar{x}_k + Pg_k T_k f_k \bar{x}_k = 0. \tag{A_k}$$

The first Fourier coefficient for (A_k) is

$$\hat{\bar{x}}_{k1} + G_k(i\omega)\nu_k(\omega)F_k(a)\hat{\bar{x}}_{k1} = 0,$$

where $a = |\hat{\bar{x}}_{k1}|$. Assume $a \neq 0$. Then

$$F_k(a) + (G_k(i\omega)\nu_k(\omega))^{-1} = 0. \tag{B_k}$$

The criterion we need for nonexistence will involve finding a number $\sigma_k(\omega)$ so that the expression on the left side of (B_k) is bounded away from zero by $\sigma_k(\omega)$. Equation (B_k) may be solved graphically by writing it in the form

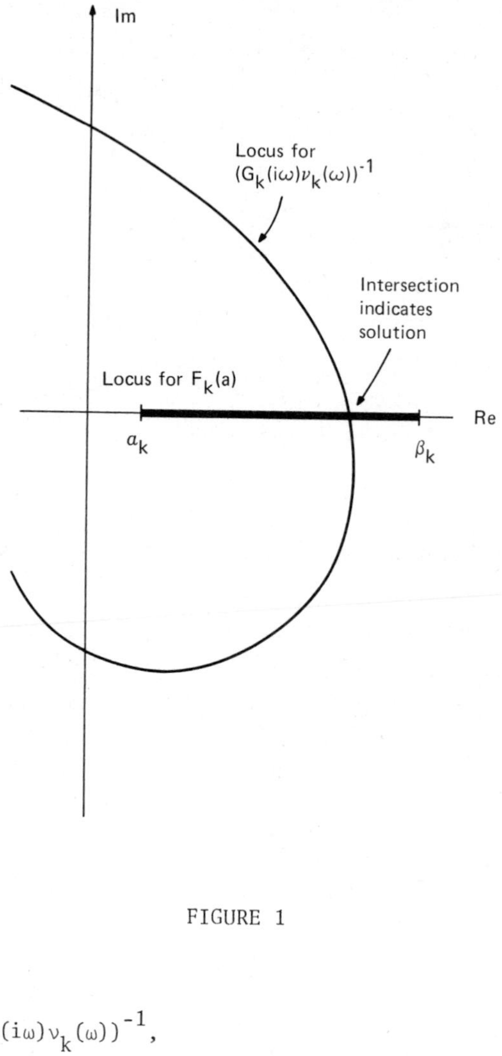

FIGURE 1

$$F_k(a) = -(G_k(i\omega)\nu_k(\omega))^{-1},$$

and graphing the curves represented by the left and right hand sides in the same complex plane. Solutions will be given by the crossings of these curves (see Figure 1).

4. CRITERIA

The criteria for existence and nonexistence of periodic solutions of $(\sum_k)$ are given in the steps of this section. As a preliminary step, we study equation (B_k) for $k = 1, \ldots, \ell$. We find an interval $[b,c]$ so that if

$b \le \omega \le c$, then at most one of the equations (B_k) (and hence at most one of the describing function approximations) has a solution for some $a > 0$. We now perform each of the following steps. If one of the steps cannot be done, then the criterion is inconclusive.

STEP 1. Define a test matrix $R(\omega)$ as follows: Let

$$\rho_k(\omega) = \inf_{n \text{ odd}, |n|>1} \left| (G_k(in\omega)\nu_k(n\omega))^{-1} + \frac{\alpha_k + \beta_k}{2} \right|$$

and ζ_k be as in assumption (iv). The k,mth component of $R(\omega)$ is

$$r_{km}(\omega) = \begin{cases} \dfrac{2\rho_k(\omega)}{\beta_k - \alpha_k} - 1 - \dfrac{2|b_{kk}|}{\zeta_k(\omega)(\beta_k - \alpha_k)}, & k = m \\[2ex] \dfrac{-2|b_{km}|}{\zeta_k(\omega)(\beta_k - \alpha_k)}, & k \ne m \end{cases}$$

Frequently, ρ_k is a rapidly increasing function of ω (see Remark 1), so that $R(\omega)$ is diagonally dominant for moderate sizes of ω.

STEP 2. Find continuous vectors $d(\omega)$ and $e(\omega)$ with positive components so that $R(\omega)e(\omega) = d(\omega)$ for $b \le \omega \le c$. This can always be done if $R(\omega)$ is an M - matrix (see [6]) or if $R(\omega)$ is diagonally dominant. (In the latter case, let $e \equiv (1, \ldots, 1)^T$ and calculate d.) The choices for e and d can often be optimized to obtain the best possible estimates on the periodic solution (see [10]).

STEP 3. Calculate

$$\eta_k(\omega) = \frac{\beta_k - \alpha_k}{2} e_k(\omega) + \sum_{m=1}^{\ell} \frac{|b_{km}|}{\zeta_k(\omega)} d_m(\omega)$$

and

$$\sigma_k(\omega) = \eta_k(\omega)/d_k(\omega).$$

STEP 4.

Case 1. Determine that

$$|(G_k(i\omega)\nu_k(\omega))^{-1} + F_k(a)| > \sigma_k(\omega)$$

for $a > 0$, $b \le \omega \le c$, and $k = 1, \ldots, \ell$.

This step can be done graphically as indicated in Step 5 below.

THEOREM 1. If Steps 1 - 3 and Step 4, Case 1, can be accomplished, then the interconnected system $(\sum_k)$ has no solution $x \in H_\ell(\omega)$ with $x \neq 0$ and $b \leq \omega \leq c$.

The proof of this result is similar to that in [10].

STEP 4.

Case 2. Determine that

$$|(G_k(i\omega)\nu_k(\omega))^{-1} + F_k(a)| \geq \sigma_k(\omega)$$

for $a > 0$, $b \leq \omega \leq c$ and $k = 2, \ldots, \ell$, and that

$$(G_1(i\omega_0)\nu_1(\omega_0))^{-1} + F_1(a_0) = 0$$

for some $\omega_0 \in (b,c)$ and some $a_0 > 0$. If Step 4, Case 2, is satisfied, proceed to Steps 5 and 6.

STEP 5. Find numbers a_1, a_2, ω_1, ω_2 so that $0 < a_1 < a_0 < a_2$ and $0 < b < \omega_1 < \omega_0 < \omega_2 < c$, and

$$|(G_1(i\omega)\nu_1(\omega))^{-1} + F_1(a)| \geq \sigma_1(\omega)$$

for (a,ω) on the boundary of the rectangle $[a_1, a_2] \times [\omega_1, \omega_2]$.

This step may be accomplished graphically as indicated in Figure 2. (See also [10].)

STEP 6. Determine that the function

$$J(a, \omega) = (G_1(i\omega)\nu_1(\omega))^{-1} + F_1(a)$$

is one-to-one on the rectangle

$$[a_1, a_2] \times [\omega_1, \omega_2].$$

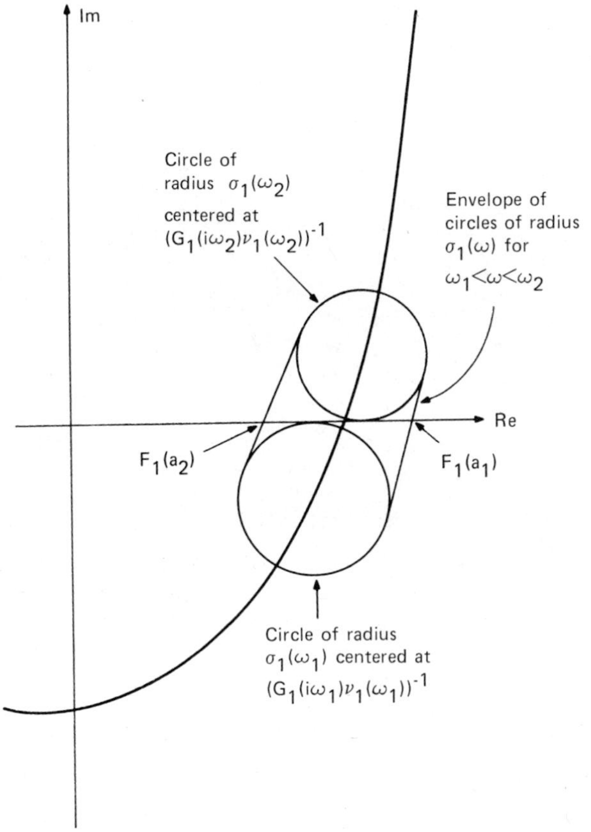

FIGURE 2

Under our assumptions, it is sufficient to determine that $F_1(a)$ is one-to-one for $a_1 \le a \le a_2$ and Im $(G_1(i\omega)\nu_1(\omega))^{-1}$ is one-to-one for $\omega_1 \le \omega \le \omega_2$.

THEOREM 2. If Steps 1 - 6 can be done, with Case 2 of Step 4, then the interconnected system $(\sum_k)$ has a solution $x \in H_\ell(\omega)$ with $x \neq 0$ and $\omega_1 \le \omega \le \omega_2$. In addition, $0 < a_1 \le a \equiv |\hat{x}_1| \le a_2$ and

$$\|x_k\|_\omega^2 \le a^2(d_k(\omega)^2 + e_k(\omega)^2)/d_1(\omega)^2$$

for $k = 1, \ldots, \ell$.

The proof of this proposition is similar to that in [10].

5. EXAMPLE

To illustrate the nature of the results, consider the system

$$x_1'''(t) + x_1''(t) + x_1'(t) + f_1(x_1(t - \pi/4)) = 0.2x_2(t)$$

$$x_2'''(t) + x_2''(t) + 2x_2'(t) + x_2(t) + f_2(x_2(t - \pi/4)) = 0.2x_1(t)$$

where ' = d/dt, and for j = 1, 2, f_j satisfies

$$f_j(u) = -f_j(-u)$$

and

$$0 \le f_j(u) - f_j(v) \le u - v$$

for $u \ge v$. By applying the procedure described in this paper, we can determine that this system has a periodic solution of period $2\pi/\omega$ with $0.68 \le \omega \le 0.77$ and another with $1.2 \le \omega \le 1.22$. There are no other solutions $x \in H_2(\omega)$, $x \ne 0$, for other values of $\omega > 0.5$.

REFERENCES

1. A. R. Bergen and R. L. Franks, Justification of the describing function method, *SIAM J. Control and Opt.* 9 (1979), 568-587.

2. L. Cesari, Functional analysis and periodic solutions of nonlinear differential equations, *Contributions to Differential Eqns.* 1 (1962), 149-187.

3. A. Gelb and W. E. VanderVelde, *Multiple-Input Describing Functions and Nonlinear System Design*, McGraw-Hill, New York, 1968.

4. W. S. Loud, Nonsymmetric periodic solutions of certain second-order nonlinear differential equations, *J. Differential Equations* 5 (1969), 352-368.

5. A. I. Mees and A. R. Bergen, Describing functions revisited, *IEEE Trans. Automatic Contr.* AC-20 (1975), 473-478.

6. A. N. Michel and R. K. Miller, *Qualitative Analysis of Large Scale Dynamical Systems*, Academic Press, New York, 1977.

7. R. K. Miller and A. N. Michel, On existence of periodic motions in nonlinear control systems with periodic inputs, *SIAM J. Control and Opt.* 18 (1980), 585-598.

8. D. D. Siljak, *Large Scale Dynamic Systems: Stability and Structure*, North Holland, New York, 1978.

9. S. J. Skar, R. K. Miller and A. N. Michel, Periodic solutions of systems of ordinary differential equations, *Differential Equations*, Academic Press, New York, 1980, 51-64.

10. S. J. Skar, R. K. Miller and A. N. Michel, On existence and nonexistence of limit cycles in interconnected systems, *IEEE Trans. Automatic Contr.*, AC-26 (1981), 1153-1169.

11. M. Urabe, Galerkin's procedure for nonlinear periodic systems, *Arch. Rat. Mech. and Anal.* 20 (1965), 120-152.

DIFFERENTIABILITY PROPERTIES OF PSEUDOPARABOLIC POINT CONTROL PROBLEMS

L. W. White

Department of Mathematics and Energy Resources Center
The University of Oklahoma
Norman, Oklahoma

1. INTRODUCTION

In this paper, we study the following problem: Let Ω be a nonempty bounded open subset of $\mathbb{R}^p$, p = 2 or 3, with a smooth boundary Γ, and let $Q = \Omega \times (0,T)$, $\sum = \Gamma \times (0,T)$, and $a \in \Omega$. Consider the pseudoparabolic problem

$$\begin{cases} My_t + Ly = v(t)\phi(x - a) & \text{in } Q \\ y(x,0;\ v) = 0 & \text{in } \Omega \\ y(x,t;\ v) = 0 & \text{on } \sum \end{cases} \tag{1}$$

where M = M(x) and L = L(x) are second order symmetric uniformly elliptic operators. The function ϕ may be an "approximate identity" with the properties:

$$\begin{cases} \phi \in C_0^\infty(\mathbb{R}^p), \quad \text{supp } \phi(x - a) \subset B(a,\varepsilon) \subset \Omega \text{ where } B(a,\varepsilon) \\ = \{y \in \mathbb{R}^p \mid \|y - a\| \le \varepsilon\}, \quad \phi \ge 0, \quad \int_\Omega \phi(x - a)dx = 1, \end{cases} \tag{2}$$

or ϕ may be the Dirac measure at a, $\delta(x - a)$. Together with the equation (1), we study the optimization problem

$$\begin{cases} \text{minimize} \quad J(v) = \int_0^T v^2(t)dt + \int_\Omega (y(x,T;\ v) - z(x))^2 dx \\ \text{subject to} \quad v \in L^2(0,T) \quad \text{with } z \in L^2(\Omega). \end{cases} \tag{3}$$

The differential equation (1) arises in the modelling of various physical systems such as flow of fluid in fissured strata [2] and the flow of second order fluids [6]. We refer to the work of Carroll and Showalter [3] for an extensive bibliography concerning these equations.

The control problem embodied in (1) and (3) is studied in [7, 8]. There the existence of a unique solution u_a is established. Furthermore, it is shown that the function from Ω into $\mathbb{R}$ defined by $a \to j(a) = J(u_a)$ is continuous from Ω to $\mathbb{R}$. Here, we determine differentiability properties of this function. More specifically, for the case of the Dirac measure we show that for $\Omega \subseteq \mathbb{R}^2$ and $z \in H^{\frac{1}{2}}(\Omega)$, the function $a \to j(a)$ is differentiable. In the approximate identity case, the differentiability properties are independent of the space dimension and the smoothness of z. Section 2 considers the case of an approximate identity and Section 3 treats the Dirac function case.

2. THE CASE OF AN APPROXIMATE IDENTITY

We begin with the equations that characterize the solution of the control problem (1) and (3), cf. [7, 8].

PROPOSITION 1. The control problem (1) and (3) where ϕ satisfies (2) has a unique solution characterized by the system

$$\begin{cases} My_t + Ly = u_a(t)\phi(x - a) \quad \text{in} \quad Q \\ y(\cdot,0;\ u_0) = 0 \quad \text{in} \quad \Omega \\ y(x,t;\ u_a) = 0 \quad \text{on} \quad \Sigma, \end{cases} \tag{4}$$

$$\begin{cases} -Mq_t + Lg = 0 \quad \text{in} \quad Q \\ q(\cdot,T;\ u_a) = M^{-1}(y(\cdot,T;u_a) - z(\cdot)) \quad \text{in} \quad \Omega \\ q(x,t;\ u_a) = 0 \quad \text{on} \quad \Sigma \end{cases} \tag{5}$$

and

$$u_a(t) + \int_\Omega q(x,t;\ u_a)\phi(x - a)dx = 0 \quad \text{a.e. in} \quad (0,T). \tag{6}$$

As a function of $a \in \Omega$, we have

$$j(a) = J(u_a) = \|u_a\|^2_{L^2(0,T)} + \|y(\cdot,T;\ u_a) - z(\cdot)\|^2_{L^2(\Omega)}. \tag{7}$$

We calculate the gradient of j to obtain

$$\nabla j(a) = (j_{a_1}(a),\ j_{a_2}(a),\ j_{a_3}(a)), \tag{8}$$

where

$$j_{a_i}(a) = 2(u_a,\ \delta_{a_i}u_a)_{L^2(0,T)} + 2(y(T;\ u_a) - z,\ \delta_{a_i}y(T;\ u_a))_{L^2(\Omega)} \tag{9}$$

for i = 1,2,3, and show that equation (9) makes sense.

We consider j_{a_1}, the other derivatives being similar. Set $\eta_1 = \delta_{a_1}y$, $\zeta_1 = \delta_{a_1}q$, $\phi_1 = \dfrac{\partial\phi}{\partial a_1}$, and $w_1 = \delta_{a_1}u_a$. Taking the variation of equations (4) - (6), we have

$$\begin{cases} M\dfrac{\partial\eta_1}{\partial t} + L\eta_1 = w_1\phi(x - a) - u_a\phi_1(x - a) & \text{in } Q \\ \eta_1(0) = 0 & \text{in } \Omega \\ \eta_1|_\Sigma = 0, \end{cases} \tag{10}$$

$$\begin{cases} -M\dfrac{\partial\zeta_1}{\partial t} + L\eta_1 = 0 & \text{in } Q \\ \zeta_1(T) = M^{-1}\eta_1(T) & \text{in } \Omega \\ \zeta_1|_\Sigma = 0 \end{cases} \tag{11}$$

and

$$w_1(t) + \int_\Omega \zeta_1(x,t)\phi(x - a)dx - \int_\Omega q(x,t)\phi_1(x - a)dx = 0. \tag{12}$$

Multiplying (10) by q and integrating, we have

$$(y(T; u_a) - z, \ \eta_1(T))_{L^2(\Omega)} =$$

$$\int_0^T \{w_1(t)\int_\Omega q(x,t)\phi(x - a)dx - u_a(t)\int_\Omega q(x,t)\phi_1(x - a)dx\}dt.$$

Thus, we may rewrite equation (9) for i = 1 as

$$j_{a_1}(a) = 4\int_0^T w_1(t)u_a(t)dt - 2\int_0^T u_a(t)\int_\Omega q(x,t)\phi_1(x - a)dxdt. \tag{13}$$

LEMMA 2. Equation (13) defines $j_{a_1}(a)$ if the system (10) - (12) has a unique solution.

We approach the problem of proving the existence and uniqueness of a solution of (10) - (12) by considering the following quadratic control problem.

$$\begin{cases} M\dfrac{\partial\eta_1(v)}{\partial t} + L\eta_1(v) = v(t)\phi(x - a) - u_a(t)\phi_1(x - a) \quad \text{in } Q \\ \eta_1(0,v) = 0 \quad \text{in } \Omega \\ \eta_1(v)|_\Sigma = 0 \end{cases} \tag{14}$$

$$\begin{cases} \text{minimize } \{\|v\|^2_{L^2(0,T)} + \|\eta_1(T;v)\|^2_{L^2(\Omega)} \\ \qquad\qquad - 2(v, \int_\Omega q(x,t)\phi_1(x - a)dx)_{L^2(0,T)}\} \\ \text{subject to } v \in L^2(0,T). \end{cases} \tag{15}$$

REMARK 3. Note that since ϕ is smooth, the solution of (14) has trace at time T in $L^2(\Omega)$ for any $v \in L^2(0,T)$. That is, $\eta_1(\cdot,T;\ v) \in L^2(\Omega)$ for any $v \in L^2(0,T)$.

The functional in (15) makes sense, and the following is a standard result.

LEMMA 4. There exists a unique solution w_1 to problem (15).

By taking the variation of (15) at w_1 and introducing equation (11), we obtain equation (12). Hence, we have proved the following.

LEMMA 5. There exists a unique solution to the system (10) - (12).

Lemmas 2 and 5 imply the following.

THEOREM 6. The partial derivative $j_{a_1}(a)$ is given by equation (13).

REMARK 7. Note that if ϕ satisfies (2), then the set Ω may be taken to be in $\mathbb{R}^p$. By inspecting the previous arguments, we can see that further differentiability is possible, depending on the differentiability of ϕ. In particular, if ϕ is infinitely differentiable, then so is j.

3. THE DELTA FUNCTION CASE

We now study the problem for $\phi = \delta$. That is,

$$\begin{cases} My_t + Ly = v(t)\delta(x - a) \quad \text{in} \quad Q \\ y(0) = 0 \quad \text{in} \quad \Omega \\ y|_\Sigma = 0, \end{cases} \tag{16}$$

where Ω is in $\mathbb{R}^2$ and Γ is smooth.

REMARK 8. Recall that for $\Omega \subset \mathbb{R}^p$, $H^n(\Omega) \subset C^0(\bar{\Omega})$ if $n > \frac{p}{2}$, [1]. For $p = 2$, we see that $H^{3/2}(\Omega) \subset C^0(\bar{\Omega})$ and $\delta \in (H^{3/2}(\Omega))^*$. Furthermore, by interpolation, it follows that $y \in H^1(0,T; H^{1/2}(\Omega))$ so that the trace $y(\cdot,T; v) \in H^{1/2}(\Omega)$ for each v in $L^2(0,T)$.

From the above remark, it is clear that the minimization problem (3) makes sense. In [7] it is shown that there exists a unique solution u_a in $L^2(0,T)$, in fact in $C^\infty(0,T)$.

PROPOSITION 9. There exists a unique solution u_a for the problem given by (16) and (3) that is characterized by the system

$$\begin{cases} My_t + Ly = u_a(t)\delta(x - a) \quad \text{in} \quad Q \\ y(0) = 0 \quad \text{in} \quad \Omega \\ y|_\Sigma = 0, \end{cases} \tag{17}$$

$$\begin{cases} -Mq_t + Lq = 0 \quad \text{in} \quad Q \\ q(T) = M^{-1}(y(T; u_a) - z) \text{ in } \Omega \\ q|_\Sigma = 0 \end{cases} \tag{18}$$

and

$$u_a(t) + q(a,t; u_a) = 0 \quad \text{in} \quad (0,T). \tag{19}$$

As in the previous section, we (formally) calculate $j_{a_1}(a)$ and the variation of equations (17) - (19) to obtain the system of equations

$$\begin{cases} M \frac{\partial \eta_1}{\partial t} + L\eta_1 = w_1(t)\delta(x - a) - u_a(t)\delta_1(x - a) \quad \text{in} \quad Q \\ \eta_1(0) = 0 \quad \text{in} \quad \Omega \\ \eta_1|_\Sigma = 0, \end{cases} \tag{20}$$

$$\begin{cases} -M \frac{\partial \zeta_1}{\partial t} + L\zeta_1 = 0 \quad \text{in} \quad Q \\ \zeta_1(T) = M^{-1}\eta_1(T) \quad \text{in} \quad \Omega \\ \zeta_1|_\Sigma = 0, \end{cases} \tag{21}$$

$$w_1(t) + \zeta_1(a,t) = -q_{x_1}(a,t) \quad \text{in} \quad (0,T), \tag{22}$$

and

$$j_{a_1}(a) = -2\int_0^T u_a(t)\zeta_1(a,t)dt + 4\int_0^T w_1(t)q(a,t)dt. \tag{23}$$

We seek to provide the proper setting for these equations. Because of the irregularity involved, we prove existence of a solution of the system (20) - (22) by transposition [4,5].

We begin with some observations concerning the regularity of the solution of (17) - (19) that follow from interpolation and results in [5].

LEMMA 10. The solution $y(u_a)$ of equation (17) belongs to $H^1(0,T; H^{1/2}(\Omega))$. The solution q of equation (18) belongs to $H^k(0,T; H_0^1(\Omega) \cap H^{5/2}(\Omega))$ for $k \geq 0$ if $z \in H^{1/2}(\Omega)$.

REMARK 11. The map $t \to q(\cdot,t)$ is an infinitely differentiable map of (0,T) into $H_0^1(\Omega) \cap H^{5/2}(\Omega)$. Hence, $t \to q(a,t)$ is continuous and in $L^2(0,T)$. Further, with $q_{x_1}(\cdot,t) \in H^{3/2}(\Omega)$ for each t, we see that $t \to q_{x_1}(a,t)$ is continuous and in $L^2(0,T)$.

For equations (20) - (22) with the variation w_1 in $L^2(0,T)$ and with δ_1 belonging to $H^{-5/2}(\Omega)$, the right side of equation (20) is in $L^2(0,T;H^{-5/2}(\Omega))$. Thus, we seek a solution η_1 in $H^1(0,T; H^{-1/2}(\Omega))$.

REMARK 12. In this case, we only have $\eta_1(T)$ in $H^{-1/2}(\Omega)$. Hence, the method of demonstrating the existence of a solution to the variational equations that is used in Section 2 is not applicable here.

However, we note that if $\eta_1(\cdot,T)$ is in $H^{-1/2}(\Omega)$, the solution ζ_1 of equation (21) belongs to $H^p(0,1; H_0^1(\Omega) \cap H^{3/2}(\Omega))$. Accordingly, for each $a \in \Omega$, $\zeta_1(a,t)$ is defined and is a continuous function of t in [0,T].

LEMMA 13. If there exists a solution to the system of equations (20) - (22) with $\zeta_1(a,t)$ in $L^2(0,T)$, then formula (23) has meaning.

We prove the existence of a solution to (20) - (22) by transposition. To this end, we consider the system

$$\begin{cases} -M\psi_t + L\psi = \theta \quad \text{in} \quad Q \\ \psi(T) = M^{-1}\alpha(T) \quad \text{in} \quad \Omega \\ \psi|_\Sigma = 0, \end{cases} \tag{24}$$

$$\begin{cases} M\alpha_t + L\alpha = \beta - \psi(a,t)\delta(x - a) \quad \text{in} \quad Q \\ \alpha(0) = 0 \quad \text{in} \quad \Omega \\ \alpha|_\Sigma = 0, \end{cases} \tag{25}$$

where $\theta \in L^2(0,T; H^{1/2}(\Omega))$ and $\beta \in L^2(0,T; H^{-3/2}(\Omega))$.

Multiplying equation (20) by ψ and using equation (22), we integrate to obtain

$$\int_\Omega \eta_1(x,T)\alpha(x,T)dx + \int_0^T\int_\Omega \eta_1(x,t)\theta(x,t)dxdt$$

$$= \int_0^T (q_{x_1}(a,t) - \zeta_1(a,t))\psi(a,t)dt \tag{26}$$

$$- \int_0^T u_a(t)\psi_{x_1}(a,t)dt.$$

Similarly, multiplying equation (21) by α and integrating, we find that

$$\int_\Omega \eta_1(x,T)\alpha(x,T)dx = \int_0^T\int_\Omega \zeta_1(x,t)\beta(x,t)dxdt - \int_0^T \psi(a,t)\zeta_1(a,t)dt. \tag{27}$$

Combining equations (26) and (27), we have

$$\int_0^T\int_\Omega \zeta_1(x,t)\beta(x,t)dxdt + \int_0^T\int_\Omega \eta_1(x,t)\theta(x,t)dxdt$$

$$= \int_0^T q_x(a,t)\psi(a,t)dt - \int_0^T u_a(t)\psi_{x_1}(a,t)dt. \tag{28}$$

LEMMA 14. If for every pair (θ,β) in $L^2(0,T;\ H^{1/2}(\Omega)) \times L^2(0,T;\ H^{-3/2}(\Omega))$ there exists a unique solution of (24) and (25), then the solution (ζ_1,η_1) in $L^2(0,T;\ H^{3/2}(\Omega)) \times L^2(0,T;\ H^{-1/2}(\Omega))$ of (20) - (22) is defined by equation (28).

We now show that the system of equations (24) and (25) has a unique solution. Thus, we consider the problem

$$\begin{cases} M\alpha_t(v) + L\alpha(v) = \beta - v(t)\delta(x - a) \quad \text{in} \quad Q \\ \alpha(0) = 0 \quad \text{in} \quad \Omega \\ \alpha|_\Sigma = 0. \end{cases} \tag{29}$$

With $\beta \in L^2(0,T; H^{-3/2}(\Omega))$ given, the equation (29) defines $\alpha \in H^1(0,T; H^{1/2}(\Omega))$, cf. [7], by interpolation [5]. Hence, it follows that the trace $\alpha(\cdot,T)$ belongs to $H^{1/2}(\Omega)$ (see [5]) and, as in the previous section, we introduce the minimization problem

$$\begin{cases} \text{minimize} \quad \|v\|^2_{L^2(0,T)} + \|\alpha(T;\, v)\|^2_{L^2(\Omega)} + 2(\theta,\, \alpha(v))_{L^2(Q)} \\ \\ \text{subject to} \quad v \in L^2(0,T). \end{cases} \tag{30}$$

Clearly, there exists a unique solution u to problem (30), see [4, 7]. Again, a characterization can be obtained by taking the variation at u of the functional in (30). We have

$$(u,v)_{L^2(0,T)} + (\alpha(T;\, u),\, (\delta\alpha)(T))_{L^2(\Omega)} + (\theta,\, (\delta\alpha))_{L^2(Q)} = 0, \tag{31}$$

where the variations satisfy

$$\begin{cases} M(\delta\alpha)_t + L(\delta\alpha) = -v(t)\delta(x - a) \quad \text{in} \quad Q \\ (\delta\alpha)(0) = 0 \quad \text{in} \quad \Omega \\ (\delta\alpha)|_\Sigma = 0. \end{cases} \tag{32}$$

We again introduce the adjoint equation (cf. (24))

$$\begin{cases} -M\psi_t + L\psi = \theta \quad \text{in} \quad Q \\ \psi(T) = M^{-1}\alpha(T;\, u) \quad \text{in} \quad \Omega \\ \psi|_\Sigma = 0, \end{cases}$$

and note that, with $\theta \in L^2(0,T; H^{1/2}(\Omega))$, the solution ψ of (24) belongs to $H^1(0,T; H^1_0(\Omega) \cap H^{5/2}(\Omega))$. Multiplying (32) by ψ and integrating, we see that

$$\int_0^T\int_\Omega \psi(M(\delta\alpha)_t + L(\delta\alpha))dxdt = -\int_0^T v(t)\psi(a,t)dt$$

so that

$$(\alpha(T;\ u),\ (\delta\alpha)(T))_{L^2(\Omega)} + (\theta,\ (\delta\alpha))_{L^2(Q)} = -\int_0^T v(t)\psi(a,t)dt.$$

Hence, we see that

$$(u - \psi(a,\cdot),\ v)_{L^2(0,T)} = 0$$

for all $v \in L^2(0,T)$, and we have

$$u(t) = \psi(a,t) \tag{33}$$

almost everywhere in $[0,T]$. The characterizing equations then are again given by (25) and (24),

$$\begin{cases} M\alpha_t + L\alpha = \beta - \psi(a,t)\delta(x - a) & \text{in } Q \\ \alpha(0) = 0 & \text{in } \Omega \\ \alpha|_\Sigma = 0, \end{cases}$$

and

$$\begin{cases} -M\psi_t + L\psi = \theta & \text{in } Q \\ \psi(T) = M^{-1}\alpha(T) & \text{in } \Omega \\ \psi|_\Sigma = 0, \end{cases}$$

respectively, and we have shown that the system of equations (24) and (25) has a solution.

If $\theta = 0$ and $\beta = 0$, we have, by multiplying (25) by ω and integrating, that

$$\|\alpha(T)\|_{L^2(\Omega)} + \|\psi(a,\cdot)\|^2_{L^2(0,T)} = 0,$$

so that $\psi = 0$ and $\alpha = 0$.

PROPOSITION 15. If $\beta \in L^2(0,T;\ H^{-3/2}(\Omega))$ and $\theta \in L^2(0,T;\ H^{1/2}(\Omega))$, then there exists a unique solution (α,ψ) of (24) and (25) with $\psi \in H^1(0,T;\ H^1_0(\Omega) \cap H^{5/2}(\Omega))$ and $\alpha \in H^1(0,T;\ H^{1/2}(\Omega))$.

From Proposition 15 and Lemma 14, we deduce the following.

COROLLARY 16. There exists a solution ζ_1 such that $\zeta_1(a,\cdot)$ belongs to $L^2(0,T)$ and, in fact, to $C(0,T)$.

Thus, from Lemma 13 we conclude the following.

THEOREM 17. Let $\Omega \subset \mathbb{R}^2$ and $z \in H^{1/2}(\Omega)$. Then $j_{a_1}(a)$ is well-defined and is given by equation (23).

REMARK 18. An analogous argument holds for $j_{a_2}(a)$, and thus $\nabla j(a)$ is defined for each $a \in \Omega$.

ACKNOWLEDGEMENTS

The author would like to thank Professor J. L. Lions for his interest and comments concerning this work.

This research was supported in part by a National Science Foundation Grant No. MCS-7902037 and the Institut National de Recherche en Informatique et en Automatique.

REFERENCES

1. Adams, R. A., *Sobolev Spaces*, Academic Press, New York, 1975.

2. Barenblatt, G. I., Iu. P. Zheltov, and I. N. Kochina, Basic concepts in the theory of seepage of homogeneous liquids in fissured rocks, *J. Appl. Math. Mech.* 24 (1960), 852-864.

3. Carroll, R. W. and R. E. Showalter, *Singular and Degenerate Cauchy Problems*, Academic Press, New York, 1976.

4. J. L. Lions, *Optimal Control of Systems Governed by Partial Differential Equations*, translated by S. K. Mitter, Springer-Verlag, New York, 1971.

5. Lions, J. L. and E. Magenes, *Non-Homogeneous Boundary Value Problems and Applications I*, translated by P. Kenneth, Springer-Verlag, New York, 1972.

6. Ting, T. W., Certain non-steady flows of second order fluids, *Arch. Rat. Mech. Anal.* 14 (1963), 1-26.

7. White, L. W., Point control of pseudoparabolic problems, *J. Differential Equations*, 42(1981), 366-374.

8. White, L. W., Point control: approximations of parabolic problems and pseudoparabolic problems, *Appl. Anal.*, to appear.